Robust Parameter Estimation with Sensor Arrays in Complex Electromagnetic Environments

Robust Parameter Estimation with Sensor Arrays in Complex Electromagnetic Environments

Guest Editors

Jianfeng Li
Ding Wang

MDPI

Basel • Beijing • Wuhan • Barcelona • Belgrade • Novi Sad • Cluj • Manchester

Guest Editors
Jianfeng Li
College of Electronic
Information Engineering
Nanjing University of
Aeronautics and Astronautics
Nanjing
China

Ding Wang
College of Information
System Engineering
PLA Strategic Support Force
Information Engineering
University
Zhengzhou
China

Editorial Office
MDPI AG
Grosspeteranlage 5
4052 Basel, Switzerland

This is a reprint of the Special Issue, published open access by the journal *Sensors* (ISSN 1424-8220), freely accessible at: https://www.mdpi.com/journal/sensors/special_issues/RPESACEE.

For citation purposes, cite each article independently as indicated on the article page online and as indicated below:

Lastname, A.A.; Lastname, B.B. Article Title. *Journal Name* **Year**, *Volume Number*, Page Range.

ISBN 978-3-7258-4393-0 (Hbk)
ISBN 978-3-7258-4394-7 (PDF)
https://doi.org/10.3390/books978-3-7258-4394-7

Contents

sensors

MDPI

Communication

Direction of Arrival Estimation of Coherent Wideband Sources Using Nested Array

Yawei Tang, Weiming Deng, Jianfeng Li * and Xiaofei Zhang

College of Electronic and Information Engineering, Nanjing University of Aeronautics and Astronautics, Nanjing 211106, China; ywtang@nuaa.edu.cn (Y.T.); dengweiming0202@126.com (W.D.); zhangxiaofei@nuaa.edu.cn (X.Z.)

* Correspondence: lijianfeng@nuaa.edu.cn

Abstract: Due to their ability to achieve higher DOA estimation accuracy and larger degrees of freedom (DOF) using a fixed number of antennas, sparse arrays, etc., nested and coprime arrays have attracted a lot of attention in relation to research into direction of arrival (DOA) estimation. However, the usage of the sparse array is based on the assumption that the signals are independent of each other, which is hard to guarantee in practice due to the complex propagation environment. To address the challenge of sparse arrays struggling to handle coherent wideband signals, we propose the following method. Firstly, we exploit the coherent signal subspace method (CSSM) to focus the wideband signals on the reference frequency and assist in the decorrelation process, which can be implemented without any pre-estimations. Then, we virtualize the covariance matrix of sparse array due to the decorrelation operation. Next, an enhanced spatial smoothing algorithm is applied to make full use of the information available in the data covariance matrix, as well as to improve the decorrelation effect, after which stage the multiple signal classification (MUSIC) algorithm is used to obtain DOA estimations. In the simulation, with reference to the root mean square error (RMSE) that varies in tandem with the signal-to-noise ratio (SNR), the algorithm achieves satisfactory results compared to other state-of-the-art algorithms, including sparse arrays using the traditional incoherent signal subspace method (ISSM), the coherent signal subspace method (CSSM), spatial smoothing algorithms, etc. Furthermore, the proposed method is also validated via real data tests, and the error value is only 0.2 degrees in real data tests, which is lower than those of the other methods in real data tests.

Keywords: direction-of-arrival (DOA); sparse array; initial-estimation-free CSSM; enhanced spatial smoothing

1. Introduction

In recent years, DOA estimations [1–3] have been widely applied in many fields, such as radar, sonar, wireless communication [4–7], etc. Therefore, research into DOA estimation has attracted more attention from researchers [8–12]. In general, the spectral estimation accuracy is positively correlated with the number of antennas, while the sparse arrays can obtain a large number of continuous elements in the virtual array via the virtualization method, thereby obtaining a higher accuracy of DOA estimation. To improve the accuracy of DOA estimation and increase the DOF with a fixed number of antennas, sparse arrays have gradually replaced uniform line arrays as the most prevalent array geometry. Examples of these arrays include the minimum redundancy array (MRA) [13], the coprime array and the nested array. Compared with MRA, coprime array [14–16] and nested array [17,18] are more feasible in terms of engineering and they can obtain a large number of continuous elements in the virtual array via the virtualization method, thereby obtaining a higher accuracy of DOA estimation and a larger DOF. However, nowadays, research based on sparse arrays is mainly used for incoherent signals, as the performance is poor at handling coherent wideband signals. In reality, it can be challenging to evade coherent signals. As a result, the resolution of coherent wideband signals in sparse arrays is the primary focus of this paper.

Currently, incoherent narrowband signal DOA estimation techniques have been extensively used. However, due to the complex propagation environment [19,20], coherent wideband signals have become an important signal type in real life, and the coherent signal becomes an important factor affecting the performance of the algorithm. Traditional incoherent narrowband DOA estimation methods are inadequate for handling coherent wideband signals. Therefore, research into DOA estimation methods for coherent wideband signals holds significant practical importance. Many algorithms have been developed to counteract the effect of coherence [21–26], such as the widely used algorithm known as spatial smoothing pre-processing (SSP) [27], which divides the entire array into a series of overlapping subarrays to obtain a new data covariance matrix with recovered rank. In this paper, in order to take full advantage of the covariance matrix of individual subarrays and the mutual covariance matrix of different subarrays, an enhanced spatial smoothing algorithm is used to more effectively counteract the effects of coherence.

For wideband signals, many types of studies have been conducted regarding wideband DOA estimation algorithms. Usually, the wideband signal is first decomposed into several narrowband signals with multiple frequencies via time-to-frequency conversion. Then, two widely used classes of wideband DOA estimation algorithms are used. One is the incoherent signal subspace method (ISSM) [28–30], which applies a high-precision narrowband DOA estimation algorithm (such as the MUSIC [31] algorithm) to the narrowband signals at multiple frequencies after the time–frequency conversion and averages the results to obtain the estimated values. This method can achieve high accuracy in a high SNR, but in a low SNR, the accuracy of the algorithm may be greatly affected, and this method cannot effectively deal with the problem of coherent sources [32]. Moreover, another algorithm, known as the coherent signal subspace method (CSSM) [33–37], is often proposed, which focuses the signal subspace at multiple frequencies to the signal subspace at the reference frequency by constructing a focus matrix and summing these focused covariance matrices to construct a single correlation matrix, after which a high-precision narrowband DOA estimation algorithm can be applied to this covariance matrix to obtain the final estimated value. This algorithm achieves great performance in the case of low SNR, and the process of averaging after focusing can reduce the coherence coefficient between signals. Therefore, it can achieve some decorrelation effect. However, the construction of the focusing matrix in this algorithm depends on the pre-estimated angle, and the accuracy of the pre-estimated angle will have a large impact on the accuracy of the final estimated angle. Therefore, a focusing algorithm that does not require pre-estimated values is used in this paper to avoid the main drawbacks of the traditional focusing algorithm.

In this paper, firstly, we choose a sparse array in order to obtain a higher spectral estimation accuracy and a larger DOF using a fixed number of antennas. Then, we choose the CSSM method to assist in the process of decorrelating the coherent signal. In order to avoid the final estimation result relying on the pre-estimation, we propose a focusing algorithm without pre-estimation. Then, we virtualize the covariance matrix of sparse array due to the decorrelation operation. To more effectively counteract the effects of coherence, we adopt an enhanced spatial smoothing algorithm to make full use of the information in the covariance matrix of individual subarrays and the mutual covariance matrix of different subarrays. Finally, the MUSIC algorithm is applied to obtain the final estimated values.

The paper is organized as follows: In the Section 2, the sparse array based wideband coherent signal model is described. Then, the algorithm proposed in this paper is presented in the Section 3. The simulations, actual measurement and analysis are mentioned in the Section 4. Finally, the conclusion is presented in the Section 5 of this paper.

2. Array Model

Considering that the K far-field wideband sources impinged on a two-level nested array, the received signals were coherent. The sensor positions of nested array could be expressed as $P = \{m_1 d, 1 \leq m_1 \leq M_1\} \cup \{m_2(M_1+1)d, 1 \leq m_2 \leq M_2\}$, where $d = \lambda/2$ (λ denotes the signal wavelength), and M_1 and M_2 represented the number of sensors of the each of the subarrays, the array is shown in Figure 1. The wideband sources from different DOAs $\theta_1, \theta_2, \cdots, \theta_k$ were assumed to be independent. The array output of the $m-$ **th** sensor could be represented as follows:

$$x_m(t) = \sum_{k=1}^{K} s_k(t - \tau_{km}) + n_m(t), m = 1, 2, \cdots, M_1, M_1+1, 2(M_1+1), \cdots M_2(M_1+1) \quad (1)$$

where $s_k(t)$ and $n_m(t)$ represent the $k-$ **th** source signal and the noise at the $m-$ **th** sensor, respectively, and $n_m(t)$ is the additive white Gaussian noise with zero mean and variance σ^2. τ_{km} is the propagation delay associated with the $k-$ **th** source and $m-$ **th** sensor.

$$\tau_{km} = (m-1)d \sin \theta_k / c \quad (2)$$

where c is the speed of signal propagation.

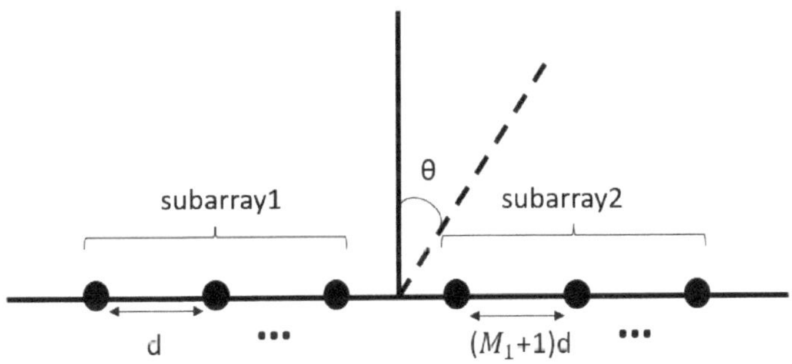

Figure 1. Two-level nested array.

We transformed the output of each sensor into the frequency domain and partitioned the frequency band into J non-overlapping narrowband blocks via $J-$ point DFT. Then, the wideband array model at J frequencies can be formulated as follows:

$$X(f_j) = A(f_j)S(f_j) + N(f_j), \ j = 1, 2, \cdots, J \quad (3)$$

where $S(f_j) = [S_1(f_j), S_2(f_j), \cdots, S_K(f_j)]^T$ and $N(f_j) = [N_1(f_j), N_2(f_j), \cdots, N_M(f_j)]^T$ represent the signal and noise at frequency f_j after DFT transformation, respectively.

$$A(f_j) = [a_{\theta_1}(f_j), a_{\theta_2}(f_j), \cdots a_{\theta_K}(f_j)], \ j = 1, 2 \cdots J \quad (4)$$

where $a_{\theta_k}(f_j) = [e^{-j2\pi f_j \tau_{k1}}, e^{-j2\pi f_j \tau_{k2}}, \cdots, e^{-j2\pi f_j \tau_{kM}}]^T$ represents the frequency-dependent array steering vectors for $k-$ **th** source angles.

The data covariance matrix can be written as follows:

$$\begin{aligned} R(f_j) &= E[X(f_j)X^H(f_j)] \\ &= A(f_j)Rss(f_j)A^H(f_j) + \sigma^2 I \end{aligned} \quad (5)$$

where $Rss(f_j) = E[S(f_j)S^H(f_j)]$ is the covariance matrix of the source signals at frequency f_j.

3. Proposed Method

3.1. Multi-Frequency Focused Decorrelation Method

The traditional CSSM algorithm was used to focus information derived from different frequency bins on a reference frequency bin, and narrowband algorithms were then used to obtain DOA estimation values. The key purpose of this algorithm was to construct a focusing matrix $T(f_j)$ to transform the array manifold at frequency f_j to the reference frequency f_0 as follows:

$$T(f_j)A(f_j) = A(f_0) \tag{6}$$

where $T(f_j)$ is the focusing matrix at frequency f_j, and it is a unitary matrix.

The data covariance matrix was generated as follows:

$$R_c = \sum_{j=1}^{J} T(f_j)R(f_j)T^H(f_j) \tag{7}$$

The focusing matrices at different frequencies are non-unique, and the methods used to construct this matrix have been extensively explored. The rotational signal subspace (RSS) algorithm is now the approach most widely used to construct unitary focusing matrices by minimizing the Frobenius norm of the array manifold mismatches.

However, for most existing methods of constructing focus matrices, initial estimations were critical, and poor initial estimations could have a significant impact on the final DOA estimations.

In order to avoid excessive dependence on initial estimations for the final accuracy, we adopted a focusing algorithm that did not require pre-estimation. It can be observed that the unitary matrix with columns that are the eigenvectors of the data covariance matrix spanned the same subspace across each frequency as the array manifold, meaning that we can use them to construct the focusing matrix. This method does not require initial estimations compared to the approach of building a focusing matrix based on the array steering vector.

Firstly, we define the focusing matrix as follows:

$$T_{auto}(f_j) = \frac{1}{\sqrt{J}}U(f_0)U^H(f_j) \tag{8}$$

where $U(f_j)$ is a unitary matrix, and its columns are the eigenvectors of the covariance matrix $R(f_j)$.

In practical cases, we multiply $T_{auto}(f_j)$ and $X(f_j)$, and we then find

$$Y(f_j) = T_{auto}(f_j)X(f_j) \tag{9}$$

The covariance matrix can be presented as follows:

$$\begin{aligned} R_y(f_j) &= E[Y(f_j)Y^H(f_j)] \\ &= T_{auto}(f_j)A(f_j)Rss(f_j)A^H(f_j){T_{auto}}^H(f_j) + \tfrac{1}{J}\sigma^2 I \end{aligned} \tag{10}$$

Summing up the matrices at J frequencies, we can write the covariance matrix as follows:

$$\begin{aligned} R_{coh} &= \sum_{j=0}^{J-1} R_y(f_j) \\ &= \tfrac{1}{J}U(f_0)\left(\sum_{j=0}^{J-1} U^H(f_j)A(f_j)Rss(f_j)A^H(f_j)U(f_j)\right)U^H(f_0) \\ &+\sigma^2 I \end{aligned} \tag{11}$$

By constructing a focusing matrix to focus all frequency components to the reference frequency and then averaging the covariance matrix, the correlation coefficients between the signals were reduced, meaning that the rank of the covariance matrix was equal to the number of sources used for decorrelation.

Finally, we constructed the focusing matrix without the need for initial estimations, focused the covariance matrices at different frequencies to the reference frequency f_0, and, ultimately, obtained the required covariance matrix.

Since the array we used was a two-level nested array, it needed to be virtualized before it could be used to perform DOA estimation. Thus, we vectorized the above covariance matrix and found

$$z = vec(R_{coh}) \tag{12}$$

We sorted z and removed redundancy to obtain $\tilde{z}$, which is the received signal of this virtual array. The number of continuous elements in the virtual array is $N = 2M_1(M_2 + 1) - 1$.

3.2. Enhanced Spatial Smoothing Method

As the received signal is coherent, subspace-based and propagator-based algorithms cannot be used directly. And one of the most famous decorrelation algorithms is the spatial smoothing algorithm. The entire array is divided into P subarrays, and L represents the number of array elements in each subarray, the subarray is shown in Figure 2, where dotted line represents subarrays not shown in the figure. The relationships between P, L and N can be expressed as follows:

$$N = L + P - 1 \tag{13}$$

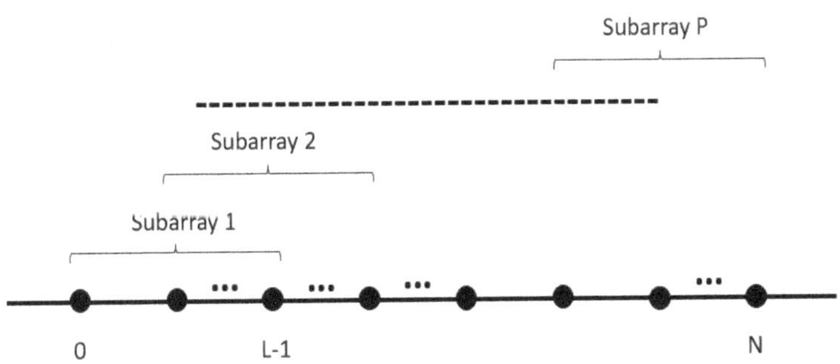

Figure 2. Overlapping subarrays used in the spatial smoothing method.

Therefore, the cross-covariance matrix R_{ij} of the $i-$ **th** subarray and the $j-$ **th** subarray can be written as follows:

$$R_{ij} = E[\tilde{z}_i \tilde{z}_j^H] \qquad i, j \in (1, 2, \cdots P) \tag{14}$$

where $\tilde{z}_p$ represents the $p-$ **th** L row of $\tilde{z}$.

And the backward cross-covariance matrix can be expressed as follows:

$$\overline{R_{ij}} = JR_{ij}^{*}J \tag{15}$$

where J denotes the $(L \times L)$ exchange matrix, and the operator $*$ represents the complex conjugate.

One of the drawbacks of many spatial smoothing methods is that they do not make full use of the information in the subspace, while the spatial smoothing method used in this paper utilizes information including the cross-covariance matrix R_{ij} of different subarrays and the covariance matrix R_{ii}/R_{jj} of a single subarray. The rank-restored data covariance matrix after enhanced spatial smoothing can be written as follows:

$$\begin{aligned} R_{ESS} &= \frac{1}{2P}\sum_{i=1}^{P}\sum_{j=1}^{P}\left\{\left(R_{ij}R_{ji}+\overline{R_{ij}R_{ji}}\right)\right. \\ &\quad \left.+\left(R_{ii}R_{jj}+\overline{R_{ii}R_{jj}}\right)\right\} \end{aligned} \tag{16}$$

Furthermore, we apply the MUSIC algorithm to the covariance matrix R_{ESS}, and the covariance matrix R_{ESS} can be divided into the signal subspace and the noise subspace, which can be written as follows:

$$R_{ESS} = U_s\Sigma_s U_s^H + U_N\Sigma_N U_N^H \tag{17}$$

where Σ_s is the diagonal matrix consisting of the maximum eigenvalues sorted in descending order, which have the same numbers as the sources. Σ_N is the diagonal matrix composed of the other eigenvalues. U_s and U_N are matrices with the eigenvectors that correspond to the eigenvalues as columns.

The MUSIC spectrum is generated as follows:

$$P_{\text{MUSIC}}(\theta) = \frac{1}{a_1^H U_N U_N^H a_1} \tag{18}$$

where $a_1 = \left[0, e^{-j2\pi f_0 d\sin\theta/c}, e^{-j2\pi f_0(2d)\sin\theta/c}, \ldots, e^{-j2\pi f_0(L-1)d\sin\theta/c}\right]^T$ is the subarray steering vector at the reference frequency f_0.

Then, DOA estimation values are obtained through a spectrum search. The main steps of the proposed method are shown in Algorithm 1.

Algorithm 1: The Proposed Method

Input: The received data $X(f_j)$
Output: DOA estimation values

1. Construct the focus matrix $T_{auto}(f_j)$ according to Equation (8);
2. Obtain the covariance matrix R_{coh} after focusing according to Equations (10) and (11);
3. Vectorize the above covariance matrix R_{coh} and obtain the received signal $\widetilde{z}$ of the virtual array according to Equation (12);
4. Apply an enhanced spatial smoothing method to obtain the rank-restored covariance matrix R_{ESS} according to Equations (14)–(16);
5. Apply the MUSIC algorithm to obtain the estimation results according to Equations (17) and (18)

4. Performance Analysis

4.1. Simulation Analysis

To prove the superiority of the proposed method, we compared it to several other methods. The four methods used are as follows: (i) Using the method in [15], which uses conventional ISSM algorithm, decorrelation was performed via the traditional spatial smoothing algorithm (named ISSM-ss); (ii) after constructing the focus matrix using the traditional CSSM algorithm, we used the enhanced spatial smoothing algorithm for decorrelation (named CSSM-enss); (iii) we then used the focusing algorithm that does not require the initial estimation to generate the focus matrix (named IEF-CSSM); and (iv) the algorithm proposed in this article was used.

In this section, the root mean square error (RMSE) was used to verify algorithm performance. The expression of RMSE is as follows:

$$\text{RMSE} = \sqrt{\frac{1}{KN}\left(\sum_{n=1}^{N}\sum_{k=1}^{K}\left(\theta_k - \hat{\theta}_{k,n}\right)^2\right)} \tag{19}$$

where K and N denote the numbers of sources and Monte Carlo trials, respectively. $\hat{\theta}_{k,n}$ represents the estimated values of $n-$ **th** Monte Carlo trial. In the following simulations, the number of Monte Carlo trials is 500.

We design two wideband sources impinging on a two-level nested array from 20° and 45°, and the number of sensors used for both subarrays is 10. The more frequency bins there are present, the slower the processing speed will be, and the accuracy of the results will raise. When the number of frequency bins reaches a certain threshold, the change in result accuracy will become negligible or even non-existent. As a result, between 280 and 320 MHz, a total of 40 frequency bins is uniformly sampled for wideband DOA estimation.

As shown in Figure 3, all four methods perform decorrelation and have spectral peaks present around the angle of the two sources, but the heights of the spectrum peaks are different due to the varying performances of the decorrelation methods. The method with the lowest peak is the IEF-CSSM method, which proves that the CSSM method is indeed helpful in decorrelation, while the method with a lower peak is the ISSM-ss method. There are some deviations in the positions of the spectral peaks due to the use of the ISSM method and the common spatial smoothing algorithm, which is not as effective as the other two methods using the enhanced spatial smoothing algorithm. Moreover, the CSSM-enss method results in the final spectral peak height being lower than that of the method proposed in this paper due to the difference in the focusing algorithm.

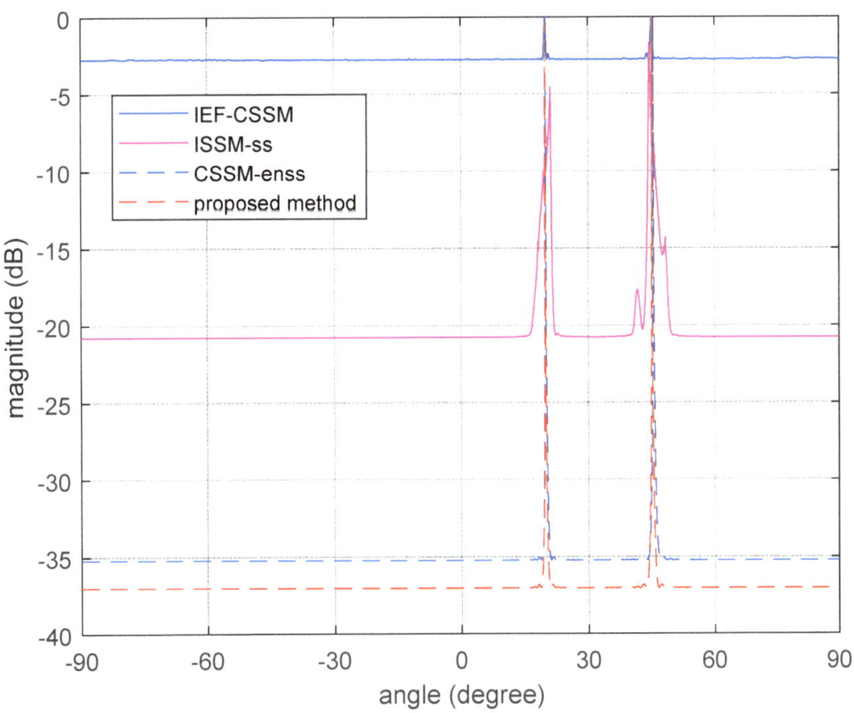

Figure 3. Comparison between the spatial spectra of different algorithms.

As shown in Figure 4, the RMSE of each method becomes smaller as the SNR gradually increases. Among the methods, the CSSM-enss method suffers from a definite impact on the final estimates because the method used to construct the focus matrix is the conventional CSSM algorithm that requires an initial estimation value, though CSSM-enss has a better performance than ISSM-ss in its decorrelation algorithm, meaning that the performance of the CSSM-enss method is slightly better than that of the ISSM-ss method. However, the IEF-CSSM method is slightly more accurate because it uses an algorithm that does not require initial estimation values, but the final performance is weaker than that of the proposed method because of the poor decorrelation method.

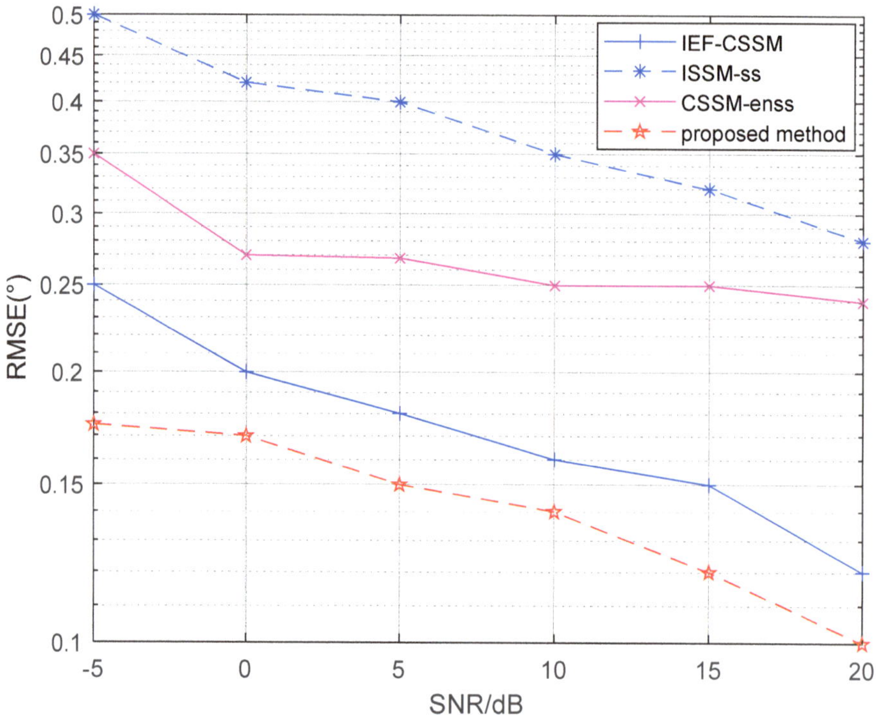

Figure 4. RMSE versus SNR.

4.2. Experimental Results

To demonstrate the effectiveness of the proposed method, we conducted a set of experiments inside of the room, in which a wideband signal with a bandwidth of 10 Mhz impinged on the sparse array from 12°. A total of 32 frequency bins were used for wideband DOA estimation, the experimental equipment and scene are shown in Figures 5 and 6, respectively. As Figure 7 shows, the performance of the IEF-CSSM method in decorrelation is weaker than those of other methods in practical testing. On the other hand, the CSSM-enss method performs better than the ISSM-ss method but is worse than the proposed method. Furthermore, in Table 1, after the processing of the proposed method, there is a peak at 12.20° in the spectrum, while the real angle is 12°, which shows that the proposed method has satisfactory accuracy in practical application. And in order to ensure that the signal is coherent, we chose to perform experiments in an indoor environment, and the final processing results prove that the proposed method has the effect of decorrelation.

Table 1. Accuracy of the four methods for real data.

Method	DOA Estimation	Error Value
IEF-CSSM	8.25°	3.75°
ISSM-ss	12.80°	0.80°
CSSM-enss	12.70°	0.70°
Proposed method	12.20°	0.20°

(a) (b)

Figure 5. (**a**) is the sparse array used in the experiment, which used a tripod and a spirit level to ensure the stability of the support frame used for observation. (**b**) is the signal generator used in the experiment.

Figure 6. Experiment scene designed to satisfy the conditions required for coherent signals; we performed experiments inside of the room.

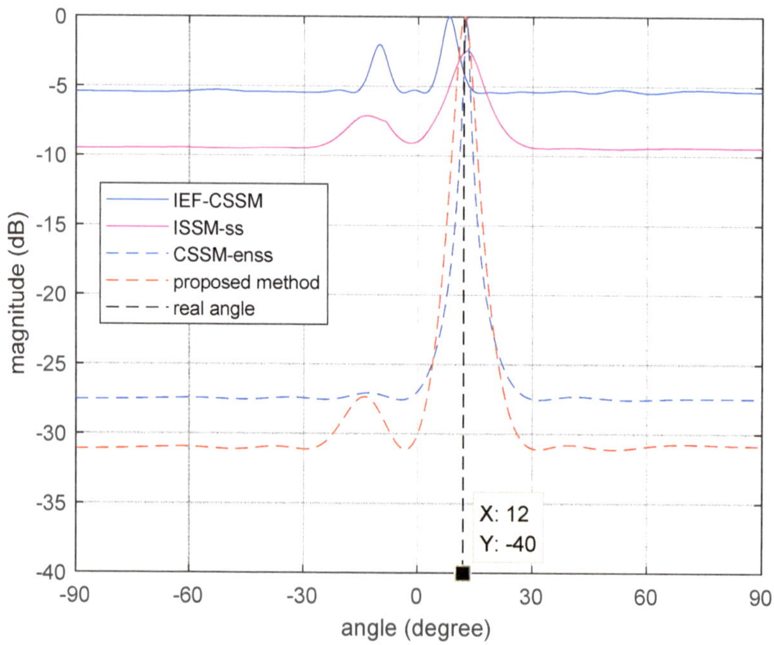

Figure 7. Spectrum of the proposed method for real data.

5. Conclusions

In this paper, we propose a DOA estimation algorithm for wideband coherent signals using the nested array. Firstly, a two-level nested array is used to ensure large DOF in the virtual array. Then, the focusing matrix is constructed without initial estimation to ensure that the initial estimation's influence on the final estimation in the conventional algorithm is avoided. Furthermore, an enhanced spatial smoothing algorithm is used for decorrelation. Finally, the MUSIC algorithm is applied to perform high-accuracy DOA estimation. The proposed method has better decorrelation and higher accuracy than the ISMM-ss, CSMM-enss and IEF-CSSM methods. In our experimental tests, the proposed method is effective for coherent wideband signal processing, even in indoor environments, and we have demonstrated the advantage of the proposed method in relation to accuracy compared to other methods. In the future, we will investigate low-rank matrix recovery and denoising using the obtained covariance matrix before angle estimation, which could help to reduce residual interference on the covariance matrix.

Author Contributions: Conceptualization, J.L.; methodology, Y.T.; software, Y.T. and W.D.; investigation, W.D.; data curation, Y.T.; writing—original draft preparation, Y.T.; writing—review and editing, J.L. and X.Z. All authors have read and agreed to the published version of the manuscript.

Funding: This work was supported in part by the National Key R&D Program of China under Grant 2020YFB1807602 and under Grant 2020YFB1807604, in part by the China University Innovation Fund for Production, education and research under Grant 2021ZYA0301, in part by the Jiangsu Planned Projects for Postdoctoral Research Funds under Grant 2020Z013, in part by the Pre-research project of SongShan Laboratory under Grant YYJC022022018, and in part by the China Postdoctoral Science Foundation under Grant 2020M681585 and under Grant 2023T160312.

Institutional Review Board Statement: Not applicable.

Informed Consent Statement: Not applicable.

Data Availability Statement: Not applicable.

Conflicts of Interest: The authors declare no conflict of interest.

References

1. Tuncer, T.E.; Friedlander, B. *Classical and Modern Direction-of-Arrival Estimation*; Academic Press: Cambridge, MA, USA, 2009.
2. Xiang, H.H.; Chen, B.X.; Yang, M.L.; Xu, S.Q. Angle Separation Learning for Coherent DOA Estimation with Deep Sparse Prior. *IEEE Commun. Lett.* **2021**, *25*, 465–469. [CrossRef]
3. Jiang, H.; Li, L.; Zhang, H.; Pan, H.; Li, X. Stepwise DOA Estimation for Coherent and Uncorrelated Mixed Signals on Coprime Array. In Proceedings of the 2022 IEEE 5th Advanced Information Management, Communicates, Electronic and Automation Control Conference (IMCEC), Chongqing, China, 16–18 December 2022; pp. 1685–1688. [CrossRef]
4. Li, J.F.; Li, P.; Li, P.; Tang, L.M.; Zhang, X.F.; Wu, Q.H. Self-Position Awareness Based on Cascade Direct Localization Over Multiple Source Data. *IEEE Trans. Intell. Transp. Syst.* **2022**, *9*. [CrossRef]
5. Garcia, N.; Wymeersch, H.; Larsson, E.G.; Haimovich, A.M.; Coulon, M. Direct Localization for Massive MIMO. *IEEE Trans. Signal Process.* **2017**, *65*, 2475–2487. [CrossRef]
6. Khabbazibasmenj, A.; Hassanien, A.; Vorobyov, S.A.; Morency, M.W. Efficient Transmit Beamspace Design for Search-Free Based DOA Estimation in MIMO Radar. *IEEE Trans. Signal Process.* **2014**, *62*, 1490–1500. [CrossRef]
7. Choi, S.; Kim, B.; Kim, J.; Kim, D.; Cho, H. Doppler Coherent Focusing DOA Method for Efficient Radar Map Generation. In Proceedings of the 5th IEEE MTT-S International Conference on Microwaves for Intelligent Mobility (ICMIM), Detroit, MI, USA, 15–16 April 2019; pp. 1–4.
8. Millhiser, J.; Sarangi, P.; Pal, P. Initialization-Free Implicit-Focusing (IF2) for Wideband Direction-of-Arrival Estimation. In Proceedings of the 47th IEEE International Conference on Acoustics, Speech and Signal Processing (ICASSP), Singapore, 22–27 May 2022; pp. 4973–4977.
9. Zhang, J.; Bao, M.; Zhang, X.P.; Chen, Z.F.; Yang, J.H. DOA Estimation for Heterogeneous Wideband Sources Based on Adaptive Space-Frequency Joint Processing. *IEEE Trans. Signal Process.* **2022**, *70*, 1657–1672. [CrossRef]
10. Yoon, Y.S.; Kaplan, L.M.; McClellan, J.H. TOPS: New DOA estimator for wideband signals. *IEEE Trans. Signal Process.* **2006**, *54*, 1977–1989. [CrossRef]
11. Yuanyuan, J.; Dan, L.; Xiaohuan, W.; Wei-Ping, Z. A Gridless Wideband DOA Estimation Based On Atomic Norm Minimization. In Proceedings of the 2020 IEEE 11th Sensor Array and Multichannel Signal Processing Workshop (SAM), Hangzhou, China, 8–11 June 2020. [CrossRef]
12. Kulhandjian, H.; Kulhandjian, M.; Kim, Y.; D'Amours, C. 2-D DOA Estimation of Coherent Wideband Signals with Auxiliary-Vector Basis. In Proceedings of the IEEE International Conference on Communications (ICC), Kansas City, MO, USA, 20–24 May 2018.
13. Zoltowski, M.D.; Silverstein, S.D.; Mathews, C.P. Beamspace Root-Music for Minimum Redundancy Linear Arrays. *IEEE Trans. Signal Process.* **1993**, *41*, 2502–2507. [CrossRef]
14. Li, J.F.; Li, D.; Jiang, D.F.; Zhang, X.F. Extended-Aperture Unitary Root MUSIC-Based DOA Estimation for Coprime Array. *IEEE Commun. Lett.* **2018**, *22*, 752–755. [CrossRef]
15. Peng, Z.; Ding, Y.T.; Ren, S.W.; Wu, H.X.; Wang, W.J. Coprime Nested Arrays for DOA Estimation: Exploiting the Nesting Property of Coprime Array. *IEEE Signal Process. Lett.* **2022**, *29*, 444–448. [CrossRef]
16. Qin, Y.H.; Liu, Y.M.; Liu, J.Y.; Yu, Z.Y. Underdetermined Wideband DOA Estimation for Off-Grid Sources with Coprime Array Using Sparse Bayesian Learning. *Sensors* **2018**, *18*, 11. [CrossRef]
17. Shi, J.P.; Hu, G.P.; Zhang, X.F.; Zhou, H. Generalized Nested Array: Optimization for Degrees of Freedom and Mutual Coupling. *IEEE Commun. Lett.* **2018**, *22*, 1208–1211. [CrossRef]
18. Ren, S.W.; Dong, W.T.; Li, X.N.; Wang, W.J.; Li, X.R. Extended Nested Arrays for Consecutive Virtual Aperture Enhancement. *IEEE Signal Process. Lett.* **2020**, *27*, 575–579. [CrossRef]
19. Shan, T.J.; Wax, M.; Kailath, T. On spatial smoothing for direction-of-arrival estimation of coherent signals. *IEEE Trans. Acoust. Speech Signal Process.* **1985**, *ASSP-33*, 806–811. [CrossRef]
20. Ziskind, I.; Wax, M. Maximum likelihood localization of multiple sources by alternating projection. *IEEE Trans. Acoust. Speech Signal Process.* **1988**, *36*, 1553–1560. [CrossRef]
21. Zoltowski, M.; Haber, F. A vector space approach to direction finding in a coherent multipath environment. *IEEE Trans. Antennas Propag.* **1986**, *AP-34*, 1069–1079. [CrossRef]
22. Pillai, S.U.; Kwon, B.H. Forward/backward spatial smoothing techniques for coherent signal identification. *IEEE Trans. Acoust. Speech Signal Process.* **1989**, *37*, 8–15. [CrossRef]
23. Pham, G.T.; Loubaton, P.; Vallet, P. Performance Analysis of Spatial Smoothing Schemes in the Context of Large Arrays. *IEEE Trans. Signal Process.* **2016**, *64*, 160–172. [CrossRef]
24. Qu, L.L.; Sun, Q.; Yang, T.H.; Zhang, L.L.; Sun, Y.P. Time-Delay Estimation for Ground Penetrating Radar Using ESPRIT With Improved Spatial Smoothing Technique. *IEEE Geosci. Remote Sens. Lett.* **2014**, *11*, 1315–1319. [CrossRef]
25. Jin, A.S.; Zhang, F. 2D DOA estimation of coherent sources based on reconstruction of Toeplitz matrix sets. In Proceedings of the Chinese Automation Congress (CAC), Shanghai, China, 6–8 November 2020; pp. 508–513.

26. Liu, X.Z.; Song, M.Y.; Yang, Y.H. An Effective DOA Estimation Method of Coherent Signals Based on Reconstruct Weighted Noise Subspace. In Proceedings of the 29th Chinese Control And Decision Conference (CCDC), Chongqing, China, 28–30 May 2017; pp. 2218–2222.
27. Evans, J.E.; Sun, D.F.; Johnson, J.R. *Application of Advanced Signal Processing Techniques to Angle of Arrival Estimation in ATC Navigation and Surveillance Systems*; Massachusetts Institute of Technology, Lincoln Laboratory: Cambridge, MA, USA, 1982.
28. Wax, M.; Shan, T.J.; Kailath, T. Spatio-temporal spectral analysis by eigenstructure methods. *IEEE Trans. Acoust. Speech Signal Process.* **1984**, *ASSP-32*, 817–827. [CrossRef]
29. Shi, J.; Zhang, Q.F.; Wang, Y. Wideband DOA Estimation based on A-Shaped Array. In Proceedings of the IEEE International Conference on Signal Processing, Communications and Computing (ICSPCC), Xiamen, China, 22–25 October 2017.
30. Yadav, A.K.; Santosh, S. Comparative Analysis of Wideband DOA Estimation Methods with Nested Array. In Proceedings of the 2016 IEEE International Conference on Recent Trends in Electronics, Information & Communication Technology (RTEICT), Bengaluru, India, 20–21 May 2016; pp. 1367–1370.
31. Schmidt, R.O. Multiple emitter location and signal parameter estimation. *IEEE Trans. Antennas Propag.* **1986**, *AP-34*, 276–280. [CrossRef]
32. Di Claudio, E.D.; Parisi, R. WAVES: Weighted average of signal subspaces for robust wideband direction finding. *IEEE Trans. Signal Process.* **2001**, *49*, 2179–2191. [CrossRef]
33. Pal, P.; Vaidyanathan, P.P. A Novel Autofocusing Approach for Estimating Directions-of-Arrival of Wideband Signals. In Proceedings of the 2009 43rd Asilomar Conference on Signals, Systems and Computers, Pacific Grove, CA, USA, 1–4 November 2009; pp. 1663–1667. [CrossRef]
34. Wang, H.; Kaveh, M. Coherent signal-subspace processing for the detection and estimation of angles of arrival of multiple wide-band sources. *IEEE Trans. Acoust. Speech Signal Process.* **1985**, *ASSP-33*, 823–831. [CrossRef]
35. Bo, L.K.; Xiong, J.Y.; Luo, L.Y. A Novel Wideband DOA Estimation Method Using Direction-Free Focusing Matrix. In Proceedings of the 3rd International Conference on Computer Science and Network Technology (ICCSNT), Dalian, China, 12–13 October 2013; pp. 1065–1069.
36. Hung, H.; Kaveh, M. Focussing matrices for coherent signal-subspace processing. *IEEE Trans. Acoust. Speech Signal Process.* **1988**, *36*, 1272–1281. [CrossRef]
37. Valaee, S.; Kabal, P. The optimal focusing subspace for coherent signal subspace processing. *IEEE Trans. Signal Process.* **1996**, *44*, 752–756. [CrossRef]

sensors

Article

Low-Complexity Joint Angle of Arrival and Time of Arrival Estimation of Multipath Signal in UWB System

Weiming Deng, Jianfeng Li *, Yawei Tang and Xiaofei Zhang

College of Electronic Information Engineering, Nanjing University of Aeronautics and Astronautics, Nanjing 211106, China; dengweiming@nuaa.edu.cn (W.D.); ywtang@nuaa.edu.cn (Y.T.); zhangxiaofei@nuaa.edu.cn (X.Z.)

* Correspondence: lijianfeng@nuaa.edu.cn

Abstract: In an ultra-wideband (UWB) system, the two-dimensional (2D) multiple signal classification (MUSIC) algorithms based on high-precision 2D spectral peak search can jointly estimate the time of arrival (TOA) and angle of arrival (AOA). However, the computational complexity of 2D-MUSIC is very high, and the corresponding data model is only based on the dual antennas. To solve these problems, a low-complexity algorithm for joint AOA and TOA estimation of the multipath ultra-wideband signal is proposed. Firstly, the dual antenna sensing data model is extended to the antenna array case. Then, based on the array-sensing data model, the proposed algorithm transforms the 2D spectral peak search of 2D MUSIC into a secondary optimization problem to extract the estimation of AOA via only 1D search. Finally, the acquired AOA estimations are brought back, and the TOA estimations are also obtained through a 1D search. Moreover, in the case of an unknown transmitted signal waveform, the proposed method can still distinguish the main path signal based on the time difference of arrival of different paths, which shows wider applications. The simulation results show that the proposed algorithm outperforms the Root-MUSIC algorithm and the estimation of signal parameters using the rotational invariance techniques (ESPRIT) algorithm, and keeps the same estimation accuracy but with greatly reduced computational complexity compared to the 2D-MUSIC algorithm.

Keywords: multipath environment; angle of arrival (AOA); time of arrival (TOA); multiple signal classification (MUSIC)

1. Introduction

Angle of arrival (AOA) and time of arrival (TOA) are important issues in array signal processing, which is widely used in radar, navigation, indoor positioning, and wireless communication [1–3]. In the outdoor environment, the general positioning accuracy of the global positioning system (GPS) is more than 10 m without aids. GPS fails to help us in areas such as close-range complex following systems or indoor positioning. However, there are many applications in these complex environments, such as daily physical activity tracking [4], intelligent transportation [5], and localization of sensor nodes with a nano unmanned Aerial Vehicle [6]. In these close-range and complex scenarios, accurate estimation of AOA and TOA is important for positioning. AOA and TOA mean to measure the angle of arrival and time of arrival between the base station and the mobile station, respectively.

Ultra-wideband (UWB) signals [7] can solve the multipath problems to be faced in these scenarios. UWB signals are widely used in low-power short-range wireless communication [8] due to their large bandwidth, low power consumption, and strong ability to distinguish multipath signals. The Saleh-Valenzuela (S-V) model [9–11] has been adopted as a UWB system channel model for small-scale fading. In the S-V model, the multipath components are modeled as several rays arriving within different clusters. The multipath components of different clusters have constant AOA and TOA, but the fading is randomly varying. A series of algorithms are thus proposed to estimate AOA and TOA using the properties of the S-V channel model.

Earlier, only TOA in the UWB system was estimated, which was divided into time-domain methods, frequency-domain methods, and deep learning methods. Ref. [12] proposes a TOA estimation method based on matched filter threshold detection. Refs. [13,14] propose TOA estimation methods for incoherent energy detection based on a low sampling rate. A two-step TOA estimation method based on energy coherent detection of the direct path component is proposed in ref. [15]. The accuracy of the time-domain methods [12–15] is limited by the bandwidth and sampling interval. To meet the requirement of high resolution, TOA algorithms from the frequency domain [16–18] have been explored. Refs. [16–18] use a spectral peak search leading to an increase in computational complexity. Refs. [19–21] propose deep learning methods for TOA estimation. Ref. [19] proposes a machine learning method based on kernel principal component analysis, which projects selected channel parameters into a nonlinear orthogonal high-dimensional space and then uses a subset of these projections to estimate TOA. A deep learning model consisting of cascaded convolutional neural networks is proposed in ref. [20] to estimate TOA. Ref. [21] uses a deep learning model of ResNet50 [22] to estimate TOA. The deep learning methods reduce the computational complexity, but it is influenced by the data in the training set, which limits the applicability.

The AOA estimation in UWB systems has also been studied. Due to the high time resolution of ultra-wideband systems, a number of joint AOA and TOA estimation algorithms have been proposed. In [23], TOA is firstly estimated and then the best linear unbiased estimation is performed to obtain AOA. Ref. [24] improves the accuracy of TOA estimation by performing rough estimation and refining estimation, and the improvement of TOA estimation leads to the improvement of AOA estimation. In [25], a high-precision TOA estimation method is proposed based on two-dimensional multiple signal classification (2D-MUSIC), and the AOA can be obtained based on the time difference of adjacent array elements. Meanwhile, a Root-MUSIC algorithm is proposed to reduce the computational complexity in [25]. In [26], the computational complexity is further reduced compared with [25], which requires a spatial-spectral search to estimate the TOA. Ref. [26] estimates the TOA of two antennas using a polynomial root method, but the estimation accuracy is slightly reduced. Ref. [27] uses discrete Fourier transform (DFT) processing to obtain a coarse estimate of TOA, and designs a compensation matrix to phase compensate the time delay vector to obtain a fine estimate of TOA. The computational complexity of ref. [27] is greatly reduced because there is no need for eigenvalue decomposition, but the accuracy of TOA and AOA estimation is seriously affected by the number of DFT points. Ref. [28] exploits the conjugate symmetry of the time delay matrix to extend the sample points as well as the number of clusters to improve the estimation accuracy and the number of sources that can be identified. However, its computational complexity increases. In [29], the AOA is firstly estimated by frequency domain waveforms, and the TOA is estimated from AOA. Unlike [23–28], its AOA accuracy is not affected by the TOA estimation accuracy, but its TOA estimation accuracy is unsatisfied. Refs. [23–29] only consider dual antennas case, which does not consider the case of multiple antennas. In [30], the joint estimation of AOA and TOA is extended to three-dimensional space positioning using an iterative approach with high computational complexity, which is only applicable to the case of a single snapshot.

Although there are some works, e.g., refs. [23–29], considering the joint estimation of AOA and TOA in UWB system, only the data model of dual antennas is considered, which is not a general case. Meanwhile, the estimation accuracy and real-time performance of the methods in refs. [23–30] need to be improved. Moreover, in case of unknown prior information of the transmitted signal waveform, they are unable to distinguish the main path signal. To solve these problems, a low-complexity algorithm for joint AOA and TOA estimation of multipath signal in the UWB system is proposed. Firstly, the proposed algorithm transforms the 2D spectral search operation into a secondary optimization problem, and the AOA estimation is extracted separately. Then, the acquired AOA is brought into the 2D function, and the TOA estimation is also obtained through a 1D search. The main advantages of the proposed algorithm are as follows: (1) The proposed method has enhanced accuracy compared to the Root-MUSIC algorithm and ESPRIT algorithm, and maintains the same accuracy as the 2D-MUSIC algorithm; (2) Compared with the 2D-MUSIC, Root-MUSIC and EPRIT algorithm, the proposed algorithm requires lower complexity; (3) In case of unknown prior information of the transmitted signal, the proposed method can still distinguish the main path signal based on the time difference of arrival (TDOA) of a different path, which shows wider applications.

This paper is structured as follows: Section 2 introduces the received data model of the reference antenna and the proposed array-sensing data model. Section 3 introduces the traditional joint AOA and TOA estimation algorithms, including the 2D-MUSIC algorithm and the Root-MUSIC algorithm, respectively. Section 4 presents the proposed low-complexity joint AOA and TOA estimation algorithm. Section 5 shows the computational complexity and estimation accuracy of different algorithms, while conclusions are made in Section 6.

Notation: This article uses bold letters to denote vectors or matrices. $()^T, ()^*, ()^H, ()^{-1}$ represent transpose, conjugate, conjugate transpose, and inverse of a matrix, respectively. $\mathbf{I}_N$ represents the $N \times N$ identity matrix, and $E(\cdot)$ denotes expectation. $\wedge$ denotes an estimated expression and ./ means point division of matrices. ∂ represents the partial derivative and $|\cdot|$ represents the absolute value. Finally, $diag(\boldsymbol{a})$ denotes the diagonal matrix with the elements of vector $\boldsymbol{a}$ as diagonal elements.

2. Data Model

2.1. Single-Antenna Sensing Data Model

Figure 1 shows the scenario of multipath propagation and antenna array structure for joint AOA and TOA estimation. The antenna array is a uniform linear array (ULA), and the number of array elements is M. The array element spacing is $d = \lambda/2$, and λ is the wavelength corresponding to the center frequency of the incident signal. Assume that the incident signal propagating through the channel produces L multipath. The form of the signal received by the single antenna is denoted as

$$y(t) = \sum_{l=1}^{L} \beta_l s(t - \tau_l) + w(\mathrm{t}) \tag{1}$$

where $s(t)$ is the transmitted signal, τ_l is the time delay value corresponding to each path, β_l is the attenuation coefficient corresponding to each path, and $w(\mathrm{t})$ is the zero-mean additive Gaussian white noise.

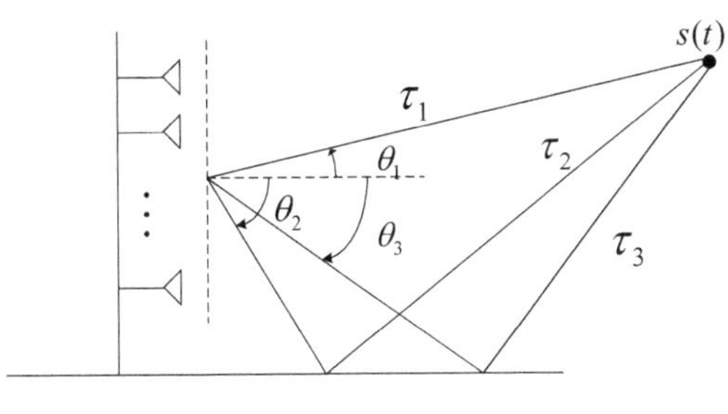

Figure 1. ULA structure for joint TOA and DOA estimation.

Then, we transform the time-domain form of the signal into the frequency-domain form as

$$Y(\omega) = \sum_{l=1}^{L} \beta_l S(\omega) e^{-j\omega\tau_l} + W(\omega) \tag{2}$$

Sampling the received signal in the frequency-domain at equal intervals of N $(N > L)$ points, with a sampling interval of $\Delta\omega = 2\pi/N$, the discrete frequency-domain form of the sampled received signal can be obtained as follows:

$$Y(\omega_n) = \sum_{l=1}^{L} \beta_l S(\omega_n) e^{-j\omega_n \tau_l} + W(\omega_n) \tag{3}$$

where $\omega_n = n\Delta\omega, n = 0, 1, \cdots, N-1$. Dividing the data into K segments (i.e., number of frequency-domain snapshots), the kth segment frequency-domain data can be written in the following concise vector form:

$$\mathbf{y}_k = \mathbf{S}\mathbf{E}_{\tau 1}\boldsymbol{\beta}_k + \mathbf{w}_k \tag{4}$$

where $\mathbf{y}_k = [Y^{(k)}(\omega_0), \cdots, Y^{(k)}(\omega_{N-1})]^T \in \mathbb{C}^{N\times 1}$ is the N-point frequency-domain equal interval sampling of the received signal in the kth segment. $\mathbf{S} = diag([S(\omega_0), \cdots, S(\omega_{N-1})])$ $\in \mathbb{C}^{N\times N}$ is the diagonal matrix, and the diagonal elements are the N-point frequency-domain equal interval sampling values of the transmitted signal $s(t)$. $\mathbf{w}_k = [W^{(k)}(\omega_0), \cdots,$ $W^{(k)}(\omega_{N-1})]^T \in \mathbb{C}^{N\times 1}$ is the frequency-domain sampling vector of Gaussian white noise. $\mathbf{E}_{\tau 1} \in \mathbb{C}^{N\times L}$ is the matrix containing the signal multipath time delay information. We refer to it as the time delay matrix. It can be expressed as $\mathbf{E}_{\tau 1} = [\mathbf{e}_{\tau_1}, \cdots, \mathbf{e}_{\tau_l}, \cdots, \mathbf{e}_{\tau_L}]$, where $\mathbf{e}_{\tau_l} = [e^{-j\omega_0\tau_l}, e^{-j\omega_1\tau_l}, \cdots, e^{-j\omega_{N-1}\tau_l}]^T$. The coefficients of the fading channel are contained in the vector which can be expressed as $\boldsymbol{\beta}_k = [\beta_1^{(k)}, \beta_2^{(k)}, \cdots, \beta_L^{(k)}]^T$.

According to the S-V channel model, we can obtain the complex-value fading coefficient of the kth segment of the lth path as

$$\beta_l^{(k)} = \alpha_l^{(k)} e^{j\phi_l^{(k)}} \tag{5}$$

where $\alpha_l^{(k)}$ is the fading amplitude of the kth segment of the lth path, and $\phi_l^{(k)}$ is an arbitrary phase variable representing a uniform distribution in the range $[0, 2\pi]$. Through the S-V channel model, we can learn that $\beta_l^{(k)}$ is a random complex-value fading coefficient.

The current literature only considers the dual-antenna model, without considering the more generalized array antenna model. Here we propose the multiple array antenna elements receiving model.

2.2. Proposed Array Antenna Sensing Data Model

Previous methods have performed joint AOA and TOA estimation based on two receiving antennas, and have ignored the intrinsic connection of the delay matrix from each antenna. Here, we extend it to the case of multiple antennas. The frequency-domain received signal $\mathbf{Y}_i \in \mathbb{C}^{N\times K}(i = 1, \cdots, M)$ of each antenna can be obtained from the data model. It can be expressed as $\mathbf{Y}_i = \mathbf{S}\mathbf{E}_{\tau i}\mathbf{B} + \mathbf{W}_i$, where $\mathbf{B} = [\boldsymbol{\beta}_1, \cdots, \boldsymbol{\beta}_k, \cdots, \boldsymbol{\beta}_K] \in \mathbb{C}^{L\times K}$ represents the coefficient of the complex-value fading-channel, and $\mathbf{W}_i = [\mathbf{w}_{i1}, \cdots, \mathbf{w}_{ik}, \cdots, \mathbf{w}_{iK}] \in \mathbb{C}^{N\times K}$ denotes the frequency-domain sampling matrix of Gaussian white noise received by each antenna. $\mathbf{E}_{\tau i} \in \mathbb{C}^{N\times L}$ is the delay matrix of each antenna. The time delay matrix of the reference array element is expressed as

$$\mathbf{E}_{\tau 1} = \begin{bmatrix} e^{-j\omega_0\tau_1} & e^{-j\omega_0\tau_2} & \cdots & e^{-j\omega_0\tau_L} \\ e^{-j\omega_1\tau_1} & e^{-j\omega_1\tau_2} & \cdots & e^{-j\omega_1\tau_L} \\ \vdots & \vdots & \ddots & \vdots \\ e^{-j\omega_{N-1}\tau_1} & e^{-j\omega_{N-1}\tau_2} & \cdots & e^{-j\omega_{N-1}\tau_L} \end{bmatrix} \tag{6}$$

The columns of the time delay matrix $\mathbf{E}_{\tau i}$ of the ith antenna are related to the columns of the time delay matrix of the reference antenna as follows:

$$\mathbf{e}_{\tau il} = \boldsymbol{\Phi}^{i-1}(\theta_l)\mathbf{e}_{\tau 1l} \tag{7}$$

where $\mathbf{e}_{\tau il}$ is the lth column of $\mathbf{E}_{\tau i}$ and $\mathbf{\Phi}(\theta_l) = diag(e^{-j\omega_0 \frac{d\sin\theta_l}{c}}, \ldots, e^{-j\omega_{N-1}\frac{d\sin\theta_l}{c}})$.

Because the general sampling bandwidth is larger than the signal bandwidth, many elements of $[S(\omega_0), \ldots, S(\omega_{N-1})]$ are zero or close to zero. Assuming non-zero at these frequencies in $\omega_{c0}, \ldots, \omega_{c(P-1)}$, the received signal matrix $\mathbf{Z} \in \mathbb{C}^{MP\times K}$ of M antenna array elements can be constructed as:

$$\begin{aligned}\mathbf{Z} = \begin{bmatrix} \mathbf{Y}_1 \\ \mathbf{Y}_2 \\ \vdots \\ \mathbf{Y}_M \end{bmatrix} &= \underbrace{\begin{bmatrix} \mathbf{SF}_{\tau 1} \\ \mathbf{SF}_{\tau 2} \\ \vdots \\ \mathbf{SF}_{\tau M} \end{bmatrix}}_{\mathbf{A}(\tau,\theta)} \mathbf{B} + \underbrace{\begin{bmatrix} \mathbf{V}_1 \\ \mathbf{V}_2 \\ \vdots \\ \mathbf{V}_M \end{bmatrix}}_{\mathbf{V}} \\ &= \mathbf{A}(\tau,\theta)\mathbf{B} + \mathbf{V}\end{aligned} \tag{8}$$

where $\mathbf{S} = diag([S(\omega_{c0}), \cdots, S(\omega_{c(P-1)})])$. $\mathbf{F}_{\tau i}$ and $\mathbf{V}_i$ are composed of rows of the matrix $\mathbf{E}_{\tau i}$ and $\mathbf{W}_i$ extracted according to $\omega_{c0}, \ldots, \omega_{c(P-1)}$, where $\mathbf{F}_{\tau i} \in \mathbb{C}^{P\times L}$ and $\mathbf{V}_i \in \mathbb{C}^{P\times K}$. τ is the time delay of the multipath signal arriving at the reference array, and θ is the angle of incidence of the multipath signal arriving at the array.

3. Traditional Based Joint AOA and TOA Algorithm

3.1. 2D-MUSIC Method

The covariance matrix of the received signal matrix $\mathbf{Z}$ can be written as follows:

$$\hat{\mathbf{R}}_{ZZ} = \mathbf{Z}\mathbf{Z}^H/K \tag{9}$$

where $\hat{\mathbf{R}}_{ZZ} \in \mathbb{C}^{MP\times MP}$. The 2D-MUSIC [31] space spectral function is constructed as follows:

$$P_{2D-MUSIC}(\tau,\theta) = \frac{1}{\mathbf{a}(\tau,\theta)^H \hat{\mathbf{U}}_N \hat{\mathbf{U}}_N^H \mathbf{a}(\tau,\theta)} \tag{10}$$

where $\mathbf{a}(\tau,\theta)$ is the column vector in matrix $\mathbf{A}(\tau,\theta)$, and $\hat{\mathbf{U}}_N \in \mathbb{C}^{MP\times(MP-L)}$ represents the signal subspace composed of eigenvectors corresponding to the smaller $MP-L$ eigenvalues. Let τ and θ vary and perform a 2D spectral peak search on Equation (10). $\hat{\tau}$ and $\hat{\theta}$ corresponding to the extreme spectral peak are the joint AOA and TOA estimates, where the shortest TOA corresponds to the main path signal.

3.2. Root-MUSIC Method

The method is a joint AOA and TOA estimation algorithm proposed in ref. [25] for dual-antenna. According to the previous data model, the reception data model of the dual-antenna can be obtained

$$\begin{aligned}\mathbf{Y}_1 &= \mathbf{SE}_{\tau 1}\mathbf{B} + \mathbf{W}_1 \\ \mathbf{Y}_2 &= \mathbf{SE}_{\tau 2}\mathbf{B} + \mathbf{W}_2\end{aligned} \tag{11}$$

The received signal model of the dual antenna is

$$\mathbf{Z} = \begin{bmatrix} \mathbf{Y}_1 \\ \mathbf{Y}_2 \end{bmatrix} = \begin{bmatrix} \mathbf{SE}_{\tau 1} \\ \mathbf{SE}_{\tau 2} \end{bmatrix}\mathbf{B} + \begin{bmatrix} \mathbf{W}_1 \\ \mathbf{W}_2 \end{bmatrix} \tag{12}$$

where the TOA information contained in the time delay matrix $\mathbf{E}_{\tau 1}$ of the first array element is $\tau_l(l = 1, 2, \ldots, L)$. The TOA information contained in the time delay matrix $\mathbf{E}_{\tau 2}$ of the second array element is $\varsigma_l(l = 1, 2, \ldots, L)$.

The eigenvalue decomposition of the covariance matrix $\hat{\mathbf{R}}_1 = \mathbf{Y}_1\mathbf{Y}_1^H/K$ is performed to obtain the signal subspace and noise subspace.

$$\hat{\mathbf{R}}_1 = \hat{\mathbf{E}}_S\hat{\mathbf{\Lambda}}_S\hat{\mathbf{E}}_S^H + \hat{\mathbf{E}}_N\hat{\mathbf{\Lambda}}_N\hat{\mathbf{E}}_N^H \tag{13}$$

where $\hat{\mathbf{\Lambda}}_S$ is an $L \times L$ diagonal matrix whose diagonal elements contain L maximum eigenvalues. $\hat{\mathbf{\Lambda}}_N$ represents the diagonal matrix whose diagonal elements contain $N-L$ minimum eigenvalues. $\hat{\mathbf{E}}_S$ is a matrix consisting of the eigenvectors corresponding to the L largest eigenvalues of $\hat{\mathbf{R}}_1$. $\hat{\mathbf{E}}_N$ stands for the matrix composed of the remaining eigenvectors. Then, we construct the one-dimensional spectral peak function as follows

$$P_{MUSIC}(\tau) = \frac{1}{[\mathbf{Sb}(\tau)]^H \hat{\mathbf{E}}_N \hat{\mathbf{E}}_N^H [\mathbf{Sb}(\tau)]} \tag{14}$$

By utilizing the Root-MUSIC [32] idea, we use polynomial rooting instead of spectral peak search. Let $z = e^{-j\Delta\omega\tau}$, $\mathbf{p}(z) = \left[1, z, \ldots, z^{N-1}\right]$, and $\mathbf{p}(z)^H = \mathbf{p}(z^{-1})^T = \left[1, z^{-1}, \ldots, z^{-(N-1)}\right]$. Equation (14) is transformed into the following form

$$\hat{f}(z) = z^{N-1}[\mathbf{Sp}(z)]^H \mathbf{E}_N \mathbf{E}_N^H [\mathbf{Sp}(z)] \tag{15}$$

Since the polynomial coefficients are symmetric, the roots are in the form of complex conjugate pairs, and the L roots $\hat{z}_1, \hat{z}_2, \ldots, \hat{z}_L$ closest to the unit circle within the unit circle can be selected. Then the time delay of the signal received by the first antenna is obtained

$$\hat{\tau}_{l,\,0} = -angle(\hat{z}_l)/\Delta\omega,\ (l = 1,\ 2,\ \ldots,\ L) \tag{16}$$

Let $\Delta\hat{\tau}_l = \hat{\varsigma}_l - \hat{\tau}_l$, which is the time delay difference of the adjacent array elements of the lth multipath. From Figure 1, we can obtain $\Delta\hat{\tau}_l = \frac{d \sin\theta_l}{c}$. Exploiting the relationship between τ_l and ς_l, we can obtain $\tau_l - \frac{d}{c} \le \varsigma_l \le \tau_l + \frac{d}{c}$. We perform eigenvalue decomposition of the covariance matrix $\hat{\mathbf{R}} = \mathbf{Z}\mathbf{Z}^H/K$ to obtain the noise subspace $\hat{\mathbf{U}}_N \in {}^{2N\times(2N-L)}$. ς_l can be estimated using the following one-dimensional spectral peak search function

$$\hat{\varsigma}_l = \underset{\varsigma \in \left[\hat{\tau}_{l,0} - \frac{d}{c}, \hat{\tau}_{l,0} + \frac{d}{c}\right]}{\arg\max} \frac{1}{\mathbf{a}(\hat{\tau}_{l,0}, \varsigma)^H \hat{\mathbf{U}}_N \hat{\mathbf{U}}_N^H \mathbf{a}(\hat{\tau}_{l,0}, \varsigma)},\ l = 1,\ 2,\ \ldots,\ L \tag{17}$$

The TOA value of the second antenna can be obtained by the local search of Equation (17). Then τ_l can be estimated via Equation (18) by locally searching τ_l within $\tau_l \in [\hat{\tau}_{l,0} - \Delta\tau, \hat{\tau}_{l,0} + \Delta\tau]$, where is $\Delta\tau$ a small value.

$$\hat{\tau}_l = \underset{\tau \in \left[\hat{\tau}_{l,0} - \frac{d}{c}, \hat{\tau}_{l,0} + \frac{d}{c}\right]}{\arg\max} \frac{1}{\mathbf{a}(\tau, \hat{\varsigma}_l)^H \hat{\mathbf{U}}_N \hat{\mathbf{U}}_N^H \mathbf{a}(\tau, \hat{\varsigma}_l)}, l = 1, 2, \ldots, L \tag{18}$$

Then we can obtain an estimate of the angle of incidence $\hat{\theta}_l = \arcsin\left(\frac{\Delta\hat{\tau}_l c}{d}\right)$, $l = 1,\ 2,\ \ldots,\ L$. So far, we have implemented the Root-MUSIC algorithm for joint AOA and TOA estimation in UWB systems.

4. Proposed Algorithm

4.1. Theoretical Derivation

The denominator of Equation (10) can be rewritten as:

$$\begin{aligned} f(\tau,\theta) &= \mathbf{a}(\tau,\theta)^H \hat{\mathbf{U}}_N \hat{\mathbf{U}}_N^H \mathbf{a}(\tau,\theta) \\ &= [\mathbf{Se}_{\tau 1l}]^H \mathbf{Q}(\theta) [\mathbf{Se}_{\tau 1l}] \end{aligned} \tag{19}$$

$$\mathbf{Q}(\theta) = \begin{bmatrix} \mathbf{I} \\ \mathbf{\Phi}(\theta) \\ \vdots \\ \mathbf{\Phi}^{M-1}(\theta) \end{bmatrix}^H \mathbf{U}_N \mathbf{U}_N^H \begin{bmatrix} \mathbf{I} \\ \mathbf{\Phi}(\theta) \\ \vdots \\ \mathbf{\Phi}^{M-1}(\theta) \end{bmatrix} \tag{20}$$

Then, Equation (19) can be written as follows:

$$f(\tau,\theta) = |S(\omega_{c0})|^2 (\frac{e^{-j\omega_{c0}\tau}}{S^*(\omega_{c0})}[\mathbf{S}\mathbf{e}_{\tau 1l}]^H \mathbf{Q}(\theta) * [\mathbf{S}\mathbf{e}_{\tau 1l}] \frac{e^{j\omega_{c0}\tau}}{S(\omega_{c0})}) \tag{21}$$

where $S^*(\omega_{c0})$ represents the conjugate of $S(\omega_{c0})$. Since the constants do not affect the search of the spectral peak function, Equation (21) can be written as

$$f(\tau,\theta) = \frac{e^{-j\omega_{c0}\tau}}{S^*(\omega_{c0})}[\mathbf{S}\mathbf{e}_{\tau 1l}]^H \mathbf{Q}(\theta)[\mathbf{S}\mathbf{e}_{\tau 1l}] \frac{e^{j\omega_{c0}\tau}}{\mathbf{S}(\omega_{c0})} \tag{22}$$

Define $\mathbf{e}(\tau) = \mathbf{S}\mathbf{e}_{\tau 1l} \frac{e^{j\omega_{c0}\tau}}{S(\omega_{c0})} = \begin{bmatrix} 1 & \frac{S(\omega_{c1})}{S(\omega_{c0})} e^{-j\omega_1 \tau} & \dots & \frac{S(\omega_{c(P-1)})}{S(\omega_{c0})} e^{-j\omega_{(P-1)}\tau} \end{bmatrix}^T$, and rewrite Equation (22) as

$$f(\tau,\theta) = \mathbf{e}(\tau)^H \mathbf{Q}(\theta)\mathbf{e}(\tau) \tag{23}$$

Equation (23) is an objective function, and considering the use of $\mathbf{e}_1^H \mathbf{e}(\tau) = 1$ to eliminate the tame solution of $\mathbf{e}(\tau) = 0$, where $\mathbf{e}_1 = [1,0,\dots,0]^T \in \mathbb{R}^{P\times 1}$, Equation (23) can be reformulated as:

$$\min_{\theta,\tau} e(\tau)^H \mathbf{Q}(\theta) e(\tau) || s.t. e_1^H e(\tau) = 1 \tag{24}$$

The cost function is expressed as:

$$L(\theta,\tau) = \mathbf{e}(\tau)^H \mathbf{Q}(\theta)\mathbf{e}(\tau) - \lambda(\mathbf{e}_1^H \mathbf{e}(\tau) - 1) \tag{25}$$

where λ is a constant. Then, we give the derivative of $L(\theta,\tau)$

$$\frac{\partial}{\partial \mathbf{e}(\tau)} L(\theta,\tau) = 2\mathbf{Q}(\theta)\mathbf{e}(\tau) - \lambda \mathbf{e}_1 \tag{26}$$

According to Equation (26), we obtain $\mathbf{e}(\tau) = \mu \mathbf{Q}(\theta)^{-1}\mathbf{e}_1$, where μ is a constant. Since $\mathbf{e}_1^H \mathbf{e}(\tau) = 1$, combined with $\mathbf{e}(\tau) = \mu \mathbf{Q}(\theta)^{-1}\mathbf{e}_1$, we obtain $\mu = 1/\mathbf{e}_1^H \mathbf{Q}(\theta)^{-1}\mathbf{e}_1$. We bring μ into the $\mathbf{e}(\tau) = \mu \mathbf{Q}(\theta)^{-1}\mathbf{e}_1$ to obtain $\hat{\mathbf{e}}(\tau) = \frac{\mathbf{Q}(\theta)^{-1}\mathbf{e}_1}{\mathbf{e}_1^H \mathbf{Q}(\theta)^{-1}\mathbf{e}_1}$.

By bringing $\hat{\mathbf{e}}(\tau)$ into $\min_{\theta} \mathbf{e}(\tau)^H \mathbf{Q}(\theta)\mathbf{e}(\tau)$, we obtain the estimate $\hat{\theta}_l$ $(l = 1,2,\dots L)$ as:

$$\hat{\theta}_l = \arg\min_{\theta} \frac{1}{\mathbf{e}_1^H \mathbf{Q}(\theta)^{-1}\mathbf{e}_1} = \arg\max_{\theta} \mathbf{e}_1^H \mathbf{Q}(\theta)^{-1}\mathbf{e}_1 \tag{27}$$

The AOA is obtained by finding the angle corresponding to the peak of the (1, 1)th element of $\mathbf{Q}(\theta)^{-1}$. Then, the acquired AOA is brought into the 2D function, and the TOA estimation is also obtained through a 1D search. The 2D-MUSIC space spectral function can be rewritten as :

$$P_{1D-MUSIC}(\tau) = \frac{1}{\mathbf{a}(\tau,\hat{\theta})^H \mathbf{U}_N \mathbf{U}_N^H \mathbf{a}(\tau,\hat{\theta})} \tag{28}$$

The advantage of this proposed method is that the accuracy is close to that of the 2D search, and the computational complexity is greatly reduced compared to the 2D-MUSIC algorithm.

4.2. Separation of Multipath Signals under Unknown Waveforms

In addition, in case of unknown prior information of the transmitted signal, the proposed method can distinguish the main path signal based on TDOA. With the obtained incidence angle $\hat{\theta}_l$ $(l = 1,2,\ldots L)$ and $\hat{\mathbf{e}}(\tau) = \frac{\mathbf{Q}(\theta)^{-1}\mathbf{e}_1}{\mathbf{e}_1^H\mathbf{Q}(\theta)^{-1}\mathbf{e}_1}$, we obtain $\hat{\mathbf{e}}(\tau_l)$ $(l = 1,2,\ldots,L)$ $\in \mathbb{C}^{P\times 1}$ as:

$$\hat{\mathbf{e}}(\tau_l) = \left[\frac{S(\omega_{c0})}{S(\omega_{c0})}e^{-j\omega_0\tau_l},\ldots,\frac{S(\omega_{c(P-1)})}{S(\omega_{c0})}e^{-j\omega_{(P-1)}\tau_l}\right]^T \tag{29}$$

Let $\hat{\mathbf{e}}(\tau_l) = \begin{bmatrix}\mathbf{e}_{1l}\\ \mathbf{E}_{Al}\end{bmatrix} = \begin{bmatrix}\mathbf{E}_{Bl}\\ \mathbf{e}_{2l}\end{bmatrix}$, where $\mathbf{e}_{1l}$ and $\mathbf{e}_{2l}$ are the first and last rows of $\mathbf{e}(\tau_l)$, respectively. $\mathbf{E}_{Al}$ and $\mathbf{E}_{Bl}$ are the rest. At this point, we can obtain

$$\mathbf{E}_{Al}./\mathbf{E}_{Bl} = \left[\frac{S(\omega_{c1})}{S(\omega_{c0})}e^{-j\Delta\omega\tau_l},\ldots,\frac{S(\omega_{c(P-1)})}{S(\omega_{c(P-2)})}e^{-j\Delta\omega\tau_l}\right]^T \tag{30}$$

A simple example of judgment is given below. Suppose now that the incidence angles of all paths are obtained. At this point, we can obtain

$$\begin{aligned}
\mathbf{Q}_1 &= \mathbf{E}_{A1}./\mathbf{E}_{B1}\\
&= \left[\frac{S(\omega_{c1})}{S(\omega_{c0})}e^{-j\Delta\omega\tau_1},\ldots,\frac{S(\omega_{c(P-1)})}{S(\omega_{c(P-2)})}e^{-j\Delta\omega\tau_1}\right]^T\\
\mathbf{Q}_2 &= \mathbf{E}_{A2}./\mathbf{E}_{B2}\\
&= \left[\frac{S(\omega_{c1})}{S(\omega_{c0})}e^{-j\Delta\omega\tau_2},\ldots,\frac{S(\omega_{c(P-1)})}{S(\omega_{c(P-2)})}e^{-j\Delta\omega\tau_2}\right]^T\\
&\vdots\\
\mathbf{Q}_L &= \mathbf{E}_{AL}./\mathbf{E}_{BL}\\
&= \left[\frac{S(\omega_{c1})}{S(\omega_{c0})}e^{-j\Delta\omega\tau_L},\ldots,\frac{S(\omega_{c(P-1)})}{S(\omega_{c(P-2)})}e^{-j\Delta\omega\tau_L}\right]^T
\end{aligned} \tag{31}$$

$$\mathbf{Q}_a./\mathbf{Q}_b = \left[e^{-j\Delta\omega(\tau_a-\tau_b)},\ldots,e^{-j\Delta\omega(\tau_a-\tau_b)}\right]^T (a,b\in l) \tag{32}$$

This way, the main path signal can be distinguished by observing the positive or negative TDOA between all paths without relying on the sampled value of the transmitted signal.

5. Computational Complexity and Estimation Performance

5.1. Computational Complexity

The computational complexity of the proposed algorithm is $O\{2M^3P^3 + (K-L+N_2L)M^2P^2 + N_1M^2P^3 + N_1MP^3 + ((N_1+N_2L)M^2 - (N_1+N_2L)M + 2N_1)P^3/2 + N_2LMP^2 + N_2LMP\}$. The computational complexity of the 2D-MUSIC [26] algorithm is $O\{2M^3P^3 + (K-L+N_1N_2)M^2P^2 + N_1N_2MP^2 + N_1N_2M(M-1)P^3/2 + N_1N_2MP\}$. The computational complexity of the Root-MUSIC [25] algorithm is $O\{(10+(2N_2L+1)(M^2(M+1)^21-4)/4)P^3 + (L+K-7)P^2 + 2N_2L((M-1)(M+2)/2 + M(M+1)(2M+1)(1-2L)/6 - (1-2L))P^2\}$. The computational complexity of the Esprit [25] algorithm is $O\{(2+(N_2L+1)(M^2(M+1)^2-4)/4)P^3 + (L-3+K)P^2 + N_2L((M-1)(M+2)/2 + M(M+1)(2M+1)(1-2L)/6 - (1-L))P^2 + (3-2L+3L^2)P + (2L^3-3L^2+L-1)\}$. Here, M is the number of array elements, K is the number of frequency domain snapshots, L is the number of multipath, P is the number of sampled points, N_1 is the number of angle search points, and N_2 is the number of time delay search points.

Figure 2 shows a comparison of the computational complexity of the proposed method and the 2D-MUSIC algorithm with the variation of the number of array elements. With $K = 100$, $M = [2,4,6]$, $L = 3$, $P = 64$, $N_1 = 1001$, and $N_2 = 5001$. It can be seen from Figure 2 that the proposed method obtains lower computational complexity than other methods with different values of M. The computational complexity of the proposed algorithm is reduced by two orders of magnitude compared to the 2D-MUSIC and by one order of magnitude compared to the Esprit and Root-MUSIC. Thus, the proposed method provides an appreciable contribution to reducing the computational complexity, especially compared with other joint AOA and TOA estimation methods of the same type.

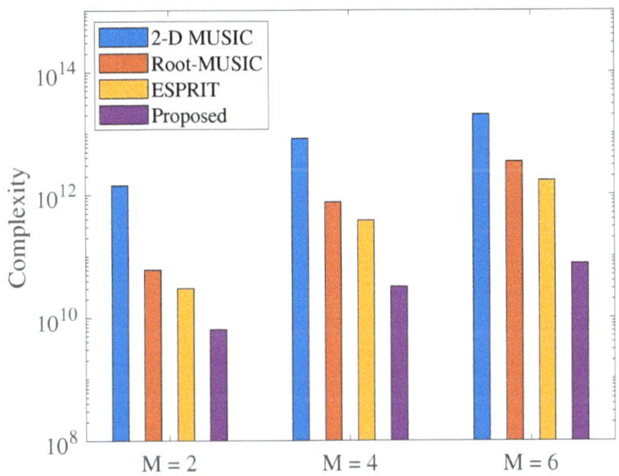

Figure 2. Computational complexity comparison.

5.2. Simulation Parameters

To evaluate the performance of the proposed algorithm, we perform Monte Carlo simulations. The signal-to-noise ratio (SNR) and root mean square error (RMSE) [33] are defined, respectively, as

$$\mathrm{SNR} = 10\lg \frac{\|y(t)\|^2}{\|w(t)\|^2} \tag{33}$$

$$\mathrm{RMSE} = \frac{1}{L}\sum_{l=1}^{L}\sqrt{\frac{1}{N}\sum_{n=1}^{N}\left(\delta_l - \hat{\delta}_{l,n}\right)^2} \tag{34}$$

where $y(t)$ is the time-domain signal received by the antenna, $w(t)$ is additive Gaussian white noise, L is the number of multipath, and M is the Monte Carlo count. $\hat{\delta}_{l,n}$ stands for the estimation of δ_l of the nth Monte Carlo trial. δ_l means the estimation of AOA and TOA.

In this section, the frequency range of the incident UWB signal is set from 3.5 GHz to 4.5 GHz, the bandwidth is 1 GHz, the sampling frequency is 10 GHz, and the number of sampling points is 2048. After that, we apply the fast Fourier transform on the sampled signal and select the number of points containing the signal as $N = 64$. The frequency-domain snapshot number is $K = 100$. The number of receiving antennas is $M = 2$. We assume that the AOA of the multipath signal is $[20°, 40°, 60°]$ and TOA of the multipath signal is $[4.5/c, 6.5/c, 8.5/c]$, where c is the speed of light. Figures 3–5 simulate the RMSE performance of AOA, TOA, and TDOA under different SNR varying from 0 dB to 25 dB. In Figure 5, except for the proposed algorithm (unknown transmitted signal) which estimates the TDOA with the unknown transmitted signal, all other algorithms estimate TDOA with the known transmitted signal. Figures 6–8 simulate the RMSE performance of AOA, TOA, and TDOA at different snapshot numbers varying from $K = 100$ to $K = 400$ and SNR = 5 dB, respectively. In Figure 8, except for the proposed algorithm (unknown transmitted signal) which estimates the TDOA with the unknown transmitted signal, all other algorithms estimate TDOA with the known transmitted signal.

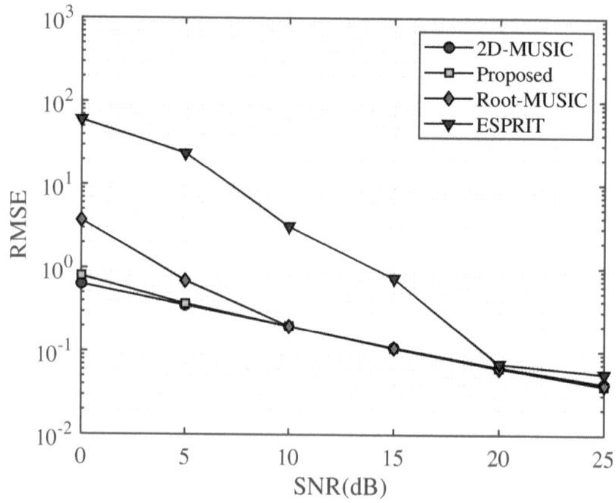

Figure 3. RMSE comparison of AOA under different SNR.

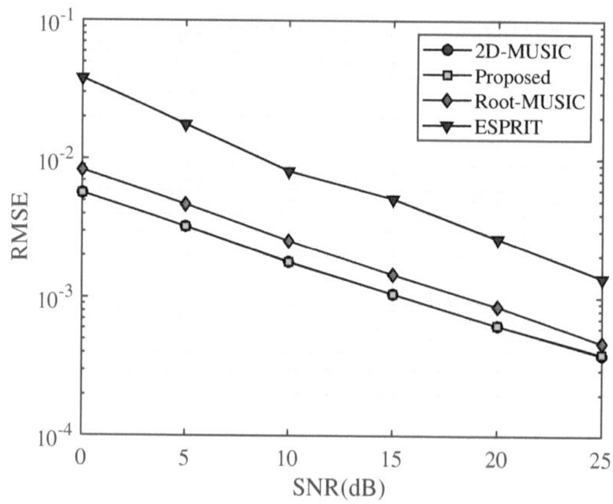

Figure 4. RMSE comparison of TOA under different SNR.

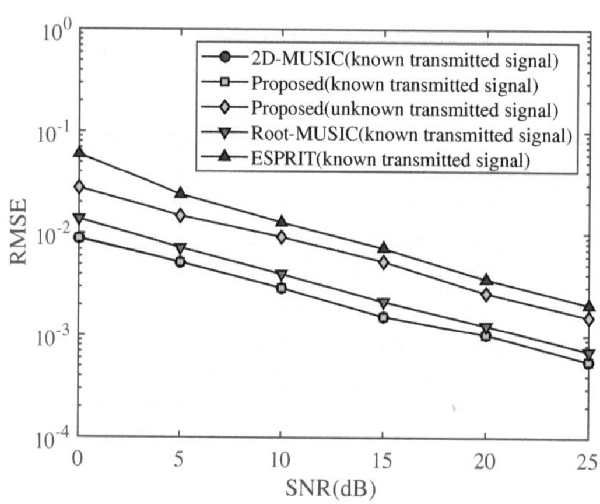

Figure 5. RMSE comparison of TDOA under different SNR.

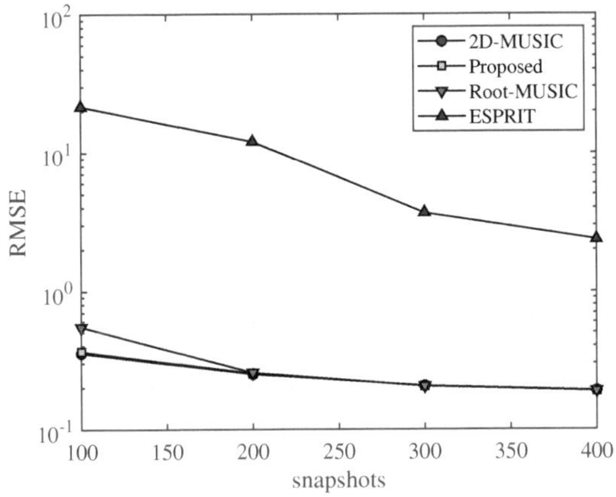

Figure 6. RMSE comparison of AOA under number of snapshots.

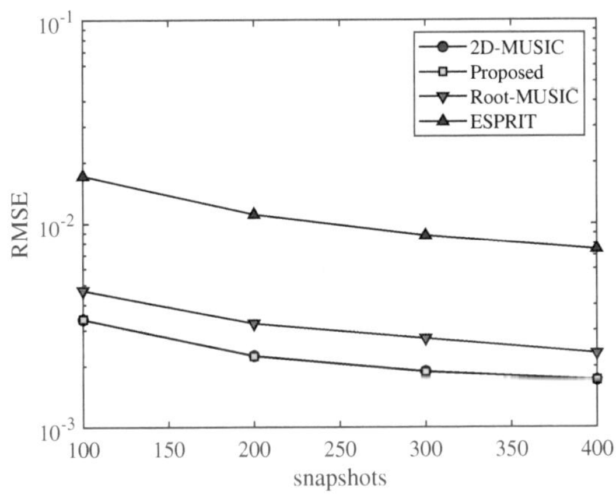

Figure 7. RMSE comparison of TOA under number of snapshots.

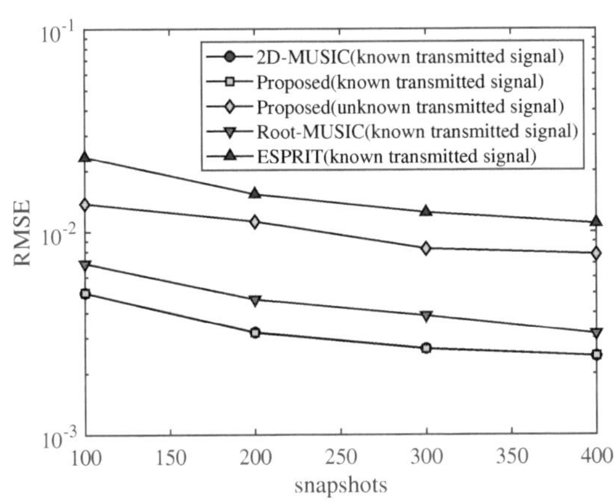

Figure 8. RMSE comparison of TDOA under number of snapshots.

5.3. Results and Discussion

In this paper, the proposed algorithm is compared with the same type of joint AOA and TOA estimation algorithms, i.e., 2D-MUSIC in ref. [26], Root-MUSIC in ref. [25], and ESPRIT in ref. [25], respectively. From Figures 3 and 4, it can be seen that the RMSE of AOA and TOA gradually decrease with the increase in the SNR. Meanwhile, it can be seen that the RMSE curves of AOA and TOA of the proposed algorithm basically overlap with those of 2D-MUSIC and are lower than Root-MUSIC and ESPRIT, indicating that the accuracy of AOA and TOA estimation of the proposed algorithm is basically consistent with 2D-MUSIC and better than Root-MUSIC and ESPRIT algorithms. Therefore, compared with other algorithms, the proposed algorithm has certain advantages in AOA and TOA estimation accuracy for the same SNR.

In order to show the ability of the proposed algorithm to discriminate the main path when the prior information of the transmitted signal is unknown, Figure 5 shows the TDOA estimation performance of the proposed algorithm in the case of unknown transmitted signal, and the rest of the compared algorithms estimate the TDOA in the case of known transmitted signal. The trend of the RMSE curve of the proposed algorithm (with known transmit signal) is similar to that of the RMSE curve of TOA shown in Figure 4. The accuracy of TDOA estimation for the proposed algorithm (known transmitted signal) is basically the same as 2D-MUSIC and better than Root-MUSIC and ESPRIT. In Figure 5, it can be seen that the TDOA estimation accuracy of the proposed algorithm (unknown transmitted signal) is better than the ESPRIT algorithm and slightly inferior to the Root-MUSIC algorithm, 2D-MUSIC algorithm, and the proposed algorithm (known transmitted signal). Although the TDOA estimation accuracy of the proposed algorithm (unknown transmitted signal) is general, it is applicable to a wide range of scenarios because it does not require a priori information of the known transmitted signal.

The number of snapshots is also an important factor affecting the estimation accuracy. In Figures 6–8, simulation results show that the performance of RMSE improves as the number of snapshots increases. It can be seen in Figures 6 and 7 that for the same number of snapshots, the AOA and TOA estimation accuracy of the proposed algorithm is basically the same as that of the 2D-MUSIC algorithm, which is better than other algorithms. It can be seen in Figure 8 that for the same number of snapshots, the TDOA estimation performance of the proposed algorithm (known transmitted signal) is basically the same as that of the 2D-MUSIC algorithm, and the TDOA performance of the proposed algorithm (unknown transmitted signal) is slightly inferior to that of the Root-MUSIC algorithm and better than that of the ESPRIT algorithm. Therefore, compared with other algorithms, the proposed algorithm has certain advantages in AOA, TOA, and TDOA estimation accuracy for the same number of snapshots, and the proposed algorithm is applicable to a wide range of scenarios.

6. Conclusions

In this paper, we proposed a high-precision and low-complexity joint AOA and TOA estimation algorithm. We extend the received data model of the dual antennas to the received data model of the multiple antennas. To reduce the computational complexity of the 2D-MUSIC joint AOA and TOA, the angular search can be extracted separately by converting the 2D spectral peak search into a secondary optimization problem. Then, the 2D search is converted into a 1D search based on the acquired AOA. As a result, the computational complexity is reduced, and the estimation accuracy of the proposed algorithm remains basically the same as that of the 2D-MUSIC algorithm. Meanwhile, in case of unknown prior information of the transmitted signal, the proposed method can distinguish the main path signal based on TDOA, which shows wider applications. Through simulation experiments, the following conclusions can be obtained:

(1) The computational efficiency of the proposed algorithm is significantly higher than that of the 2D-MUSIC algorithm, the Root-MUSIC algorithm, and the ESPRIT algorithm.
(2) At each SNR, the accuracy of AOA, TOA, and TDOA estimation of the proposed algorithm is basically the same as that of 2D-MUSIC and better than that of Root-MUSIC and ESPRIT.
(3) At each snapshot number, the accuracy of AOA, TOA, and TDOA estimation of the proposed algorithm is basically the same as that of 2D-MUSIC and better than that of Root-MUSIC and ESPRIT.
(4) In case of unknown prior information of the transmitted signal, the proposed method can still distinguish the main path signal, while the other methods fail.

The current methods can only handle a single signal. However, when multiple signals, especially those with multipath interference, are present, it becomes challenging to address the issue. Therefore, future research should focus on tackling the complexities associated with multiple signals. Additional investigation and exploration are necessary to thoroughly examine this aspect.

Author Contributions: Conceptualization, J.L.; methodology, W.D.; software, W.D. and Y.T.; investigation, Y.T.; data curation, W.D.; writing—original draft preparation, W.D.; writing—review and editing, J.L. and X.Z. All authors have read and agreed to the published version of the manuscript

Funding: This work was supported in part by the National Key R&D Program of China under Grant 2020YFB1807602 and under Grant 2020YFB1807600, in part by the China University Innovation Fund for Production, education, and research under Grant 2021ZYA0301, in part by the Jiangsu Planned Projects for Postdoctoral Research Funds under Grant 2020Z013, in part by the Pre-research project of SongShan Laboratory under Grant YYJC022022018, and in part by the China Postdoctoral Science Foundation under Grant 2020M681585.

Data Availability Statement: Not applicable.

Conflicts of Interest: The authors declare no conflict of interest.

Abbreviations

The following abbreviations are used in this manuscript.

AOA	Angle of Arrival
TOA	Time of Arrival
MUSIC	Multiple Signal Classification
2D	Two-Dimensional
Esprit	estimation of signal parameters using rotational invariance techniques
GPS	Global Positioning System
UWB	Ultra-Wideband
S-V	Saleh-Valenzuela
2D-MUSIC	Two-Dimensional Multiple Signal Classification
TDOA	Time Difference of Arrival
1D	One-Dimensional
ULA	Uniform Linear Array
SNR	Signal-to-Noise Ratio
RMSE	Root Mean Square Error

References

1. Zhang, Q.; Li, J.; Li, Y.; Li, P. A DOA Tracking Method Based on Offset Compensation Using Nested Array. *IEEE Trans. Circuits Syst. II Express Briefs* **2022**, *69*, 1917–1921. [CrossRef]
2. Vlachos, E.; Alexandropoulos, C.G.; Thompson, J. Massive MIMO Channel Estimation for Millimeter Wave Systems via Matrix Completion. *IEEE Signal Process. Lett.* **2022**, *25*, 1675–1679. [CrossRef]
3. Mao, Z.; Liu, S.; Zhang, Y.D.; Han, L.; Huang, Y. Joint DoA-Range Estimation Using Space-Frequency Virtual Difference Coarray. *IEEE Trans. Signal Process.* **2022**, *70*, 2576–2592. [CrossRef]
4. Bharadwaj, R.; Koul, S.K. Wearable UWB Technology for Daily Physical Activity Tracking, Detection, and Classification. *IEEE Sens. J.* **2022**, *22*, 20684–20694. [CrossRef]

5. Ding, Y.; Jiang, D.; Liu, Y.; Zhang, D.; He, T. SmartLOC: Indoor Localization with Smartphone Anchors for On-Demand Delivery. In Proceedings of the ACM on Interactive, Mobile, Wearable and Ubiquitous Technologies (IMWUT), New York, NY, USA, 4–6 December 2021; pp: 1–24. [CrossRef]
6. Niculescu, V.; Palossi, D.; Magno, M.; Benini, L. Energy-Efficient, Precise UWB-Based 3-D Localization of Sensor Nodes With a Nano-UAV. *IEEE Internet Things J.* **2023**, *10*, 5760–5777. [CrossRef]
7. Yaacoub, T.; Dobre, O.A.; Youssef, R.; Radoi, E. Optimal Selection of Fourier Coefficients for Compressed Sensing-Based UWB Channel Estimation. *IEEE Wirel. Commun. Lett.* **2017**, *6*, 466–469. [CrossRef]
8. Allebes, E.; Singh, G.; He, Y. A 3-to-10 GHz 180 pJ/b IEEE802.15.4z/4a IR-UWB Coherent Polar Transmitter in 28 nm CMOS with Asynchronous Amplitude Pulse-Shaping and Injection-Locked Phase Modulation. In Proceedings of the 2021 IEEE International Solid-State Circuits Conference (ISSCC), San Francisco, CA, USA, 13–22 February 2021; pp. 304–306. [CrossRef]
9. Saleh, A.A.M.; Valenzuela, R.A. A statistical model for indoor multipath propagation. *IEEE J. Sel. Areas Commun.* **1987**, *5*, 128–137. [CrossRef]
10. Feng, R.; Wang, C.X.; Huang, J.; Gao, X.Q.; Salous, S.; Haas, H. Classification and Comparison of Massive MIMO Propagation Channel Models. *IEEE Internet Things J.* **2022**, *9*, 23452–23471. [CrossRef]
11. Hollosi, G. Distribution of ultra wideband (UWB) receive timestamps in dense indoor environment based on the Saleh-Valenzuela channel model. In Proceedings of the 2022 14th International Conference on Communications (COMM), Bucharest, Romania, 16–18 June 2022; pp. 1–5. [CrossRef]
12. Wu, S.-H.; Zhang, Q.-Y.; Zhang, N.-T. TOA Estimation Based on Match-Filtering Detection for UWB Wireless Sensor Networks. *J. Softw.* **2009**, *20*, 3010–3022. [CrossRef]
13. Liu, W.; Ding, H.; Huang, X.; Liu, Z. TOA Estimation in IR UWB Ranging with Energy Detection Receiver Using Received Signal Characteristics. *IEEE Commun. Lett.* **2012**, *16*, 738–741. [CrossRef]
14. Liu, X.J.; Wang, Z.B.; Zheng, L. Simple energy envelope TOA estimation algorithm. *Comput. Eng. Appl.* **2012**, 78–80. [CrossRef]
15. Zhang, T.T.; Zhang, Q.Y.; Zhang, N.T. A Two-Step TOA Estimation Method Based on Energy Detection for IR-UWB Sensor Networks. In Proceedings of the 2009 Seventh Annual Communication Networks and Services Research Conference, Moncton, NB, Canada, 11–13 May 2009; pp. 139–145. [CrossRef]
16. Pourkhaatoun, M.; Zekavat, S.A. Concatenated Spectrum Multi-Band TOA Estimation. In Proceedings of the 2011 IEEE 22nd International Symposium on Personal, Indoor and Mobile Radio Communications (PIMRC), Toronto, ON, Canada, 11–14 September 2011; pp. 1192–1196. [CrossRef]
17. He, Z.M.; Ma, Y.; Tafazolli, R. Improved High Resolution TOA Estimation for OFDM-WLAN Based Indoor Ranging. *IEEE Wirel. Commun. Lett.* **2013**, *2*, 163–166. [CrossRef]
18. Xu, H.L.; Yang, L.Q. ToA Estimation for MB-OFDM UWB by Suppressing Energy Leakage. In Proceedings of the 2010—MILCOM 2010 Military Communications Conference (MilCOM), San Jose, CA, USA, 31 October–3 November 2010; pp. 2381–2386. [CrossRef]
19. Savic, V.; Larsson, E.G.; Ferrer-Coll, J.; Stenumgaard, P. Kernel Methods for Accurate UWB-Based Ranging with Reduced Complexity. *IEEE Trans. Wirel. Commun.* **2016**, *15*, 1783–1793. [CrossRef]
20. Abbasi, A.; Liu, H. Novel Cascade CNN Algorithm for UWB Signal Denoising, Compressing, and ToA Estimation. In Proceedings of the 2021 IEEE 11th Annual Computing and Communication Workshop and Conference (CCWC), Las Vegas, NV, USA, 27–30 January 2021; pp. 0721–0725. [CrossRef]
21. Abbasi, A.; Liu, H. OneAI—Novel Multipurpose Deep Learning Algorithms for UWB Wireless Networks. In Proceedings of the 2021 IEEE Workshop on Signal Processing Systems (SiPS), Coimbra, Portugal, 19–21 October 2021; pp. 129–133. [CrossRef]
22. He, K.; Zhang, X.; Ren, S.; Sun, J. Deep Residual Learning for Image Recognition. In Proceedings of the 2016 IEEE Conference on Computer Vision and Pattern Recognition (CVPR), Las Vegas, NV, USA, 27–30 June 2016; pp. 770–778. [CrossRef]
23. Navarro, M.; Najar, M. Joint estimation of TOA and DOA in IR-UWB. In Proceedings of the 2007 IEEE 8th Workshop on Signal Processing Advances in Wireless Communications(WPAWC), Helsinki, Finland, 17–20 June 2007; pp. 1–5. [CrossRef]
24. Navarro, M.; Najar, M. Frequency Domain Joint TOA and DOA Estimation in IR-UWB. *IEEE Trans. Wirel. Commun.* **2011**, *10*, 3174–3184. [CrossRef]
25. Wang, F.Q.; Zhang, X.F.; Wang, F. Joint Estimation of TOA and DOA in IR-UWB System Using a Successive MUSIC Algorithm. *Wirel. Pers. Commun.* **2014**, *77*, 2445–2464. [CrossRef]
26. Li, Y.; Yu, J.; He, L.Y. TOA and DOA Estimation for IR-UWB Signals: An Efficient Algorithm by Improved Root-MUSIC. *Wirel. Commun. Mob. Comput.* **2021**, *2021*, 9960273. [CrossRef]
27. Shen, C.; Guo, Y.J.; Yu, J.R.; Yang, J.B.; Xu, J.T. DFT-Based Joint TOA and DOA Estimation in Impulse Radio Ultra Wideband. *J. Data Acquis. Process.* **2022**, *37*, 1157–1168. [CrossRef]
28. Gong, L.L.; Hu, Y.; Zhang, J.Y.; Zhao, G.F. Joint TOA and DOA Estimation for UWB Systems with Antenna Array via Doubled Frequency Sample Points and Extended Number of Clusters. *Math. Probl. Eng.* **2021**, *2021*, 7521573. [CrossRef]
29. Hua, M.C.; Liu, H.C. Joint estimation of DOA and TOA for MB-OFDM UWB signals based on frequency-domain processing. *IET Radar Sonar Navig.* **2016**, *2021*, 1337–1346. [CrossRef]
30. Sun, B.; Tan, B.; Wang, W.B.; Lohan, E.S. A Comparative Study of 3D UE Positioning in 5G New Radio with a Single Station. *Sensors* **2021**, *21*, 1178. [CrossRef]

31. Xie, R.; Hu, D.; Luo, K.; Jiang, T. Performance Analysis of Joint Range-Velocity Estimator With 2D-MUSIC in OFDM Radar. *IEEE Trans. Signal Process.* **2021**, *69*, 4787–4800. [CrossRef]
32. Zhu, Y.; Zhang, W.; Yi, H.; Xu, H. Enhanced Root-MUSIC Algorithm Based on Matrix Reconstruction for Frequency Estimation. *Sensors* **2023**, *23*, 1829. [CrossRef] [PubMed]
33. Wu, Y.; Wakin, M.B.; Gerstoft, P. Gridless DOA Estimation With Multiple Frequencies. *IEEE Trans. Signal Process.* **2023**, *71*, 417–432. [CrossRef]

sensors

MDPI

Article

An Improved Unfolded Coprime Linear Array Design for DOA Estimation with No Phase Ambiguity

Pan Gong [1,*] and Xiaofei Zhang [2]

1 College of Electronic Information and Integrated Circuits, Nanjing Vocational University of Industry Technology, Nanjing 211106, China
2 College of Electronic and Information Engineering, Nanjing University of Aeronautics and Astronautics, Nanjing 210016, China; zhangxiaofei@nuaa.edu.cn
* Correspondence: gongpan@nuaa.edu.cn

Abstract: In this paper, the direction of arrival (DOA) estimation problem for the unfolded coprime linear array (UCLA) is researched. Existing common stacking subarray-based methods for the coprime array are invalid in the case of its subarrays, which have the same steering vectors of source angles. To solve the phase ambiguity problem, we reconstruct an improved unfolded coprime linear array (IUCLA) by rearranging the reference element of the prototype UCLA. Specifically, we design the multiple coprime inter pairs by introducing the third coprime integer, which can be pairwise with the other two integers. Then, the phase ambiguity problem can be solved via the multiple coprime property. Furthermore, we employ a spectral peak searching method that can exploit the whole aperture and full DOFs of the IUCLA to detect targets and achieve angle estimation. Meanwhile, the proposed method avoids extra processing in eliminating ambiguous angles, and reduces the computational complexity. Finally, the Cramer–Rao bound (CRB) and numerical simulations are provided to demonstrate the effectiveness and superiority of the proposed method.

Keywords: direction of arrival (DOA) estimation; sparse array; improved unfolded coprime linear array (IUCLA); Cramer–Rao bound (CRB); multiple signal classification (MUSIC)

1. Introduction

Direction of arrival (DOA) estimation has attracted much attention in recent years, and it has been widely applied in many fields, such as wireless communication, radar, sonar, medicine and other engineering applications [1–8]. Additionally, the recent integration of neural networks [9] into the domain of DOA marks a promising frontier in the landscape of next-generation wireless communications. Thus, a deep learning-based scheme is proposed, and the simulation results affirm the superior performance. The model's robustness is rigorously examined across various validation cases, providing conclusive evidence of its potential in real-world applications. The accurate DOA estimation and low computational complexity have become increasingly important in recent years. In past decades, numerous subspace-based DOA estimation algorithms, e.g., multiple signals classification (MUSIC) [10–17] and estimation of signal parameters via rotational invariance techniques (ESPRIT) [18–22], have been proposed for uniform linear arrays (ULA) [23,24] and have attracted much attention due to their high resolution and performance. However, the adjacent antenna element spacing is required to be less than half of a wavelength to avoid the phase ambiguity problem [25]. Additionally, it limits the array aperture and consequently influences the mutual coupling [26] between the elements and the estimation performance; therefore, a non-uniform linear array has been recently proposed to increase the degrees of freedom (DOF) of the array, known as sparse arrays. Coprime arrays [27,28] have aroused wide attention due to the improvement of DOA estimation performance. Meanwhile, due to the larger array aperture, less mutual coupling effects arise. Furthermore, a coprime linear array (CLA) consists of two ULAs, with the inter-element spacing larger than half-wavelength; therefore, higher resolution, larger array aperture and a lesser mutual coupling effect can be attained [29,30]. Recently, a subarray-based method for a coprime array to solve the ambiguity problem was developed, and it is proposed in [31]. This algorithm considers the whole array as two subarrays, and it processes the two subarrays separately. Then, the MUSIC algorithm is applied to these two subarrays, respectively. By combining the estimates and finding the nearest spectral peaks from the MUSIC spectra of two subarrays, the phase ambiguity problem can be eliminated. However, this method suffers from severe computational complexity due to the global angular searching. Then, a partial spectral search method is proposed in [32] to decrease the computational complexity, which employs the linear relationship, along with ambiguous DOA estimates, and searches through a small sector. However, some problems occur in these mentioned methods of the CLA, which are as follows:

- DOF is limited by the subarray, which has a smaller number of elements;
- Only self-information of two subarrays is exploited, but mutual information is neglected; as a result, DOA estimation performance is degraded;
- These methods need extra procedures to eliminate ambiguous angles.

To solve these problems, an unfolded coprime linear array (UCLA) is proposed, which unfolds the two subarrays in two opposite directions so that the array aperture is extended [33]. By stacking the directional matrices, both self-information and mutual information are utilized. Therefore, the ambiguity problem is suppressed. Furthermore, this method can achieve the full DOFs due to the employment of the whole array sensors. Nevertheless, this technique is not always true. When the source signals satisfy some relations, the method will be invalid. In the case of three signals, for the two subarrays, when there are two different signals that have the same steering vectors as the given DOA, the phase ambiguity still exists. Aiming to tackle this problem, Yang et al. proposed a beamforming-based technique by defining a decision variable [34] to eliminate the phase ambiguity. However, this method employs other techniques, in addition to MUSIC spectral searching, to distinguish the real DOAs, which can increase the computational complexity. Meanwhile, this method does not always work, and it usually depends on the decision variable. When the decision variable is small, phase ambiguity problems still occur.

The method in [33] can achieve DOA estimation in most cases, but it does not work in special conditions. We have demonstrated that the method will be invalid when two subarrays have the same steering vectors of the source angles. The method in [34] can achieve the DOAs' estimation; however, along with MUSIC, it needs additional techniques to eliminate the ambiguous angles, which will increase the computational complexity. We demonstrated that this method will sometimes have the problem of "false targets" when the decision variable is small. We can directly solve the problem of "false targets" by designing a multiple coprime array configuration. Furthermore, numerous simulation results are provided to verify the effectiveness of the constructed array. Moreover, compared to nested arrays, the physical aperture of coprime arrays is larger, and our designed array has a larger aperture than the original coprime array, which can achieve higher angular resolution.

Therefore, the MUSIC algorithm can be employed directly, without ambiguous angles arising. Meanwhile, we can obtain the DOAs' estimation and do not need any other technique to estimate it. The method can achieve the full DOFs of the array. Compared to original methods, we find that the proposed method can detect sources effectively, without phase ambiguity problems.

The main contributions of this paper can be summarized as follows.

(1) The proposed method can effectively eliminate the phase ambiguity problem by rearranging the reference element spacing and determining unique directional vectors of different angles.
(2) The whole array aperture of the designed array configuration is utilized to increase DOFs and improve DOA estimation performance.
(3) The proposed method can obtain unambiguous DOA estimates without additional processing, which can reduce computational complexity.

(4) CRB and numerical simulations are offered to verify the superiorities and effectiveness of the proposed method.

The paper is organized as follows: Section 2 introduces the array configuration and the data model of the received signals. In Section 3, we detail the ambiguity problem and present the proposed method; then, we analyze the DOFs and Cramer–Rao bound (CRB) of the proposed method. In Section 4, we provide extensive simulation results and demonstrate the superiority of the proposed method. Finally, we present the conclusions in Section 5.

2. Array Signal Model

As depicted in Figure 1, we consider an UCLA with $T = M_1 + M_2 - 1$ sensors. For the array, it is composed of two uniform linear subarrays. One subarray has M_1 sensors and the other one has M_2 sensors, respectively, and M_1 and M_2 are a pair of coprime integers. Meanwhile, we assume $M_1 < M_2$. Two subarrays are overlapping at the position of $(0,0)$, so we can compute the total sensors of the array as $T = M_1 + M_2 - 1$. In Figure 1, the inter-element spacing of subarray 1 can be denoted as $d_1 = M_2 d = M_2\lambda/2$, and the inter-element spacing of subarray 2 can be denoted as $d_2 = M_1 d = M_1\lambda/2$, where $d = \lambda/2$ and λ represents the wavelength.

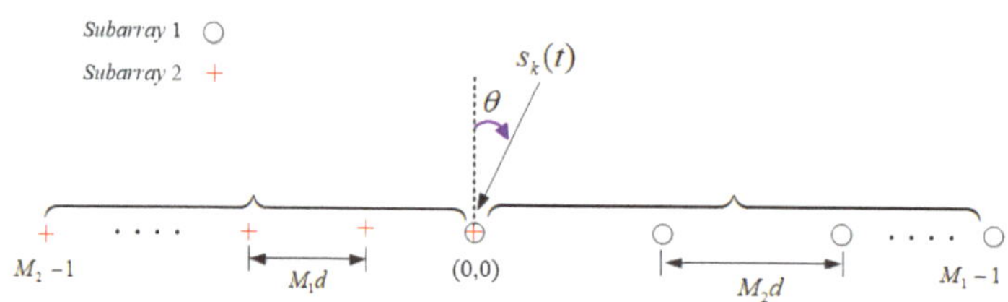

Figure 1. Unfolded coprime linear array (UCLA).

Assume that there are K narrowband sources impinging on the array, which locates at $\Theta = [\theta_1, \theta_2, \cdots, \theta_K]$, with signal powers of $\sigma_1^2, \sigma_2^2, \cdots, \sigma_K^2$. Additionally, $\theta_k \in (-\pi/2, \pi/2)$ denotes the k-th signal, wherein $K < T$ and $k \in [1, 2, \cdots, K]$. Furthermore, we suppose that these signals are far-field uncorrelated. The received signal at time t is denoted as follows [10]:

$$
\begin{aligned}
\boldsymbol{x}(t) = \begin{bmatrix} \boldsymbol{x}_1(t) \\ \boldsymbol{x}_2(t) \end{bmatrix} &= \begin{bmatrix} \boldsymbol{A}_1 \\ \boldsymbol{A}_2 \end{bmatrix} \boldsymbol{s}(t) + \begin{bmatrix} \boldsymbol{n}_1(t) \\ \boldsymbol{n}_2(t) \end{bmatrix} \\
&= \boldsymbol{A}\boldsymbol{s}(t) + \boldsymbol{n}(t)
\end{aligned}
\tag{1}
$$

where $\boldsymbol{A} = [\boldsymbol{A}_1^T, \boldsymbol{A}_2^T]^T$. $\boldsymbol{A}_1$ and $\boldsymbol{A}_2$ are steering matrices of the two subarrays, respectively. $\boldsymbol{A}_1 = [\boldsymbol{a}_1(\theta_1), \boldsymbol{a}_1(\theta_2), \cdots, \boldsymbol{a}_1(\theta_K)]$ and $\boldsymbol{A}_2 = [\boldsymbol{a}_2(\theta_1), \boldsymbol{a}_2(\theta_2), \cdots, \boldsymbol{a}_2(\theta_K)]$. $\boldsymbol{a}_1(\theta_k) = [1, e^{jM_2\pi sin\theta_k}, \cdots, e^{j(M_1-1)M_2\pi sin\theta_k}]^T$ is the steering vector for subarray 1, and $\boldsymbol{a}_2(\theta_k) = [e^{-j(M_2-1)M_1\pi sin\theta_k}, e^{-j(M_2-2)M_1\pi sin\theta_k}, \cdots, e^{-jM_1\pi sin\theta_k}, 1]^T$ is the steering vector for subarray 2, respectively. $\mathbf{s}(t) = [s_1(t), s_2(t), \cdots, s_K(t)]^T$ is the signal emitted by the k-th target at time t, where $t = 1, 2, \cdots, L$ and L represents the sampling number. $\boldsymbol{n}(t)$ stands for the additive white Gaussian noise, it obeys the normal distribution of $N(0, \sigma_k^2)$ and it is independent from source signals. $\boldsymbol{n}_1(t)$ is the noise vector of subarray 1, and $\boldsymbol{n}_2(t)$ denotes the noise vectors of subarray 2.

3. Ambiguity Problem Demonstration and Resolution

This section introduces the original coprime linear array and the cause of the phase ambiguity problem. In addition, we analyze the existing methods.

3.1. Preview

Generally, to avoid phase ambiguity problems, the inter-element spacing is usually set to be no larger than half-wavelength; however, in this way, the array aperture is restricted. For the coprime array, which is made up of two uniform linear arrays (ULA), the inter-element spacing is larger than half of a wavelength.

Figure 2 depicts the relationship between the element spacing and the number of spectral peaks. Here, peaks denote the number of DOAs. Furthermore, we consider that there is only one signal coming from $\theta_1 = 24^\circ$. Figure 2, demonstrates that when the element spacing is set to be $d = \lambda/2$, wherein λ represents the wavelength, no ambiguous angle appears. Meanwhile, when $d = 3\lambda/2$ or $d = 5\lambda/2$, ambiguous angles appear.

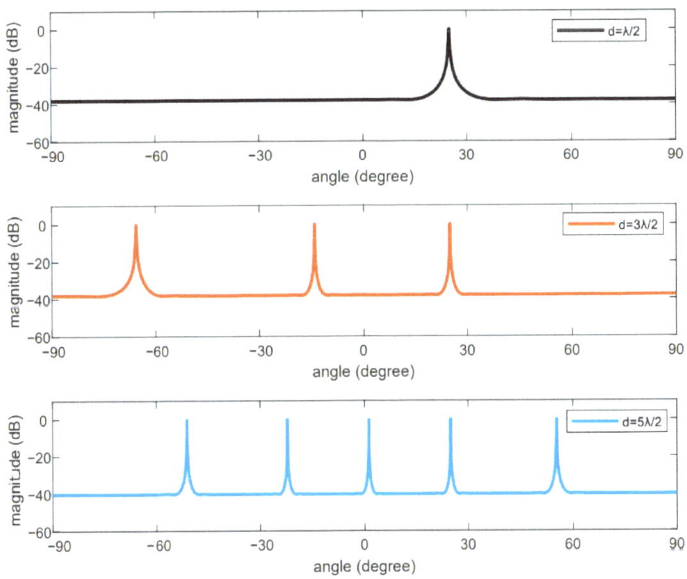

Figure 2. The relationship between the phase ambiguity problem and the inter-element spacing.

3.1.1. Case 1

In Case 1, we consider that there are two signals coming from $\Theta = [\theta_1, \theta_2]$.

We assume that there are two signals, θ_1 and θ_2, which are real angles. Then, we can obtain their corresponding directional vectors, which are $\boldsymbol{a}(\theta_1)$ and $\boldsymbol{a}(\theta_2)$.

$$\boldsymbol{a}(\theta_1) = \left[\boldsymbol{a}_1^T(\theta_1), \boldsymbol{a}_2^T(\theta_1)\right]^T \tag{2}$$

$$\boldsymbol{a}(\theta_2) = \left[\boldsymbol{a}_1^T(\theta_2), \boldsymbol{a}_2^T(\theta_2)\right]^T \tag{3}$$

We assume that θ_1' and θ_2' are ambiguous angles of θ_1 and θ_2. Therefore, we can acquire the corresponding steering vectors, which are $\boldsymbol{a}(\theta_1')$ and $\boldsymbol{a}(\theta_2')$.

$$\boldsymbol{a}(\theta_1') = \boldsymbol{a}(\theta_1) = \left[\boldsymbol{a}_1^T(\theta_1'), \boldsymbol{a}_2^T(\theta_1')\right]^T = \left[\boldsymbol{a}_1^T(\theta_1), \boldsymbol{a}_2^T(\theta_1)\right]^T \tag{4}$$

$$\boldsymbol{a}(\theta_2') = \boldsymbol{a}(\theta_2) = \left[\boldsymbol{a}_1^T(\theta_2'), \boldsymbol{a}_2^T(\theta_2')\right]^T = \left[\boldsymbol{a}_1^T(\theta_2), \boldsymbol{a}_2^T(\theta_2)\right]^T \tag{5}$$

Correspondingly, we have

$$\begin{cases} \boldsymbol{a}_1(\theta_1) = \boldsymbol{a}_1(\theta_1') \\ \boldsymbol{a}_2(\theta_1) = \boldsymbol{a}_2(\theta_1') \\ \boldsymbol{a}_1(\theta_2) = \boldsymbol{a}_1(\theta_2') \\ \boldsymbol{a}_2(\theta_2) = \boldsymbol{a}_2(\theta_2') \end{cases} \tag{6}$$

Then we have

$$\begin{cases} M_2\pi sin\theta_1 = M_2\pi sin\theta_1' + 2k_1\pi \\ M_1\pi sin\theta_1 = M_1\pi sin\theta_1' + 2k_2\pi \\ M_2\pi sin\theta_2 = M_2\pi sin\theta_2' + 2k_1\pi \\ M_1\pi sin\theta_2 = M_1\pi sin\theta_2' + 2k_2\pi \end{cases} \tag{7}$$

where $k_1 = -(M_2 - 1), \cdots, (M_2 - 1)$ and $k_2 = -(M_1 - 1), \cdots, (M_1 - 1)$.

Then we can obtain

$$\begin{cases} sin\theta_1 = sin\theta_1' + 2k_1/M_2 \\ sin\theta_1 = sin\theta_1' + 2k_2/M_1 \\ sin\theta_2 = sin\theta_2' + 2k_1/M_2 \\ sin\theta_2 = sin\theta_2' + 2k_2/M_1 \end{cases} \tag{8}$$

Then we have

$$\frac{2k_1}{M_2} = \frac{2k_2}{M_1} \tag{9}$$

Due to the coprime property of M_1 and M_2, Equation (9) is not satisfied. That is, in the case of the two signals of θ_1 and θ_2, a phase ambiguity problem does not occur.

Figure 3 depicts spectrums without ambiguity, using the method proposed in [33]. The two signals are $\theta_1 = 10°, \theta_2 = 37°$. SNR and the snapshots are set to be SNR $= 5$dB and $L = 200$, respectively. Therefore, we can draw a conclusion that the method proposed in [33] can tackle the phase ambiguity problem with Case 1.

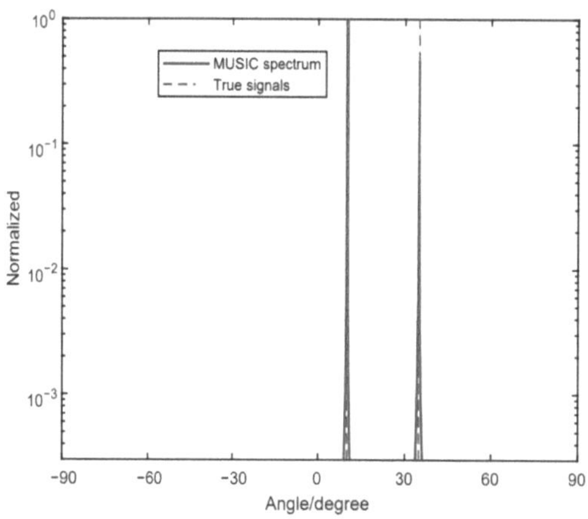

Figure 3. No ambiguous angle arises with two source signals, where $\theta_1 = 10°, \theta_2 = 37°$.

3.1.2. Case 2

In Case 2, we consider three signals with $\Theta = [\theta_1, \theta_2, \theta_3]$.

Similar to Case 1, we suppose θ_1' and θ_2' to be ambiguous angles of θ_1 and θ_2, respectively. Then, we can obtain the corresponding directional vectors of $\boldsymbol{a}(\theta_1')$ and $\boldsymbol{a}(\theta_2')$. If the relationship that the sine function of the third signal θ_3 equals to $\boldsymbol{a}(\theta_1')$ and $\boldsymbol{a}(\theta_2')$ is satisfied, the phase ambiguity problem appears. Figure 4 shows the spectrums with the method in [33], where it has three target signals, which are $\theta_1 = 10°, \theta_2 = 20°$ and $\theta_3 = 30°$. We set the number of snapshots to be $L = 200$ and SNR $= 5$dB. It is depicted clearly in Figure 4 that the method proposed in [33] still has no difficulty to resolve the three source signals. However, this method is not always accurate. When these three source signals satisfy the relationship wherein the sine function of the third signal, θ_3, equals $\boldsymbol{a}(\theta_1')$ and $\boldsymbol{a}(\theta_2')$, the method in [33] will not work; we will illustrate this in Case 3.

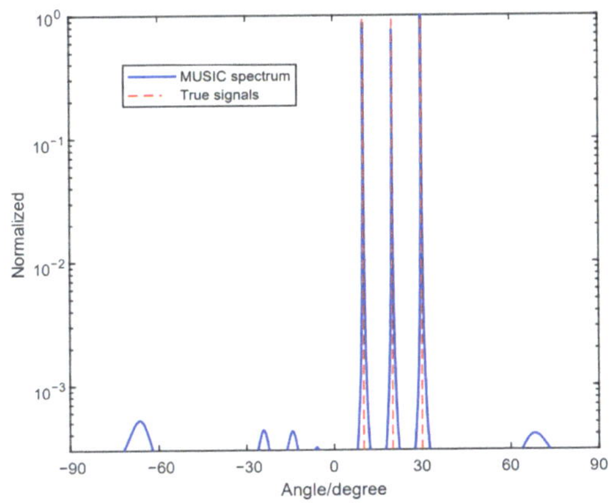

Figure 4. No ambiguous angle arises with the three given source signals, where $\theta_1 = 10°$, $\theta_2 = 20°$, $\theta_3 = 30°$.

3.1.3. Case 3

From Case 2, we notice that the method in [33] can effectively detect three signals without ambiguous angles, despite the fact that it does not always work. Thus, we illustrate Case 3 as follows.

We consider three target signals, which come from $\Theta = [\theta_1, \theta_2, \theta_3]$. Similar to Case 2, we suppose θ_1' and θ_2' to be ambiguous angles of θ_1 and θ_2, respectively. Then, we can obtain the corresponding directional vectors of $\boldsymbol{a}(\theta_1')$ and $\boldsymbol{a}(\theta_2')$. When these three signals satisfy that the sine function of θ_3 equals to $\boldsymbol{a}(\theta_1')$ and $\boldsymbol{a}(\theta_2')$, the ambiguous angle will arise. In other words, a phase ambiguity problem appears, which is invalidated in the following.

We suppose that the sine function of the third angle, θ_3, equals to $\boldsymbol{a}(\theta_1')$ and $\boldsymbol{a}(\theta_2')$. Then, it has

$$\begin{cases} sin\theta_3 = sin\theta_1 + 2(-k_1)/M_2 \\ sin\theta_3 = sin\theta_2 + 2(-k_2)/M_1 \end{cases} \tag{10}$$

Then, we can obtain the relationship among θ_1,θ_2 and θ_3 as follows:

$$\begin{cases} M_2\pi sin\theta_3 = M_2\pi sin\theta_1 + 2(-k_1)\pi \\ M_1\pi sin\theta_3 = M_1\pi sin\theta_2 + 2(-k_2)\pi \end{cases} \tag{11}$$

where $k_1 = -(M_2-1), \cdots, -1, 1, \cdots, (M_2-1)$ and $k_2 = -(M_1-1), \cdots, -1, 1, \cdots, (M_1-1)$.

It has

$$\begin{cases} \boldsymbol{a}_1(\theta_3) = \boldsymbol{a}_1(\theta_1) \\ \boldsymbol{a}_2(\theta_3) = \boldsymbol{a}_2(\theta_2) \end{cases} \tag{12}$$

so we have

$$\begin{cases} \boldsymbol{a}_1(\theta_1) + \boldsymbol{a}_1(\theta_2) - \boldsymbol{a}_1(\theta_3) = \boldsymbol{a}_1(\theta_2) \\ \boldsymbol{a}_2(\theta_1) + \boldsymbol{a}_2(\theta_2) - \boldsymbol{a}_2(\theta_3) = \boldsymbol{a}_2(\theta_1) \end{cases} \tag{13}$$

Then we define

$$\begin{cases} \boldsymbol{a}_1(\theta_4) = \boldsymbol{a}_1(\theta_1) + \boldsymbol{a}_1(\theta_2) - \boldsymbol{a}_1(\theta_3) \\ \boldsymbol{a}_2(\theta_4) = \boldsymbol{a}_2(\theta_1) + \boldsymbol{a}_2(\theta_2) - \boldsymbol{a}_2(\theta_3) \end{cases} \tag{14}$$

It has

$$\begin{cases} \boldsymbol{a}_1(\theta_4) = \boldsymbol{a}_1(\theta_2) \\ \boldsymbol{a}_2(\theta_4) = \boldsymbol{a}_2(\theta_1) \end{cases} \tag{15}$$

It is obvious that the forth angle, θ_4, arises. That is to say, a phase ambiguity problem occurs. Simulations are presented to demonstrate the analysis.

Figure 5 depicts the scenery of three signals, $\theta_1 = 12.37°$, $\theta_2 = 30°$ and $\theta_3 = 64.16°$, that come to the array. The SNR and the number of snapshots are set to be SNR $= 5$ dB and $L = 200$, respectively. It can be found that these three signals satisfy $\boldsymbol{a}_1(\theta_3) = \boldsymbol{a}_1(\theta_1)$ and $\boldsymbol{a}_2(\theta_3) = \boldsymbol{a}_2(\theta_2)$. Furthermore, we have demonstrated that the method in [33] is not effective in Case 2. Figure 5 shows that, other than these three signals, there is a forth spectrum that was detected. Aiming at solving the ambiguity problem, Yang et al. proposed a modified method by defining a decision variable [34]. Meanwhile, the method combines the beamforming technique with MUSIC, but it is not always effective. Figure 6 shows that five signals come to the array, and, by using the method in [34], the five signals can be detected successfully; however, they still have the ambiguous angle, which can increase the ineffectiveness of the DOA estimation performance. To tackle this problem, an improved unfolded coprime linear array (IUCLA) for the DOA estimation was proposed without the phase ambiguity problem. In the proposed method, we choose the third coprime integer, and, by moving the reference element, the linear combination relation of the steering vectors can be eliminated. As a result, we can use spectrum peak searching to achieve the real DOAs. Simultaneously, no additional algorithm is needed. The method is illustrated in the following part.

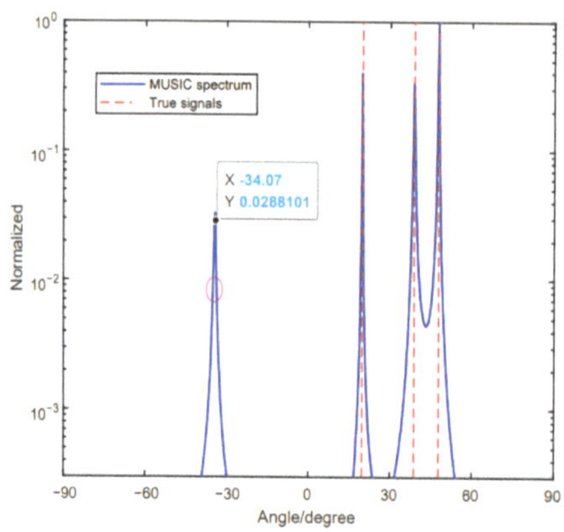

Figure 5. With the method in [33], the ambiguous angle arises with three source signals that satisfy Equation (10).

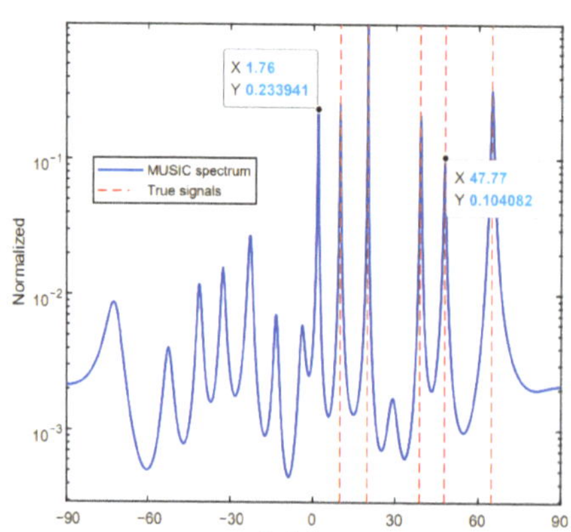

Figure 6. Using the method in [34] for the beamforming technique sometimes is not effective.

4. Proposed Method and Theoretical Performance Analysis

4.1. The Proposed Method

This section presents the proposed method to tackle the phase ambiguity problem with Case 3. In the proposed method, we consider constructing the multiple coprime arrays by introducing the third coprime integer. Specifically, Figure 7a presents an UCLA, which is composed of two subarrays, including subarray 1 with M_1 sensors and subarray 2 with M_2 sensors. Figure 7b introduces the third coprime integer, M_3, and presents a rearranged reference point instead of (0, 0). For the improved array, it still incorporates two subarrays which include M_1 and M_2 sensors, respectively. Furthermore, we can notice that the reference sensor is rearranged from (0, 0) to $(M_3d, 0)$, where M_3 can make coprime pairs with both M and N.

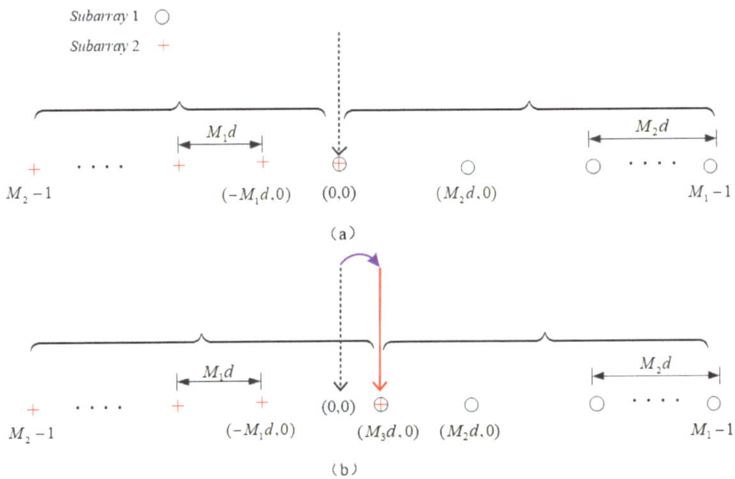

Figure 7. (**a**) The unfolded coprime linear array. (**b**) The designed and improved unfolded coprime linear array.

We employ the third coprime number, and it combines with another coprime number to form a pair of coprime integers with pairwise coprime. Multiple pairwise coprime integers are employed and the strong sidelobes can be suppressed by coprime property. Similar to the UCLA in Section 1, aside from two coprime integers, $M_1 = 5, M_2 = 7$, we employ the third coprime integers, which are $M_3 = 3$. In this way, there is not only one coprime integers pair, but also another two coprime integer pairs. Similar to our prior work, we multiply and employ the coprime property, and can eliminate the strong sidelobe.

In the rearranged array, we can denote the directional vectors of two subarrays with the k-th signal as

$$\mathbf{a}_{11}(\theta_k)= [e^{j3M_2\pi\sin\theta_k}, e^{jM_2\pi\sin\theta_k}, \cdots, e^{j(M_1-1)M_2\pi\sin\theta_k}]^T \tag{16}$$

$$\mathbf{a}_{22}(\theta_k) = [e^{-j(M_2-1)M_1\pi\sin\theta_k}, e^{-j(M_2-2)M_1\pi\sin\theta_k}, \cdots, e^{-jM_1\pi\sin\theta_k}, e^{-j3M_2\pi\sin\theta_k}]^T \tag{17}$$

Then, we can obtain the steering vectors of source signal θ_3

$$a_{11}(\theta_3) = [e^{j3M_2\pi sin\theta_3}, e^{jM_2\pi sin\theta_3}, \cdots, e^{j(M_1-1)M_2\pi sin\theta_3}]^T \tag{18}$$

$$a_{22}(\theta_3) = [e^{-j(M_2-1)M_1\pi sin\theta_3}, e^{-j(M_2-2)M_1\pi sin\theta_3}, \cdots, e^{-jM_1\pi sin\theta_3}, e^{-j3M_2\pi sin\theta_3}]^T \tag{19}$$

Additionally, $a(\theta_k)$ denotes the directional vector of the total array.

$$a(\theta_k) = \left[a_{11}^T(\theta_k), a_{22}^T(\theta_k)\right]^T \tag{20}$$

Similar to Equation (1), we can obtain the received signal

$$\begin{aligned} x_{iu}(t) = \begin{bmatrix} x_{iu1}(t) \\ x_{iu2}(t) \end{bmatrix} &= \begin{bmatrix} A_{11} \\ A_{22} \end{bmatrix} s(t) + \begin{bmatrix} n_{11}(t) \\ n_{22}(t) \end{bmatrix} \\ &= A_{iu}s(t) + n_{iu}(t) \end{aligned} \tag{21}$$

where $A_{iu} = [A_{11}^T, A_{22}^T]^T$. A_{11} and A_{22} are steering matrices of the two subarrays for IUCLA, respectively. $A_{11} = [a_{11}(\theta_1), a_{11}(\theta_2), \cdots, a_{11}(\theta_K)]$ and $A_{22} = [a_{22}(\theta_1), a_{22}(\theta_2), \cdots, a_{22}(\theta_K)]$, where $a_{11}(\theta_k)$ and $a_{22}(\theta_k)$ are denoted as Equations (16) and (17). $n_{iu}(t) = [n_{11}^T, n_{22}^T]^T$ is the total array noise vector.

The corresponding total covariance matrix can be computed with L snapshots

$$\hat{R}_{iu} = (1/L)\sum_{l=1}^{L} X_{iu}X_{iu}^H \tag{22}$$

The eigenvalue decomposition result of the total covariance matrix, $\hat{R}_{iu}$, can be denoted as

$$\hat{R}_{iu} = \hat{E}_{siu}\hat{D}_{siu}\hat{E}_{siu}^H + \hat{E}_{niu}\hat{D}_{niu}\hat{E}_{niu}^H \tag{23}$$

where $\hat{E}_{siu}$ and $\hat{E}_{niu}$ are the signal subspace and noise subspace matrices, respectively, and $\hat{D}_{siu}$ and $\hat{D}_{niu}$ include the eigenvalues.

Referring to the orthogonality between the signal subspace and the noise subspace, the spectral peak function of MUSIC can be denoted as [10]

$$f(\theta) = \frac{1}{a_{iu}^H(\theta)E_{niu}E_{niu}^H a_{iu}(\theta)} \tag{24}$$

where $a_{iu}(\theta) = \left[a_{11}^T(\theta), a_{22}^T(\theta)\right]^T$.

Referring to the derivation above, when there are three signals coming to the array and they satisfy $a_{11}(\theta_3) = a_{11}(\theta_1)$ and $a_{22}(\theta_3) = a_{22}(\theta_2)$, a phase ambiguity problem arises. In the following, we focus on proving and resolving the ambiguity problem.

Proof. Assume $a_{11}(\theta_3) = a_{11}(\theta_1)$. $a_{11}(\theta_3)$ and $a_{11}(\theta_1)$ represent the steering vectors of θ_3 and θ_1 with subarray 1, respectively. It has

$$\begin{cases} 3M_2\pi sin\theta_3 = 3M_2\pi sin\theta_1 + 2k_1\pi \\ M_2\pi sin\theta_3 = M_2\pi sin\theta_1 + 2k_1\pi \end{cases} \tag{25}$$

where $k_1 = (-M_2, M_2)$. Thus, $a_{11}(\theta_3) \neq a_{11}(\theta_1)$. □

Similarly, we can obtain that $a_{22}(\theta_3) \neq a_{22}(\theta_2)$, where $a_{22}(\theta_3)$ and $a_{22}(\theta_1)$ represent the steering vectors of θ_3 and θ_1 with subarray 2, respectively. Furthermore, we can reveal that the spectral peak function is eliminated. In this way, we can obtain the accurate DOAs' estimation without ambiguous angle θ_4, which means that the phase ambiguity problem is solved.

4.2. Theoretical Performance Analysis

4.2.1. DOF Analysis

In this part, we will provide the DOF performance of the proposed method. The method can achieve the full DOFs. Figure 8 shows that there are three signals coming to the array, with $M_1 = 3, M_2 = 2$. Furthermore, the signals are $\theta_1 = 10°, \theta_2 = 27.35°$ and $\theta_3 = 35.01°$.

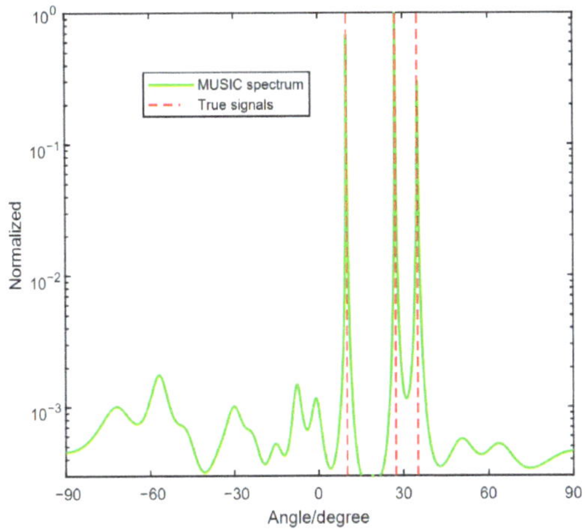

Figure 8. The reconstructed array configuration can achieve the full DOFs of three source signals.

Figure 9 shows that there are seven signals coming to the array, with $M_1 = 5$, $M_2 = 4$. Furthermore, the signals are denoted as $\theta_1 = -30°$, $\theta_2 = -10°$, $\theta_3 = 10°$, $\theta_4 = 30°$, $\theta_5 = 35°$, $\theta_6 = 40°$ and $\theta_7 = 50°$. From Figures 8 and 9, we can see that the proposed method can achieve the full DOF.

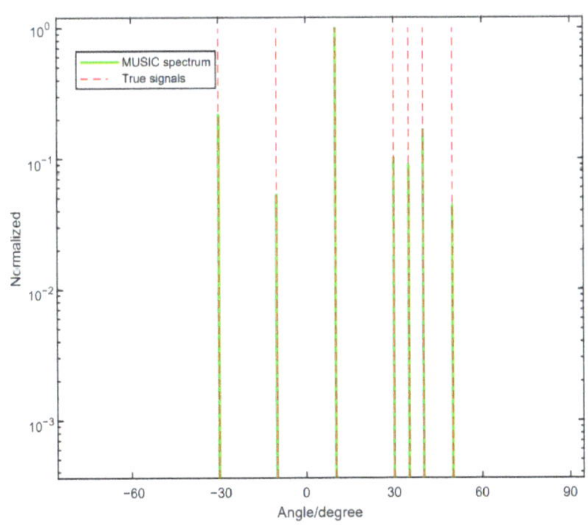

Figure 9. The reconstructed array configuration can achieve the full DOFs of seven source signals.

4.2.2. Computational Complexity Analysis

We compute the complexity of the proposed method and compare the complexity with the methods in [33,34]. The complexity of the proposed method is similar to the method in [33], which is denoted as $O(T^2L + T^3 + GT(T-K))$, where $T = M_1 + M_2 - 1$. Furthermore, $G = 180°/\tau$ is the number of the spectrum searching, where τ is the searching step, and we usually set $\tau = 0.1°$. L is the number of the snapshots. The method in [34] needs an additional algorithm to distinguish the true DOAs from the MUSIC spectrum. Thus, the complexity exceeds the method in [33] and the proposed method, which is $O(T^2L + T^3 + GT(T-K) + QT^2)$. Q is the number of searching of the additional beamforming technique. Table 1 presents the computational complexity comparison and the running time of the methods above, which are computed using MATLAB R2015b under the condition of Intel (R) Xeon (R) CPU E430 @3.10 GHz and 8 GB random access memory, where $L = 200$, $K = 3$, $(\theta_1, \phi_1) = (20°, 38.88°, 47.90°)$, $M_1 = 5, M_2 = 4$, which clearly shows that the proposed method is similar to the method in [33] and outperforms the method in [34]. Figure 10 depicts the complexity comparison versus the number of sensors with subarray 2, where the number of subarray 1 is $M_1 = 5$. As the proposed method does not need an additional procedure to identify the targets, it shows clearly that its complexity is much lower than the method in [34], and is close to the method in [33].

Table 1. Comparison of computational complexity.

Algorithms	Computational Complexity	Running Time
Method in [33]	$O(T^2L + T^3 + GT(T - K))$	4.1106 ms
Method in [34]	$O(T^2L + T^3 + GT(T - K) + QT^2)$	6.0122 ms
The proposed	$O(T^2L + T^3 + GT(T - K))$	4.1245 ms

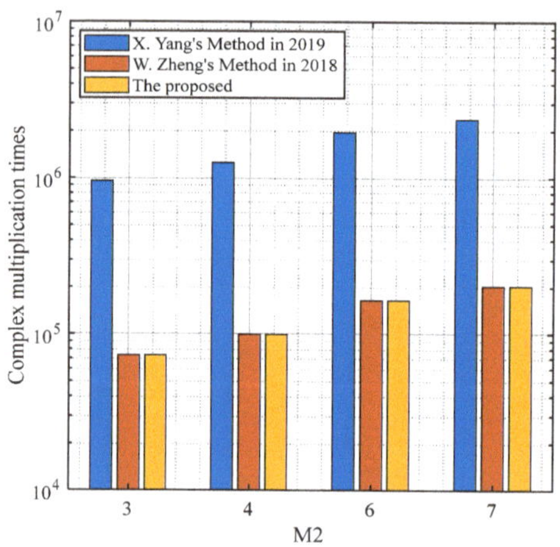

Figure 10. The complexity versus the number of sensors [33,34].

4.2.3. Advantages

In this part, we summarize the advantages of the improved array of IUCLA for DOA estimation.

(1) The proposed method can effectively eliminate the phase ambiguity problem by rearranging the reference element spacing and breaking the spectral function with the directional matrix.

(2) The improved array treats the array as a whole so that it can achieve the full DOFs; hence, it improves the estimation performance. Nevertheless, the method in [31] has a great loss in DOF, which can acquire, at most, $M_1 - 1$ DOFs, where we assume $M_1 < M_2$.

(3) CRB, as the lower bound for the unbiased estimation, is provided as a standard to measure the estimation performance. Furthermore, it is proven that the designed array can achieve the lower CRB.

(4) The designed array can achieve DOA estimation with an excellent DOA estimation performance compared to other algorithms, and it does not need additional algorithms.

(5) The proposed method performs a significant complexity decrease compared to the method in [33] due to the fact that the method does not use an additional technique, and it has a slight degrading of estimation performance compared to the method in [34].

4.2.4. Cramer–Rao Bound

CRB is the lower bound for unbiased estimation. We provide the CRB as a standard to measure the estimation performance [35,36]. In this section, we derive the CRB of the designed IUCLA.

First, we construct the steering matrix of IUCLA as

$$\boldsymbol{A}_\alpha = \begin{bmatrix} \boldsymbol{A}_{11} \\ \boldsymbol{A}_\chi \end{bmatrix} \tag{26}$$

where $\boldsymbol{A}_\chi$ represents a sub-matrix containing the second row to the last one of $\boldsymbol{A}_{22}$, since these two subarrays of IUCLA share the same element at the original point.

Referring to [36], we can obtain the CRB as

$$\boldsymbol{CRB} = \frac{\sigma_n^2}{2L}\left\{Re\left[\boldsymbol{D}^H\left[\boldsymbol{I} - \boldsymbol{A}_\alpha(\boldsymbol{A}_\alpha^H\boldsymbol{A}_\alpha)^{-1}\boldsymbol{A}_\alpha^H\right]\boldsymbol{D} \oplus \boldsymbol{R}_s\right]\right\}^{-1} \tag{27}$$

where $\boldsymbol{R}_s = \frac{1}{L}\sum_{t=1}^{L} \boldsymbol{s}(t)\boldsymbol{s}^H(t)$, $\boldsymbol{D} = \left[\frac{\partial \boldsymbol{a}_{\alpha,1}}{\partial\theta_1}, \frac{\partial \boldsymbol{a}_{\alpha,2}}{\partial\theta_2}, \ldots, \frac{\partial \boldsymbol{a}_{\alpha,K}}{\partial\theta_K}\right]$ and $\boldsymbol{a}_{\alpha,k}$ denotes the k-th column of $\boldsymbol{A}_\alpha$.

5. Simulation Results and Discussion

In the simulation section, we validate the reliability of the proposed method compared to the methods in [33,34], where we employ an UCLA. Subarray 1 is with $M_1 = 5$ sensors and subarray 2 is with $M_2 = 7$ sensors. Furthermore, we present the comparison of different methods and arrays, including UCLA, CLA and the designed IUCLA.

5.1. Reliability Comparison Analysis Based on Different Methods

Example 1: We assume that three signals, $\theta_1 = 12.37°, \theta_2 = 30°$ and $\theta_3 = 64.16°$, will be coming to the array. It can be noticed that these signals satisfy the Equation (10). In the simulation, we set SNR = -5 dB and the number of snapshots to $L = 200$. We provide the simulation result of the method in [33] using the proposed method. From Figure 11a, we can see that using the designed method has no ambiguity when three sources satisfy Equation (10), whereas the ambiguity-free method still has an ambiguity problem. In conclusion, the proposed method can tackle the phase ambiguity problem in Case 2.

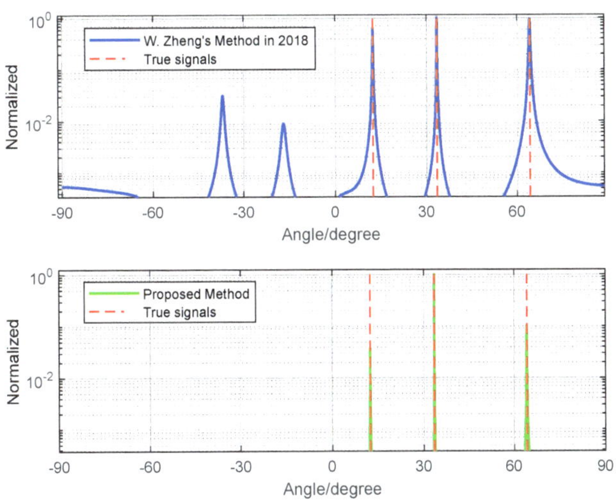

(a)

Figure 11. *Cont.*

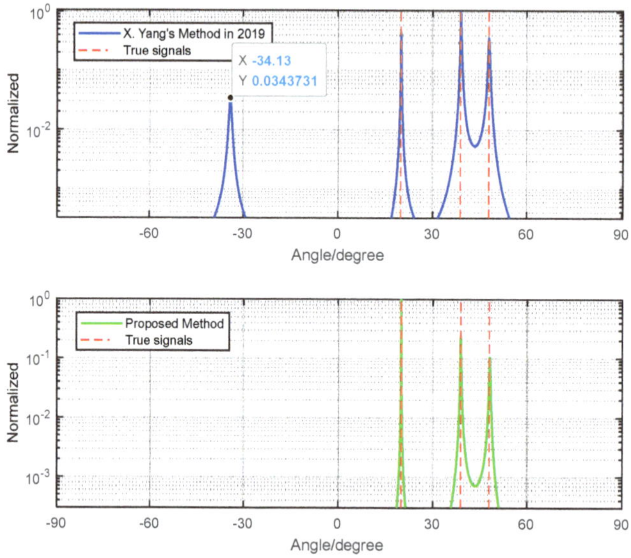

(**b**)

Figure 11. (**a**) Comparison of the proposed method to the method in [33] and (**b**) comparison of the proposed method to the method in [34].

Example 2: We assume three signals with angles of $\theta_1 = 20°, \theta_2 = 38.88°$ and $\theta_3 = 47.90°$. It can be noticed that these signals satisfy Equation (10). In the simulation, we set SNR = -5 dB and the number of snapshots to $L = 200$. We provide the simulation results of the method in [34], using the proposed method. From Figure 11b, we can see that the method in [34] is not that effective, and the designed array can effectively solve the phase ambiguity problem.

5.2. Estimation Properties Analysis of the Proposed Method

For the same simulation scenario mentioned above, we utilize the Root Mean Square Error (RMSE) to validate the estimation accuracy of the proposed method, which is defined as [33]

$$RMSE = \sqrt{\sum_{p=1}^{Q} \sum_{k=1}^{K} \left(\hat{\theta}_{k,p} - \theta_k\right)^2 / PK} \tag{28}$$

where P is the number of Monte Carlo simulations and $\hat{\theta}_{k,p}$ stands for the estimate of the p-th trial for the k-th theoretical angle θ_k. In the rest of our paper, we set $P = 1000$. Figures 12 and 13 show the RMSE versus the SNR and the number of snapshots, respectively. The CRB is provided. It is observed that the proposed method improves with the increase in SNR and the number of snapshots, owing to its robustness against noise. Compared to ESPRIT and PM algorithms, it is obvious that the estimation accuracy of the proposed method performs the superior estimation behavior.

5.3. CRB Comparison Analysis Based on Different Arrays

We compare the estimation performance of several array configurations, including UCLA, conventional ULA and our designed IUCLA. In the simulation, for fair comparison, we assume that all arrays have the same number of sensors. We can conclude from Figures 14 and 15 that the CRB and estimation performance of the proposed method are superior to the other two arrays.

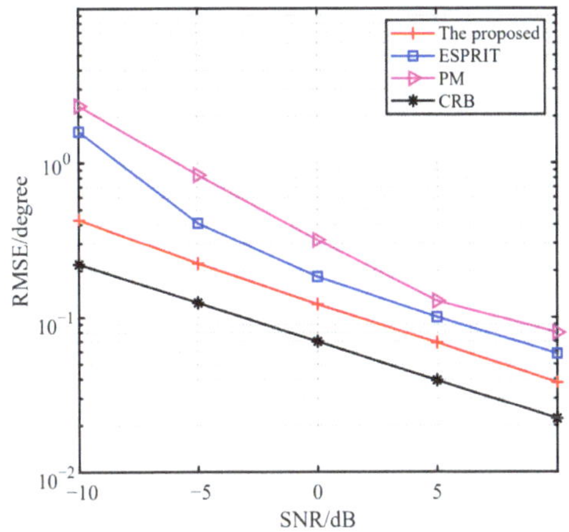

Figure 12. The RMSE versus the SNR of the proposed method.

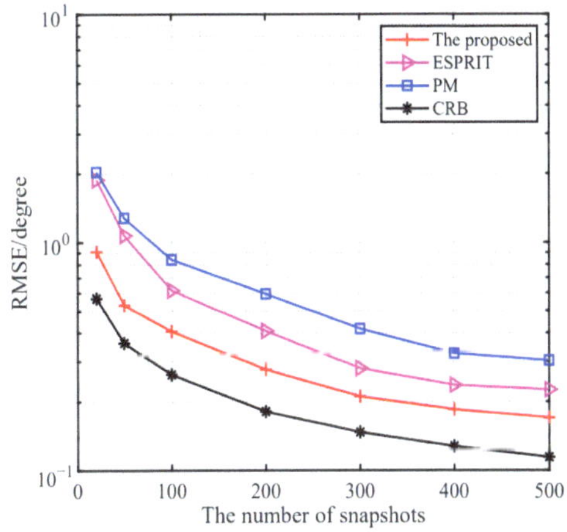

Figure 13. The RMSE versus the snapshot of the proposed method.

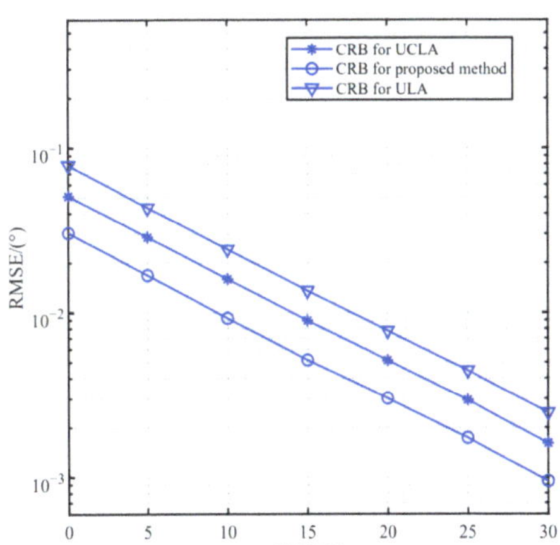

Figure 14. The RMSE versus the SNR based on different arrays.

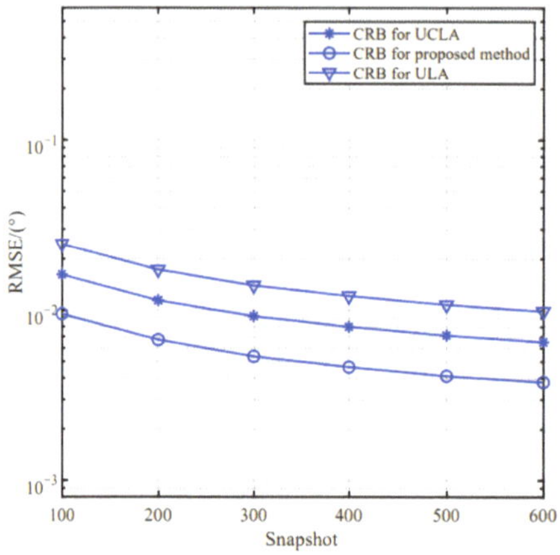

Figure 15. The RMSE versus the snapshot based on different arrays.

6. Conclusions

In this paper, an IUCLA was proposed to address the ambiguity problem of the coprime array by rearranging the reference sensor to UCLA to reshape the directional vectors of subarrays. Subsequently, we introduce the third coprime integer. It combines another coprime number to form a pair of coprime integers with pairwise coprime. Multiple pairwise coprime integers are employed, and the strong sidelobes can be suppressed by coprime property. In this process, we do not need any other technique to eliminate the ambiguous problem. The proposed method not only improves the DOA estimation performance and the number of DOFs, but it also reduces the computational complexity. CRB analysis and extensive simulation results have demonstrated the superiorities and effectiveness of our method.

Author Contributions: The main idea was proposed by P.G. and X.Z. X.Z. conceived the experiments and provided many valuable suggestions. P.G. conducted experiments and collected data. P.G. wrote the manuscript, and all of the authors participated in amending the manuscript. All authors have read and agreed to the published version of the manuscript.

Funding: This work is supported by the 2024 Jiangsu Autonomous Driving Technology Engineering Center Open Fund Project (No. ZK24-06-01), the Natural Science Research of Jiangsu Higher Education Institutions of China (No. 22KJD140004 and No. 24KJB510010), the University-level Scientific Research (No. YK20-02-11, YK20-02-12 and No. YK22-02-04) and the Jiangsu Wind Power Engineering Center Project of China (No. ZK22-03-09).

Data Availability Statement: Data are contained within the article.

Conflicts of Interest: The authors declare no conflicts of interest.

References

1. Gorcin, A.; Arslan, H. A two-antenna single RF front-end DOA estimation system for wireless communications signals. *IEEE Trans. Antennas Propag.* **2014**, *62*, 5321–5333. [CrossRef]
2. Krim, H.; Viberg, M. Two decades of array signal processing research: The parametric approach. *IEEE Signal Process. Mag.* **1996**, *13*, 67–94. [CrossRef]
3. Chen, P.; Chen, Z.; Zheng, B.; Wang, X. Efficient DOA estimation method for reconfigurable intelligent surfaces aided UAV swarm. *IEEE Trans. Signal Process.* **2022**, *70*, 743–755. [CrossRef]
4. Mei, R.; Tian, Y.; Huang, Y.; Wang, Z. 2D-DOA Estimation in Switching UCA Using Deep Learning-Based Covariance Matrix Completion. *Sensors* **2022**, *22*, 3754. [CrossRef] [PubMed]
5. Zhao, F.; Hu, G.; Zhou, H.; Guo, S. Research on Underdetermined DOA Estimation Method with Unknown Number of Sources Based on Improved CNN. *Sensors* **2023**, *23*, 3100. [CrossRef]
6. Qiu, S.; Ma, X.; Zhang, R.; Han, Y.; Sheng, W. A dual-resolution unitary ESPRIT method for DOA estimation based on sparse co-prime MIMO radar. *Signal Process.* **2023**, *202*, 108753. [CrossRef]

7. Ma, T.; Du, J.; Shao, H. A Nyström-Based Low-Complexity Algorithm with Improved Effective Array Aperture for Coherent DOA Estimation in Monostatic MIMO Radar. *Remote Sens.* **2022**, *14*, 2646. [CrossRef]
8. Hamid, U.; Wyne, S.; Butt, N.R. Joint Model-Order and Robust DOA Estimation for Underwater Sensor Arrays. *Sensors* **2023**, *23*, 5731. [CrossRef]
9. Mylonakis, C.M.; Velanas, P.; Lazaridis, P.I.; Sarigiannidis, P.; Goudos, S.K.; Zaharis, Z.D. 3D Direction of Arrival Estimation: An Innovative Deep Neural Network Approach. In Proceedings of the 2024 13th International Conference on Modern Circuits and Systems Technologies, Sofia, Bulgaria, 26–28 June 2024. [CrossRef]
10. Schmidt, R. Multiple Emitter Location and Signal Parameter Estimation. *IEEE Trans. Antennas Propag.* **1986**, *34*, 276–280. [CrossRef]
11. Sun, F.; Gao, B.; Chen, L.; Lan, P. A low-complexity ESPRIT-based DOA estimation method for co-prime linear arrays. *Sensors* **2016**, *16*, 1367. [CrossRef]
12. Wang, Y.Y.; Chen, J.T.; Fang, W.H. TST-MUSIC for joint DOA-delay estimation. *IEEE Trans. Signal Process.* **2001**, *49*, 721–729. [CrossRef]
13. Zhang, Y.; Ng, B.P. MUSIC-like DOA estimation without estimating the number of sources. *IEEE Trans. Signal Process.* **2009**, *58*, 1668–1676. [CrossRef]
14. Vallet, P.; Mestre, X.; Loubaton, P. Performance analysis of an improved MUSIC DOA estimator. *IEEE Trans. Signal Process.* **2015**, *63*, 6407–6422. [CrossRef]
15. Merkofer, J.P.; Revach, G.; Shlezinger, N.; Routtenberg, T.; van Sloun, R.J.G. DA-MUSIC: Data-driven DOA estimation via deep augmented MUSIC algorithm. *IEEE Trans. Veh. Technol.* **2023**, *73*, 2771–2785. [CrossRef]
16. Lin, J.-D.; Fang, W.-H.; Wang, Y.-Y.; Chen, J.-T. FSF MUSIC for joint DOA and frequency estimation and its performance analysis. *IEEE Trans. Signal Process.* **2006**, *54*, 4529–4542. [CrossRef]
17. Qian, C.; Huang, L.; So, H.C. Improved unitary root-MUSIC for DOA estimation based on pseudo-noise resampling. *IEEE Signal Process. Lett.* **2013**, *21*, 140–144. [CrossRef]
18. Chen, F.J.; Kwong, S.; Kok, C.W. ESPRIT-like two-dimensional DOA estimation for coherent signals. *IEEE Trans. Aerosp. Electron. Syst.* **2010**, *46*, 1477–1484. [CrossRef]
19. Herzog, A.; Habets, E.A.P. Eigenbeam-ESPRIT for DOA-vector estimation. *IEEE Signal Process. Lett.* **2019**, *26*, 572–576. [CrossRef]
20. Lin, J.; Ma, X.; Yan, S.; Hao, C. Time-frequency multi-invariance ESPRIT for DOA estimation. *IEEE Antennas Wirel. Propag. Lett.* **2015**, *15*, 770–773. [CrossRef]
21. Li, J.; Jiang, D.; Zhang, X. DOA estimation based on combined unitary ESPRIT for coprime MIMO radar. *IEEE Commun. Lett.* **2016**, *21*, 96–99. [CrossRef]
22. Wu, N.; Qu, Z.; Si, W.; Jiao, S. DOA and polarization estimation using an electromagnetic vector sensor uniform circular array based on the ESPRIT algorithm. *Sensors* **2016**, *16*, 2109. [CrossRef] [PubMed]
23. Bai, L.; Peng, C.Y.; Biswas, S. Association of DOA estimation from two ULAs. *IEEE Trans. Instrum. Meas.* **2008**, *57*, 1094–1101. [CrossRef]
24. Wu, J.X.; Wang, T.; Suo, Z.Y.; Bao, Z. DOA estimation for ULA by spectral Capon rooting method. *Electron. Lett.* **2009**, *45*, 84–85. [CrossRef]
25. Ballal, T.; Bleakley, C.J. Phase-difference ambiguity resolution for a single-frequency signal. *IEEE Signal Process. Lett.* **2008**, *15*, 853–856. [CrossRef]
26. Liao, B.; Zhang, Z.G.; Chan, S.C. DOA estimation and tracking of ULAs with mutual coupling. *IEEE Trans. Aerosp. Electron. Syst.* **2012**, *48*, 891–905. [CrossRef]
27. Pal, P.; Vaidyanathan, P.P. Nested arrays: A novel approach to array processing with enhanced degrees of freedom. *IEEE Trans. Signal Process.* **2010**, *58*, 4167–4181. [CrossRef]
28. Pal, P.; Vaidyanathan, P.P. Coprime Sampling and the MUSIC Algorithm. In Proceedings of the IEEE Digital Signal Proc. Workshop and IEEE Signal Proceedings Education Workshop (DSP/SPE), Sedona, AZ, USA, 4–7 January 2011; pp. 289–294.
29. Qin, S.; Zhang, Y.D.; Amin, M.G. Generalized coprime array configurations for direction-of-arrival estimation. *IEEE Trans. Signal Process.* **2015**, *63*, 1377–1390. [CrossRef]
30. Raza, A.; Liu, W.; Shen, Q. Thinned coprime array for second-order difference co-array generation with reduced mutual coupling. *IEEE Trans. Signal Process.* **2019**, *67*, 2052–2065. [CrossRef]
31. Zhou, C.; Shi, Z.; Gu, Y.; Shen, X. DECOM: DOA Estimation with Combined MUSIC for Coprime Array. In Proceedings of the IEEE International Conference on Wireless Communications and Signal Processing (WCSP), Hangzhou, China, 24–26 October 2013; pp. 1–5.
32. Sun, F.; Lan, P.; Gao, B. Partial spectral search-based DOA estimation method for co-prime linear arrays. *Electron. Lett.* **2015**, *51*, 2053–2055. [CrossRef]
33. Zheng, W.; Zhang, X.; Gong, P.; Zhai, H. DOA Estimation for Coprime Linear Arrays: An Ambiguity-Free Method Involving Full DOFs. *IEEE Commun. Lett.* **2017**, *22*, 562–565. [CrossRef]
34. Yang, X.; Wang, Y.; Chargé, P. Modified DOA estimation with an unfolded co-prime linear array. *IEEE Commun. Lett.* **2019**, *23*, 859–862. [CrossRef]

35. Stoica, P.; Nehorai, A. Performance study of conditional and unconditional direction-of-arrival estimation. *IEEE Trans. Acoust. Speech Signal Process.* **1990**, *38*, 1783–1795. [CrossRef]
36. Stoica, P.; Gershman, A.B. Maximum-likelihood DOA estimation by data-supported grid search. *IEEE Signal Process. Lett.* **1999**, *6*, 273–275. [CrossRef]

sensors

MDPI

Article

A Circularly Polarized Non-Resonant Slotted Waveguide Antenna Array for Wide-Angle Scanning

Guodong Han * and Weihang Liu

The 54th Research Institute of China Electronics Technology Group Corporation, Shijiazhuang 050081, China; liu_weihang@126.com
* Correspondence: hgd_cetc@163.com

Abstract: A compact circularly polarized non-resonant slotted waveguide antenna array is proposed with the aim of achieving wide-angle scanning, circular polarization, and low side-lobe levels. The designed antenna demonstrates a scanning range of +11° to +13° in the frequency domain and a beam scanning range of −45° to +45° in the phase domain. This design exhibits significant advantages for low-cost two-dimensional electronic scanning circularly polarized arrays. It employs a compact element that reduces the aperture area by 50% compared to traditional circular polarization cavities. Additionally, the staggered array method is employed to achieve an element spacing of 0.57λ within the azimuth plane. Isolation gaps were introduced into the array to enhance the circular polarization performance of non-resonant arrays. The Taylor synthesis method was employed to reduce the side-lobe levels. A prototype was designed, fabricated, and measured. The results indicate superior radiation efficiency, favorable VSWR levels, and an axis ratio maintenance below 3 dB across the scanning range. The proposed antenna and methodology effectively broaden the beam scanning angle of circularly polarized slotted waveguide array antennas.

Keywords: slotted waveguide array; circular polarization; wide-angle scanning; non-resonant array; phased array antenna; radar sensor

1. Introduction

The slotted waveguide antenna offers notable advantages, such as a high power capacity, low loss, and low side-lobe levels; thus, it is extensively utilized in radar applications. A circularly-polarized antenna possesses the capability to receive multiple polarized waves, and the circularly polarized wave has smaller energy losses when penetrating rain or fog areas [1,2]. The slotted waveguide antenna, when applied in phased array systems, has significant advantages in complex electromagnetic environments, and it can achieve many beneficial characteristics, such as fast beam tracking, beamforming, etc. [3–6]. However, to date, there has been limited research attention directed toward the circularly polarized phased slotted waveguide antenna. Based on our research (1), the dimensions of the waveguide and polarizer are the primary limiting factors regarding the beam scanning range of the phased slotted waveguide antenna, as these large structures result in a notably narrow scanning range. Consequently, this article introduces a novel approach: a compact circularly polarized non-resonant slotted waveguide, aimed at achieving a broader scanning angle.

Typically, a phased slotted waveguide array performs the phase dimensional beam scanning along the waveguide arrangement direction. This array commonly operates under resonant conditions, radiating linearly polarized electromagnetic waves [7–11]. The radiation elements typically consist of a series of slots positioned along the narrow wall of the waveguide. According to the basic phased array theory, the maximum achievable phase scanning angle satisfies the following:

$$\theta_{max} = arcsin(\frac{\lambda}{d} - 1) \quad (1)$$

where d is the element distance. In the slotted waveguide array, d represents the distance between adjacent linear waveguides. The grating lobe will manifest in the visual space if the scanning angle exceeds θ_{max}.

Due to the limitation of the element distance, it is challenging to design a phased scanning slotted waveguide array with multiple polarizations. However, the regular fixed-beam slotted waveguide antenna arrays offer numerous valuable electric characteristics, such as wide-band polarization [12], dual-band, dual polarization [13–15], and circular polarization. Circularly polarized slotted waveguide arrays can be categorized into three distinct types based on their circular polarizer structures: the compound slots, the circular polarization patch or dipole, and the circular polarization (CP) cavity. Compound slots typically consist of a pair of slots [16,17] or some complex slots [18] that are carved into the wall of the waveguide. This type of CP element has the lowest profile and simplest configuration. However, achieving identical phase excitation requires the compound slots to be spaced at λ_g, necessitating the use of dielectric-filled waveguides or SIW to reduce the waveguide's wavelength [19]. While dielectrics offer advantages in miniaturization, they also introduce additional electromagnetic attenuation losses, and many dielectric materials can compromise structural integrity. Another option is the ridge gap waveguide [20], but it also faces difficulties in miniaturization. The CP patch, dipole, and CP cavity are loaded on the waveguide slots, serving as polarizers. The CP patch or dipole has a flexible configuration, offering an excellent circular polarization performance (e.g., high efficiency [21], broadband CP [22]), and patches are easily combinable with metamaterials and other novel technologies [23–25]. However, the compact patches and dipoles loaded on the waveguides need dielectric layers, and compactly arranged patches may induce more mutual coupling effects, limiting the power capacity. On the other hand, CP cavities are typically open-ended waveguides with specific configurations [26,27]. The cavity aperture can significantly reduce the mutual coupling between elements and enhance the isolation performance, being particularly beneficial in non-resonant and phased arrays with complex aperture distributions. Cavities also offer greater robustness compared to patches. Nevertheless, cavities pose challenges in miniaturization: the conventional CP cavity's aperture typically exceeds 0.6λ, according to Equation (1); this dimension limits the widest scanning angle to ±40°. Figure 1 clearly illustrates these limitations. The study in [28] introduced a prospective approach to miniaturization employing a four-ridged cavity and a center-set slot to reduce the aperture of the element. However, this study lacked comprehensive details regarding the antenna and primarily concentrated on a small resonant array. Drawing from the insights of ref. [28], we analyzed its construction and operational principles, expanding its applications to non-resonant arrays and phased arrays. Meanwhile, we addressed several challenges in integrating this compact structure into complex arrays. Additionally, the cavity's CP frequency band is narrower than the bands of some CP patches [29].

The miniaturization of the waveguide slots poses another challenge in phased array antennas. As discussed above, the broad wall slots in phased arrays result in a very limited scanning angle. While reducing the width of the waveguide using ridge waveguides [8,9] is an option, designing them with circular polarizers is not easy. The asymmetric ridge waveguide presents a potential solution [30]; however, existing analyses have primarily focused on resonant arrays. To our knowledge, there is a dearth of research addressing phase scanning slotted waveguide arrays generating circularly polarized waves. The non-resonant slotted waveguide array offers a low-cost solution for two-dimensional beam scan arrays, integrating phase and frequency scanning within the same array. This is particularly suitable for applications insensitive to frequency variations, such as meteorological radar. Excited by traveling waves, the non-resonant slotted waveguide has the capabilities for frequency scanning [31]. The frequency scanning direction is along the waveguide. With this capability, the non-resonant slotted waveguide array can offer a two-dimensional beam scan in both the phase domain and frequency domain [32]. Furthermore, compared to resonant arrays, the non-resonant array can accommodate a wider waveguide length range, enabling a wider bandwidth.

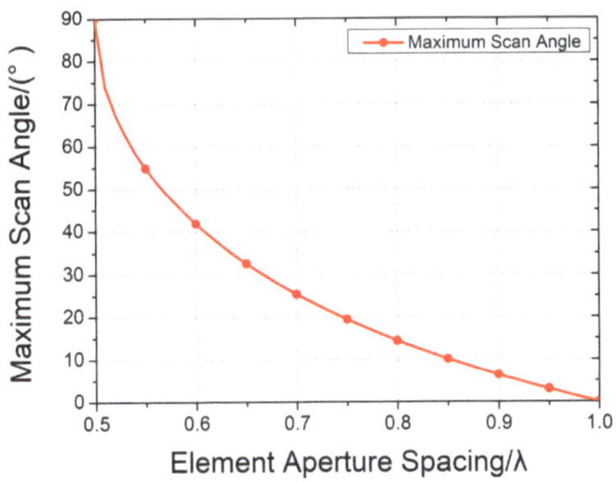

Figure 1. The relationship curve between the element aperture spacing and maximum scan angle.

To address the aforementioned problem, this article proposes an improved antenna structure aimed at achieving a wide beam scanning angle and integrating frequency domain and phase domain scanning array antennas. Our work introduced improved waveguide feeding slots, a circular polarizer, and staggered array design methods; a phase scanning angle of ±45° and a frequency domain scanning range of +11° to +13° were achieved in this study.

The key contributions of this study are as follows:

1. A center-set waveguide slot element on the centerline of the broad wall is proposed to set up a waveguide circular polarizer with four ridges;
2. A compact circular polarized slotted waveguide planar array is presented that can realize a wide phase scanning range.

The rest of this article is organized as follows. Section 2 introduces the design of the compact antenna element, including the circular polarizer and the center-set waveguide slot. Section 3 introduces the slotted waveguide array designed using this compact circularly polarized element, along with the technical details concerning beam scanning in the array design. Section 4 presents the simulation and the measured results. Finally, the conclusions are presented in Section 5.

2. Configuration of the Antenna Element

The proposed slotted antenna element configuration is depicted in Figure 2. Comprising two components, namely, the circular polarizer and the slotted ridge waveguide, the element functions as follows. A single-ridge waveguide with a straight slot positioned along the centerline of the broad wall serves as a feed waveguide. A metallic cylinder placed near the slot disturbs the electromagnetic field distribution within the waveguide. The slotted ridge waveguide emits linearly polarized waves through the straight slot on the waveguide. The circular polarizer, configured as an open-ended quad-ridge waveguide, is mounted on the slot at a 45° angle.

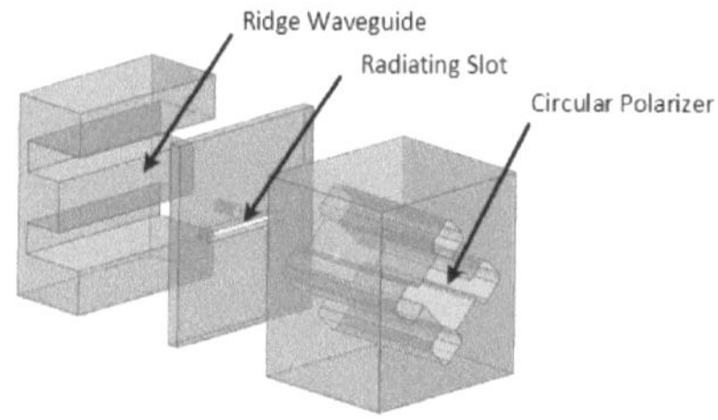

Figure 2. Configuration of the proposed antenna element.

2.1. Design of Circular Polarizer

The circular polarizer is implemented using an open-ended four-ridge waveguide. Excited by the slot, the circular polarizer reflects linearly polarized waves fully from the cavity walls, generating two orthogonal electric field components: the TE_{10} mode and the TE_{01} mode. In properly adjusting the polarizer's profile, an equal amplitude and a 90° phase gradient between these two propagating modes can be achieved. Consequently, upon reaching the radiating aperture interface, the combination of these two modes yields circularly polarized waves. The primary objective of the waveguide circular polarizer is to propagate the TE_{10} and TE_{01} modes while suppressing the TE_{11} mode within the operating frequency band.

A cross-section of the circular polarizer is depicted in Figure 3. The outer shape of the polarizer is a square, with four ridges positioned inside, where the opposite ridges are of equal size. The parameters of the ridges significantly impact the propagation characteristics of the electromagnetic wave modes. Specifically, the lower ridges primarily affect the TE_{10} mode, while the higher ridges influence the TE_{01} mode. Moreover, due to their close proximity, the TE_{11} mode is susceptible to influence from both the TE_{01} and TE_{10} modes. The electric field distributions of these three modes in the polarizer cross-section are shown in Figure 4.

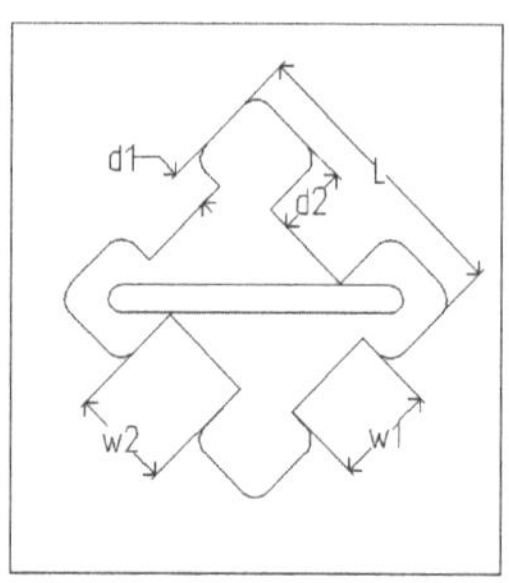

Figure 3. Top view of the circular polarizer.

The parameter values are subject to several conditions. The outer size of the polarizer is determined by the dimensions of the element radiator, which, ideally, should be minimized. The width of the ridges is contingent upon the fabrication accuracy requirements and had a negligible influence on the cutoff frequencies in this study. Consequently, the analysis focused on the ridge parameters to ascertain specific changes in the cutoff frequency. As depicted in Figure 5, $d1$ and $d2$ exert significant effects on the cutoff frequencies of the three modes. The simulated results align with the prior analysis; Figure 5a illustrates that $d1$ can alter the cutoff frequencies of the TE_{10} and TE_{11} modes, with their curves exhibiting similar trends, whereas the performance of the TE_{01} mode is minimally affected by changes in $d1$. Similarly, Figure 5b demonstrates that variations in $d2$ have a similar effect than those in $d1$, albeit with an emphasis on the TE_{10} and TE_{11} modes. Our objective is to preserve the TE_{01} and TE_{11} modes while suppressing the TE_{11} mode within the operating frequency band. The available bandwidth is indicated by the black arrow in Figure 5b. By finely adjusting these parameters, we can effectively control the propagating modes of the polarizer.

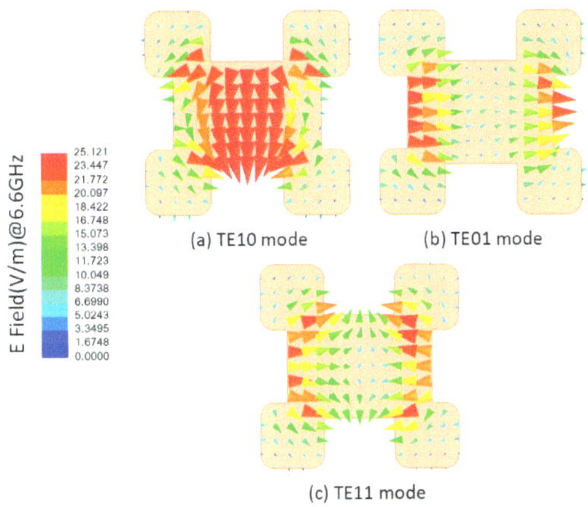

Figure 4. Electric field distributions of the three modes in a circular polarization cavity at 6.6 GHz.

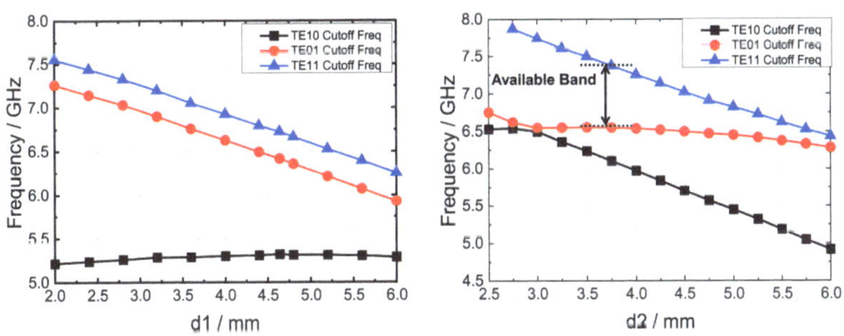

(**a**) (**b**)

Figure 5. (**a**) The relationship between the cutoff frequencies and ridge height *d*1 and (**b**) the relationship between the cutoff frequencies and ridge height *d*2.

The cutoff frequencies of each mode, as well as the waveguide wavelength, are obtained according to the parameters of the ridges. In order to satisfy the phase difference between the two modes, the height of the polarizer satisfies the following:

$$\frac{2\pi h}{\lambda_{g10}} - \frac{2\pi h}{\lambda_{g01}} = \frac{\pi}{4} \tag{2}$$

where λ_{g10} is the waveguide wavelength of the TE_{10} mode, λ_{g01} refers to the waveguide wavelength of the TE_{01} mode, and h is the height of the polarizer.

With the aforementioned analysis, the dimensional parameters of the polarizer are established. The accepted design parameters are listed in Table 1.

Table 1. The parameter values of the four-ridge cavity design.

Parameters	s1	s2	d1	d2	h
Value (mm)	7.5	7.5	2.9	5.3	31.3

2.2. *Slotted Waveguide with Metallic Cylinder Inside*

A linear center-set slotted waveguide array with a compact ridge is proposed. Longitudinal slots, serving as radiating elements, are positioned along the centerline of the waveguide's broad wall. Metallic cylinders are staggered on the upper inner wall of the waveguide, on alternating sides of the slots. The configuration of the single-slot radiator is illustrated in Figure 6.

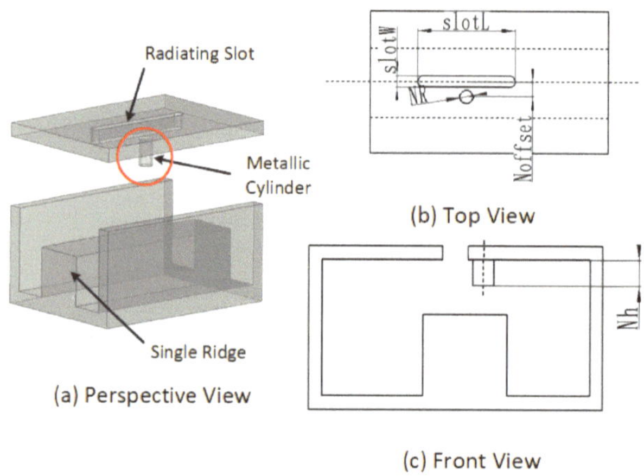

Figure 6. The configuration of the proposed feed-slotted waveguide. (**a**) Perspective view. (**b**) Top view. (**c**) Front view.

Conventional waveguide slots are typically carved at specific offset distances from the centerline of the broad wall. However, as previously discussed, the polarizer is loaded above the center of each slot at a 45° rotated angle. Consequently, conventional offset slots would lead to offset polarizers, resulting in wider antenna elements. With the proposed arrangement of the waveguide slots and polarizers, the element width can be significantly reduced. This approach allows all radiating slot elements to be arranged in a regular grid, eliminating the need for offsetting them as in traditional designs.

The introduction of special structures into the waveguide can alter its electromagnetic performance [33,34]. As depicted in Figure 6, a metallic cylinder is positioned on the inner upper wall of the waveguide at a specific offset distance from the centerline. The role of the cylinder is to disrupt the electromagnetic field distribution within the waveguide, resulting in a concentration of the electric field. In the conventional ridge waveguide, the waveguide's upper wall and the ridge's upper wall can be approximated to a plate capacitor. The capacitor's field distribution and voltage performance are influenced by the two plates' distance, which is the distance between the waveguide upper wall and the ridge upper wall in this situation. With a metallic cylinder introduced into the waveguide, the capacitor model is altered to three capacitors; the cylinder's bottom surface and the ridge's upper surface are approximated to a capacitor, and the remaining parts are approximated to other capacitors. According to capacitor theory, the closer plates have a higher voltage. Hence, the voltage at the cylinder is higher than that in other areas. Figure 7 illustrates the field distribution of the waveguide with the cylinder, showing the gathering of electric field and surface current at the cylinder position. This disruption leads to an asymmetric distribution of surface current, causing an electric potential difference at the centerline of the waveguide. Under these conditions, the center-set slot can be effectively excited by the asymmetric surface current, with the available slot positions indicated by red marks in Figure 7a. Through the preceding analysis, it is evident that both the location and scale of the cylinder influence the field distribution.

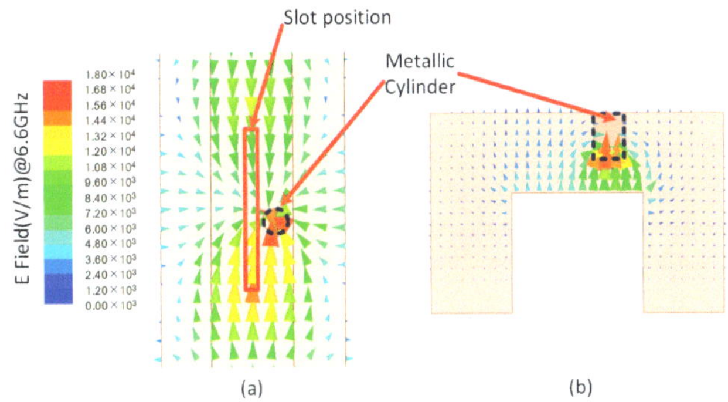

Figure 7. The field distribution of the waveguide with a cylinder. (**a**) The electric field in the upper wall of the waveguide. (**b**) The electric field in the cross-section of the waveguide.

Basic slotted waveguide antenna theory indicates that the radiation efficiency is contingent upon the excitation voltage of the slot. Leveraging these theories, we adjusted the parameters of the cylinder to analyze the impact on the radiation performance of the waveguide slot. Figure 8 depicts the simulation model of the slotted waveguide element. Figure 9 depicts the S21 results for various cylinder heights under resonant conditions. The simulation results demonstrate that longer cylinders result in more radiated energy. Furthermore, while the offset distance of the cylinder does influence radiation efficiency, its effect is not particularly significant; rather, it primarily affects the gathering position of the surface current. The radius of the cylinder impacts the intensity of the reflected wave, with thicker cylinders potentially being intersected by the slot, resulting in irregular shapes and complex field distributions. Additionally, the slot length can be adjusted to tune the resonant frequency of the element, akin to conventional waveguide slots.

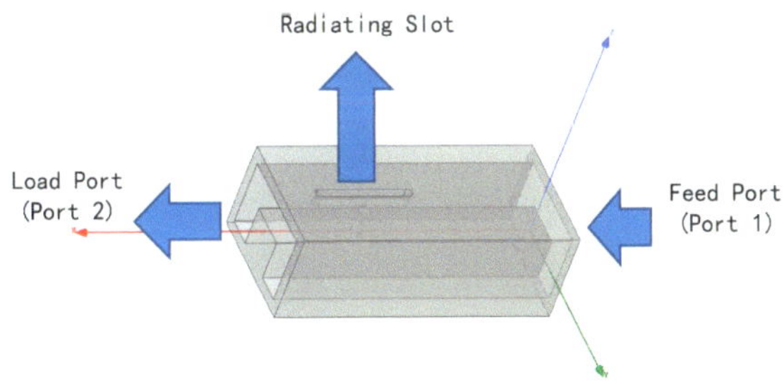

Figure 8. The simulation model of the slotted waveguide element.

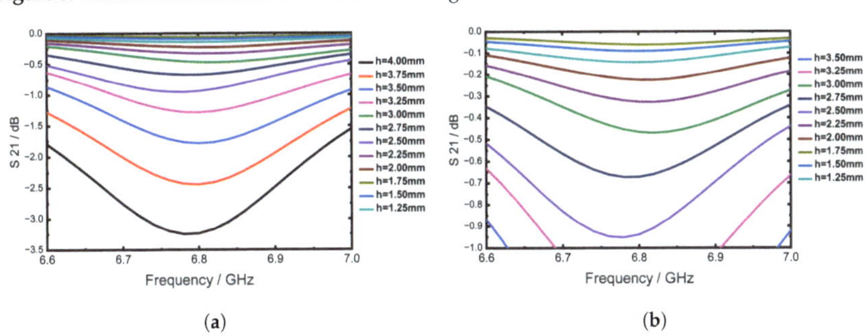

Figure 9. (**a**) The simulated S21 (dB) curves of the single resonant slot when changing the cylinder height and (**b**) the enlarged view for h = 2.50 mm to 4.00 mm.

Following the above discussion, we can manipulate the radiation efficiency by adjusting the height of the cylinder. The relation between the radiation efficiency of the slot and the cylinder height is shown in Figure 10; the cylinder height refers to *Nh* in Figure 6c. the radiation efficiency is defined as follows:

$$Eff_{rad} = 1 - (S11)^2 - (S21)^2 \quad (3)$$

where Eff_{Rad} refers to the radiation efficiency of the antenna element, S11 is the energy reflected back to the excited port, and S21 is the energy transmitted to the load or next element.

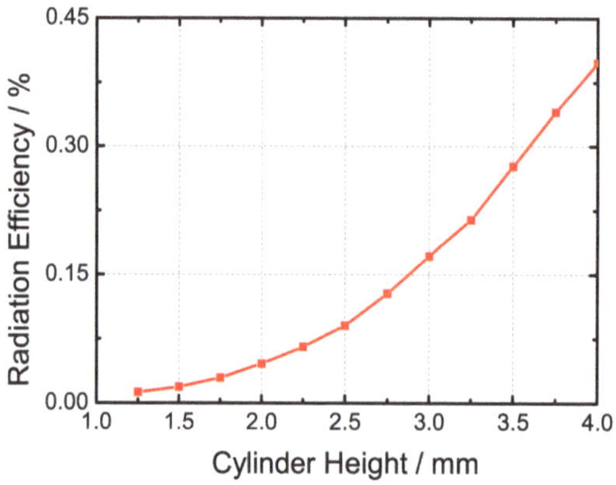

Figure 10. The relationship between the simulated radiation efficiency and the cylinder height under the resonant condition.

A method for controlling the element's radiation efficiency is introduced in this section; it lays the foundation for the follow-up array design work.

This section describes the working principle and design method of a compact circularly polarized antenna element, resulting in a miniaturized circularly polarized antenna element with an aperture size of $0.48\lambda \times 0.48\lambda$. Additionally, this section introduces the method used to control the radiation efficiency of the element, which serves as the foundation for subsequent design considerations.

3. Design of Antenna Array

3.1. Linear Array

In this section, the compact antenna element proposed in Section 2 is arranged into a linear array, with each antenna element featuring a center-set waveguide and a compact circular polarizer. They are fed by a straight-through waveguide tube, and the structure of this linear array is shown in Figure 11. The linear array is a non-resonant array that is excited by a traveling wave. In non-resonant arrays, the slots are excited with different phases, resulting in the main lobe beam deviating from the normal direction. The phase difference between the slots varies with the frequency, causing the main lobe direction to change accordingly. Thus, non-resonant arrays possess the capability of frequency scanning. In drawing on fundamental phased array antenna theory and waveguide theory, the relationship between the slots distance d and the beam angle θ satisfies the following:

$$\theta = arcsin(\frac{\lambda}{\lambda_g} - \frac{\lambda}{2d}) \quad (4)$$

The distance between the slots also impacts the resonant frequency of the linear array. When the slot distance is $0.5\lambda_g$, the non-resonant array exhibits a very high VSWR at the resonant frequency [35,36]. This phenomenon can be mitigated by increasing the slot distance. However, excessively long slot distances will result in a high grating lobe level. The suppression condition for grating lobes can also be described as in Equation (1). In conclusion, the maximum value of the slot distance is constrained by the performance of the grating lobe level, while the minimum value is determined by the VSWR performance. Additionally, the main lobe direction must be considered in the determination of the slot distance.

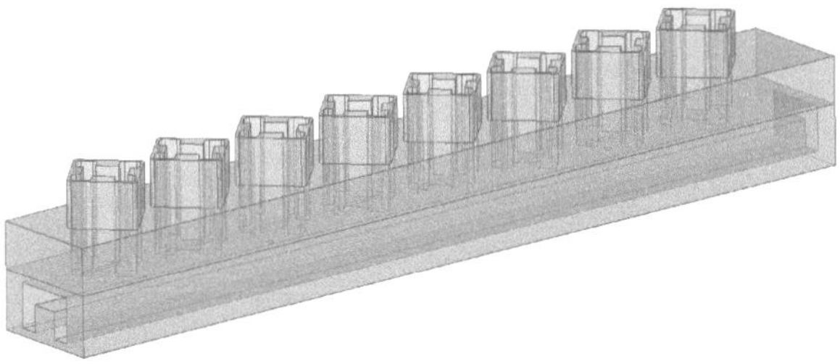

Figure 11. Configuration of an eight-element linear array.

The control of the first side-lobe level was also achieved in this work. Side-lobe control in slotted waveguide arrays is accomplished by regulating the radiation power weighting of the elements. The weighting factor is determined using the Taylor Pattern Synthesis Method. In building upon the findings from Section 2, the differential power weighting of the elements is achieved through adjustments in the height of the cylinder and the length of the slot.

3.2. Planar Array

In this section, the parallel waveguide linear array described in Section 3.1 is arranged into a two-dimensional array, and the structure of the planar array is shown in Figure 12. To reduce the width of the linear array, we adopted the staggered arrangement method to position adjacent linear arrays. However, this staggered arrangement results in an asymmetric planar array, leading to high axial ratio levels and elevated, far side-lobe levels. To address this issue, we staggered the placement of the eight-element and nine-element linear arrays. However, the staggered non-resonant array exhibits a less-than-optimal phase distribution. As depicted in Figure 13a, the phase distribution of the eight-element array does not align with that of the nine-element array. This mismatched phase distribution exacerbates the grating lobe performance during phase scanning, as indicated by the red line in Figure 14. To remedy this, we increase the excitation phase of the eight-element arrays to compensate for the phase offset induced by the staggered arrangement (Figure 13b). The magnitude of the increased phase is determined by the waveguide wavelength λ_g and the slot distance d. The precise expression for the compensated phase is given by

$$Phase_{comp} = (\frac{d}{\lambda} - \frac{1}{2})\pi \tag{5}$$

As shown in Figure 13, the phase compensation can be achieved by increasing the distance between the excite port and the first element. Figure 13a shows the situation without phase compensation; the neighboring elements exhibit non-uniform phase excitation, and the red arrow indicates the direction of the electromagnetic wave transmission. In Figure 13b, we introduce an additional distance between the first element and the feed port, as highlighted by the yellow rectangular marker; subsequently, the first element's working phase is compensated for to achieve the correct value. The compensated result is depicted by the blue line in Figure 14, and the grating lobe was suppressed effectively by applying this method.

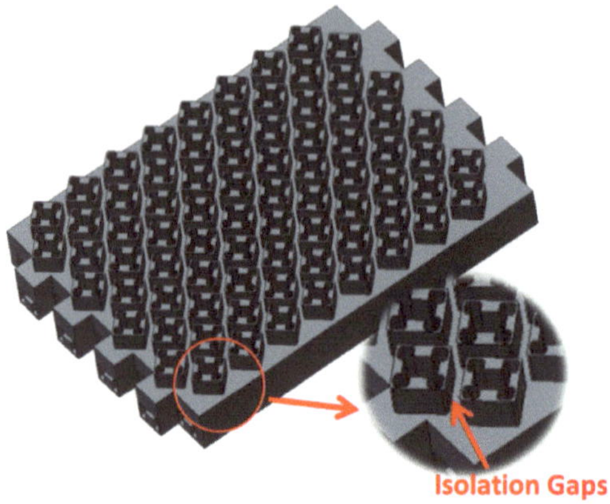

Figure 12. Configuration of the planar array.

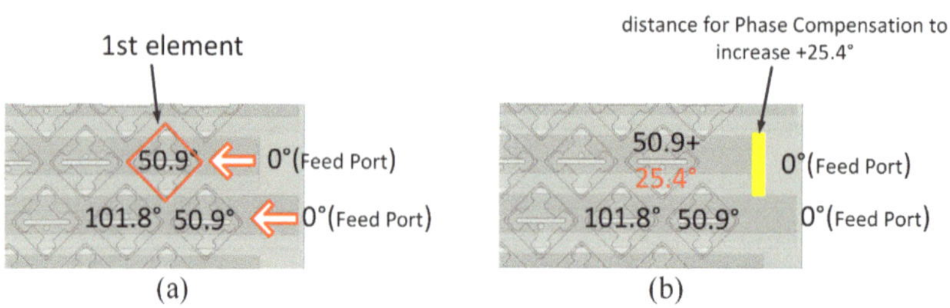

Figure 13. (**a**) The mismatching phase distribution in the planar array. (**b**) The compensation phase distribution in the planar array.

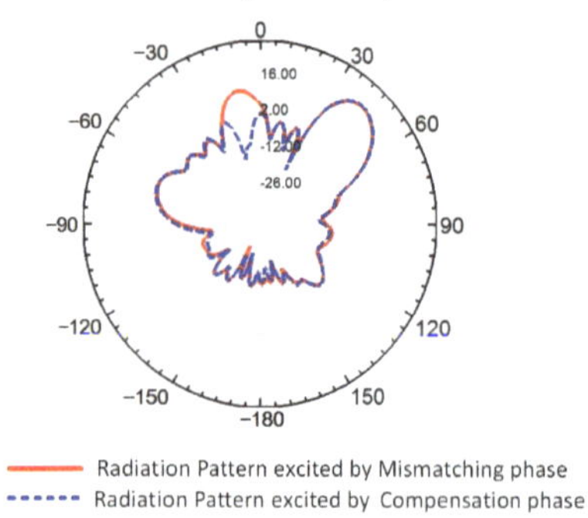

Figure 14. The red line depicts the simulated radiation pattern under the mismatching phase condition, and the blue line depicts the simulated radiation pattern under the compensation phase condition, at a 45° beam scanning angle through phase scanning.

With the introduction of the nine-element array into the planar array, the power weighting of the elements requires re-tuning. Similar to phase compensation, the radiation fields of the eight-element and nine-element arrays also need to be aligned. To address this, we utilize the near field results, as depicted in Figure 15. The radiation performance of the eight-element array serves as the reference standard. By plotting the envelope line of the near radiated electric field, we can approximate the changing curve of the near field for the eight-element linear array. The design parameters of the nine-element array are adjusted to match the curve extracted from the eight-element array. Initially, the weighting parameters of the nine-element array are defined using the pattern synthesis method, and the initial near field curve is extracted and compared with that of the eight-element array. Based on the discrepancy between the two curves, adjustments are made to the cylinder height and slot length of the nine-element array to ensure that its curve matches the standard curve. This iterative process is repeated several times throughout the design phase, resulting in the final curves depicted in Figure 15. The results demonstrate the good alignment between the two curves.

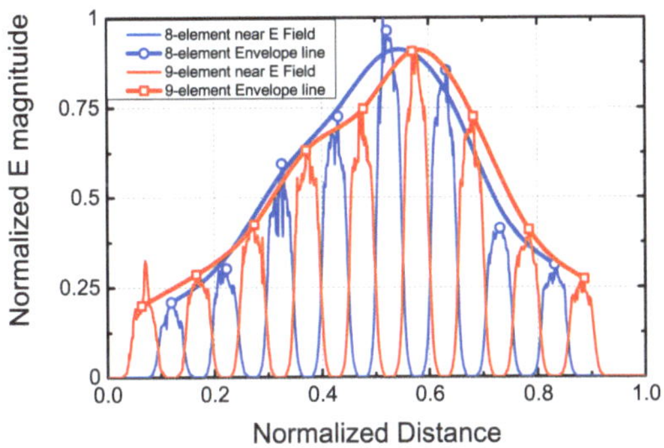

Figure 15. The near E-field curves and its envelope line of the adjacent eight- and nine-element linear arrays.

As discussed earlier, the staggered arrangement results in a complex near field distribution, exacerbated by the excitation of traveling waves, leading to the intricate phase distribution of the near field. Additionally, the compact arrangement of the radiating cavities exacerbates the coupling effects compared to normal conditions. In such scenarios, the planar array's leakage wave on the radiating surface exhibits a complex phase and field distribution, adversely affecting the axial ratio level. To mitigate the influence of the leakage wave, isolation gaps are introduced on the radiating surface. These gaps serve as channels into which the leakage wave can be transmitted, subsequently being reflected at the bottom of the gaps. By adjusting the depth of the gap, the reflected wave and transmitted wave can be neutralized within it. The configuration of the isolation gaps is illustrated in Figure 12.

With the above analyses and designs, we can obtain the final configuration of the planar array antenna. The detailed parameters are presented in Table 2.

Table 2. Detailed parameters of the planar array (unit: mm).

Slot No.	Eight-Element Array's Cylinder Height	Eight-Element Array's Slot Length	Nine-Element Array's Cylinder Height	Nine-Element Array's Slot Length
1	1.7	21.6	1.6	21.6
2	2.2	21.4	1.9	21.4
3	2.8	21.4	2.5	21.4
4	3.3	21.4	3.1	21.4
5	3.9	21.1	3.4	21.1
6	4.1	20.9	3.9	20.9
7	3.8	20.9	4.0	20.9
8	3.2	21.1	3.5	21.1
9	N/A	N/A	3.1	21.4

N/A = Not Applicable.

4. Verification and Comparison

The planar antenna was fabricated and measured in our work. During manufacturing, the antenna is divided into three layers: the waveguide layer, including the waveguide and ridge structure; the slot layer, including the slots and cylinders; and the circular polarizer layer, including the circular polarizer and isolation gap surface. The planar antenna was manufactured using CNC machining technology. During assembly, a conductive adhesive was applied to bond the waveguide layer and slot layer. This prevents the deterioration in the electrical performance caused by electromagnetic leakage resulting from the interlayer gaps; the circular polarizer layer and waveguide layer were mechanically connected using screws, which were distributed around the periphery of the prototype. The mechanical strength is the primary consideration when using screw connections, as the three layers are heavy and cannot withstand the required strength solely through adhesive bonding. The measurement was conducted using the far-field method in a 9 m × 9 m microwave anechoic chamber. The assembled antenna is depicted in Figure 16a, with dimensions of 370 mm × 240 mm × 52 mm.

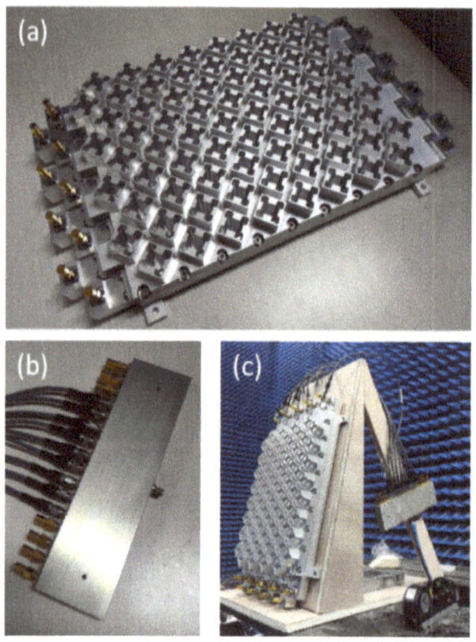

Figure 16. (**a**) The fabricated antenna. (**b**) The power divider. (**c**) The experimental setup for measurements.

The excitation method warrants mention. Although this antenna is intended for phased scan array applications, constructing a complete phased array system for measurement purposes is impractical. Instead, in this study, the phase delay was achieved using nine coaxial cables of varying lengths, ensuring approximate phase differences of 144 degrees. These cables were connected to a power divider and the antenna, as depicted in Figure 16b. The experimental setup is illustrated in Figure 16c, showing a 12° inclined support being employed to accommodate the deviated main lobe in the elevation plane.

4.1. Simulation and Measurement Results

The prototype antenna was verified, and the simulation and measurement results are presented below. The VSWR and S21 of the eight- and nine- element linear arrays were measured, respectively, abd the results are shown in Figure 17. It is observed that the measured results closely align with the simulation, indicating good agreement.

Figure 18 presents the radiation patterns of the proposed antenna. Figure 18a–c depict the elevation radiation patterns at 6.6 GHz, 6.8 GHz, and 7.0 GHz, respectively. The majority of the simulation and measurement results exhibit close agreement. However, there are slight discrepancies in the right side-lobe level (SLL) at 6.8 GHz, although all measured SLL results remained below −20 dB. Furthermore, the frequency scan angle results align well with those of the simulation. The azimuth radiation pattern results are depicted in Figure 16d–f; all measured azimuth radiation results exhibit good concordance with the simulation data. Additionally, the achieved side-lobe levels (SLLs), at approximately −14 dB, are attributed to the utilization of an equal-amplitude power divider.

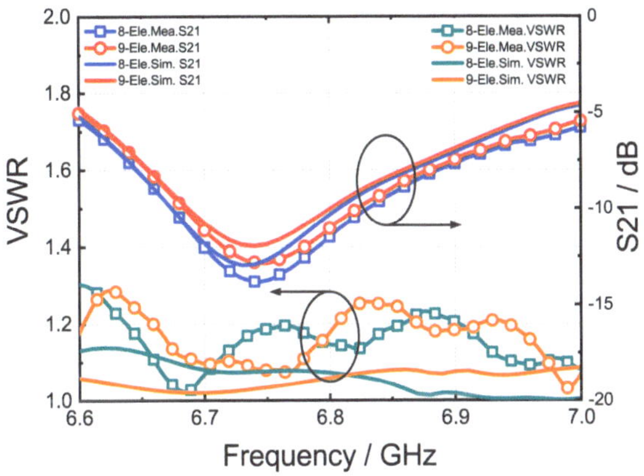

Figure 17. The simulated and measured VSWR and S21 curves within the work band.

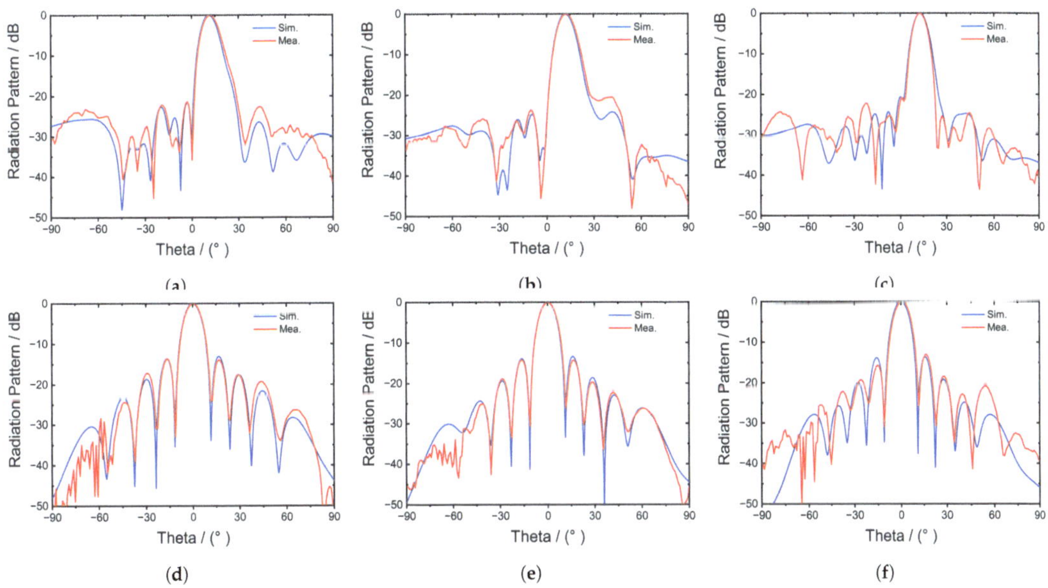

Figure 18. The measured and simulated radiation patterns in the elevation (EL.) and azimuth (Az.) planes. (**a**) El. plane at 6.6 GHz. (**b**) El. plane at 6.8 GHz. (**c**) El. plane at 7.0 GHz. (**d**) Az. plane at 6.6 GHz. (**e**) Az. plane at 6.8 GHz. (**f**) Az. plane at 7.0 GHz.

The axial ratio results are shown in Figure 19 and Table 3. A slight offset of 100–150 MHz is observed in the measured results compared to the simulation results, although the axial ratio frequency bands remain consistent. After the measurement, it was determined that the error in these results primarily arose from two sources: manufacturing and assembly errors. Specifically, the measured value of parameter $d1$ (in Section 2) ranges between 2.9 and 3.1 mm, and the depth of the circular polarizer varies within 31.4 and 31.6 mm, both of which deviate from the intended design values. It was found that the conductive adhesive method of the waveguide cavity layer and gap layer does not affect the electromagnetic performance. However, the screw connection between the circular polarizer layer and the waveguide layer has an impact on the axial ratio. This is due to the deformation of the slot layer and the circular polarizer layer at the center of the prototype. The simulation results, considering the fabrication error, are indicated in Figure 19 by the orange line. To address this problem, the circular polarizer layer and the waveguide layer should be connected using both screws and conductive adhesive, and the position of the screws needs to be redesigned to address the deformation problem after assembly. Enhancing the fabrication accuracy and applying an adhesive to the metal layers can mitigate the frequency offset.

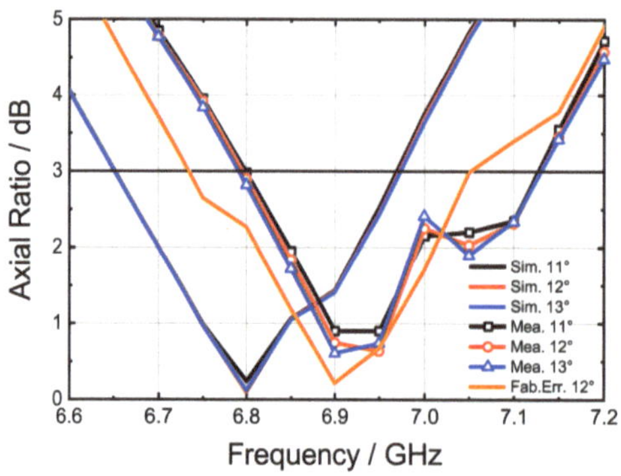

Figure 19. The measured and simulated axial ratios at 11°, 12°, and 13° at 6.6–7.2 GHz.

Table 3. The simulated axial ratio (AR) results in the azimuth plane.

Frequency/GHz	AR at 11°/dB	AR at 12°/dB	AR at 13°/dB
6.7	2.01	2.01	2.00
6.8	0.24	0.10	0.12
6.9	1.44	1.43	1.41
7.0	3.76	3.71	3.64

The advantage of the proposed antenna is the wide beam scan angle; the phase scan radiation pattern in 6.8 GHz is plotted in Figure 20. The results show that the antenna can realize beam scanning in the range of $-45°\sim+45°$ without a grating lobe; the phase mismatching effect is entirely suppressed. Table 4 shows the axial ratio result in the phased scan condition; the scanned beam can still maintain a 3 dB axial ratio under the machine error.

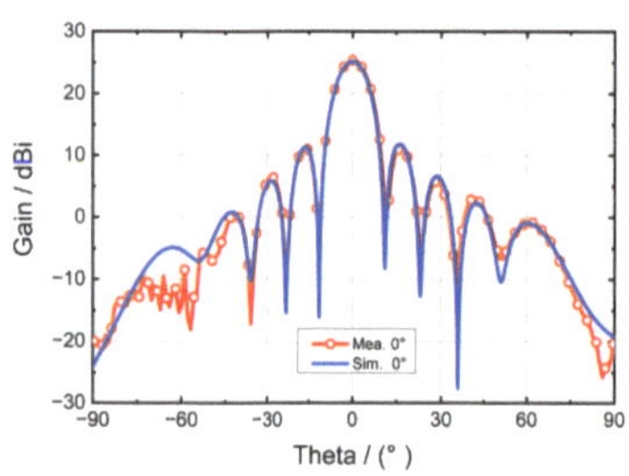

(a)

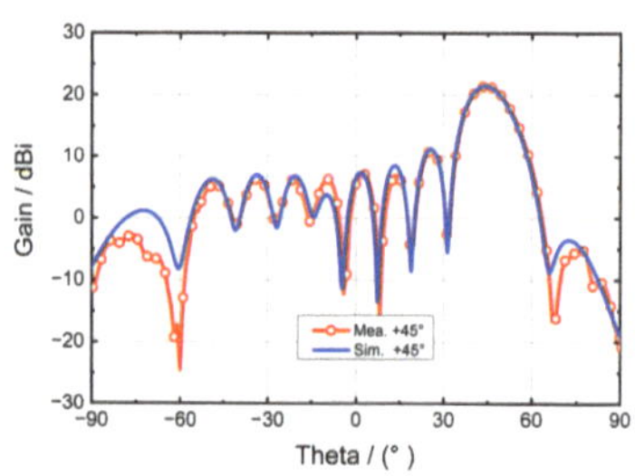

(b)

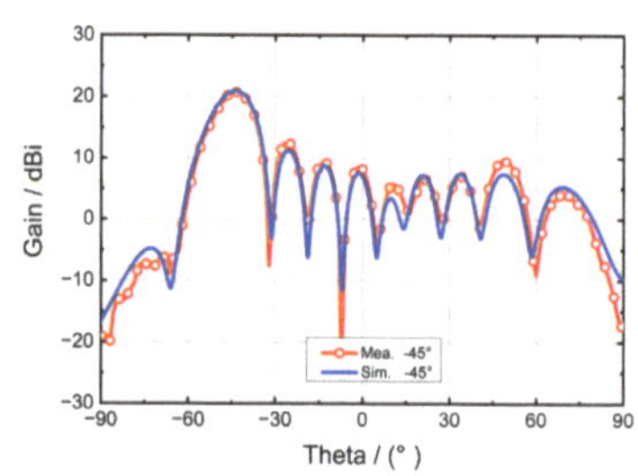

(c)

Figure 20. The measured and simulated phased scan radiation patterns in the azimuth plane: (**a**) in the normal beam direction; (**b**) in the +45° beam direction; (**c**) in the −45° beam direction.

Table 4. A comparison of the measured (Mea.) and simulated (Sim.) axial ratio results for a 6.8 GHz phase scan.

Scan Angle	−45°	0°	+45°
Axial Ratio (Sim.)	3.33 dB	0.23 dB	2.69 dB
Axial Ratio (Mea.)	2.3 dB	2.9 dB	2.2 dB

4.2. Comparison and Discussion

Table 5 compares the performance of the proposed antenna with those of several circularly polarized waveguide slot antennas reported in the literature. Our work achieved a beam scanning range of $-45°\sim+45°$ in the azimuth plane and a scanning range of $+11°\sim+13°$ in the elevation plane. Refs. [19,20] pertain to frequency scan arrays, while [22] utilized a wideband CP dipole to expand the axial ratio (AR) band. Ref. [28] employed a similar structure antenna element and proposed a fixed beam array; however, the authors did not focus on the compact array research domain. Notably, our work achieved a low side-lobe level (SLL). Comparatively, our antenna realizes a larger beam scanning range while maintaining an average or superior electrical performance in other aspects.

Table 5. Performance of the waveguide slot planar array.

Ref.	Feed Type	Polarizer Type	Number of Elements	Excitation Wave Type	Frequency Band (%)	El. Plane SLL (dB)	Gain (dBi)	Axial Ratio Band (%)	Scan Type	Scan Angle (°)
[17]	Waveguide	Slots	1 × 15	Traveling	1.33	−18 (Sim.)	21 (Sim.)	>2.7 (Sim.)	-	-
[19]	SIW	Slots	64 × 12	Traveling	8.03	−14	28	-	Freq	+36~+49 (El.)
[20]	RGW	Slots	1 × 7	Traveling	21.43	-	12.5	21	Freq	−15~+15 (El.)
[21]	Waveguide	Patches	1 × 10	Standing	2.85	−13	15.94	3.2	-	-
[22]	Waveguide	Dipoles	12 × 12	Standing	7.33	−12	28.01	14 (Sim.)	-	-
[28]	Waveguide	Waveguide	2 × 8	Standing	1.63	−13 (Sim.)	17 (Sim.)	-	Phase	-18~+18 (Az.)
This work	Waveguide	Waveguide	5 × 8 & 4 × 9	Traveling	5.89	<−20	24.3	5.0	Freq and Phase	+11~+13 (El.) −45~+45 (Az.)

SIW = substrate integrated waveguide; RGW = ridged gap waveguide; Sim. = simulation result; El. = beam scan in elevation plane; Az. = beam scan in azimuth plane.

5. Conclusions

This paper proposes a compact circularly polarized non-resonant slotted waveguide array antenna with a wide-beam scanning angle ability. We modified and improved the existing compact slotted waveguide antenna through a thorough analysis of its working principles and the optimization of design parameters, adding isolation gaps, and adapting it for use in non-resonant arrays and phased arrays. The array's performance was optimized through phase compensation and element power matching techniques. As a result, we achieved a compact slotted waveguide antenna that can work in non-resonant arrays with the capability for wide beam scanning angles. The prototype antenna is proposed, and the experimental results prove the effectiveness of the antenna design methodology. The designed antenna exhibits a phase scan capability of ±45° in the azimuth plane and a frequency scan range of +11° to +13° in the elevation plane, without the grating lobe, and it maintains a low SLL. The axial ratio remains below 3dB within this scanning range and bandwidth. Overall, the proposed antenna effectively combines circular polarization with wide-angle beam scanning capabilities in slotted waveguide array antennas.

Author Contributions: Conceptualization, G.H.; methodology, G.H. and W.L.; investigation, W.L.; resources, G.H.; visualization, W.L.; validation, W.L.; writing—original draft preparation, G.H. and W.L.; writing—review and editing, G.H. All authors have read and agreed to the published version of the manuscript.

Funding: This research received no external funding.

Institutional Review Board Statement: Not applicable.

Informed Consent Statement: Not applicable.

Data Availability Statement: The data presented in this study are available from the corresponding author upon request.

Conflicts of Interest: The authors declare no conflicts of interest.

References

1. Troughton, J.; Evans, B. Rain-induced deflection of microwave and millimetre-wave radiowaves. *Electron. Lett.* **1976**, *12*, 68–70. [CrossRef]
2. Yang, R.; Wu, Z.; Chen, Y. Study of electromagnetic wave scattering and attenuation properties due to a melting layer of precipitation. *J. Xidian Univ.* **2002**, *29*, 451–454.
3. Aljuhani, A.H.; Kanar, T.; Zihir, S.; Rebeiz, G.M. A 256-element Ku-band polarization agile SATCOM receive phased array with wide-angle scanning and high polarization purity. *IEEE Trans. Microw. Theory Tech.* **2021**, *69*, 2609–2628. [CrossRef]
4. Han, G.; Du, B.; Wu, W.; Yang, B. A novel hybrid phased array antenna for satellite communication on-the-move in Ku-band. *IEEE Trans. Antennas Propag.* **2015**, *63*, 1375–1383. [CrossRef]
5. Yang, H.; Montisci, G.; Jin, Z.; Liu, Y.; He, X.; Mazzarella, G. Improved design of low sidelobe substrate integrated waveguide longitudinal slot array. *IEEE Antennas Wirel. Propag. Lett.* **2014**, *14*, 237–240. [CrossRef]
6. Le Sage, G.P. 3D printed waveguide slot array antennas. *IEEE Access* **2016**, *4*, 1258–1265. [CrossRef]
7. Aboserwal, N.; Salazar-Cerreno, J.L.; Qamar, Z. An ultra-compact X-band dual-polarized slotted waveguide array unit cell for large E-scanning radar systems. *IEEE Access* **2020**, *8*, 210651–210662. [CrossRef]
8. Ren, Y.; Li, K.; Wang, F.; Wu, H.; Zhang, Y.; Li, L. A Dual-Polarized Phased-Array Antenna Based on Single Ridge Slotted Waveguide Array. In Proceedings of the 2019 Cross Strait Quad-Regional Radio Science and Wireless Technology Conference (CSQRWC), Taiyuan, China, 8–21 July 2019; pp. 1–3.
9. Chen, X.Q.; Wu, Q.; Tian, M.X. Study on the return loss characteristics of rectangular single ridge waveguide slot. In Proceedings of the 2013 IEEE International Conference on Applied Superconductivity and Electromagnetic Devices, Beijing, China, 25–27 October 2013; pp. 468–472.
10. Stoumpos, C.; Le Gouguec, T.; Allanic, R.; García-Vigueras, M.; Amiaud, A.C. Compact additively manufactured conformal slotted waveguide antenna array. *IEEE Antennas Wirel. Propag. Lett.* **2023**, *28*, 1843–1847. [CrossRef]
11. Chen, Q.; Yan, S.; Guo, X.; Qu, C.; Liu, J.; Huang, Z.; Yang, L.; Li, Y.; Liang, X. A 77 GHz Phased Array Antenna Based on Substrate Integrated Waveguide. *IEEE Antennas Wirel. Propag. Lett.* **2023**, *22*, 2979–2983. [CrossRef]
12. He, J.; Wu, Y.; Chen, D.; Zhang, M.; Hirokawa, J.; Liu, Q. Realization of a wideband series-fed 4× 4-element waveguide slot array in the X-band. *IEEE Access* **2021**, *9*, 83666–83675. [CrossRef]
13. Luo, H.; Xiao, Y.; Lu, X.; Sun, H. Design of a dual-polarization single-ridged waveguide slot array with enhanced bandwidth. *IEEE Antennas Wirel. Propag. Lett.* **2018**, *18*, 138–142. [CrossRef]

14. Chen, M.; Fang, X.C.; Wang, W.; Zhang, H.T.; Huang, G.L. Dual-band dual-polarized waveguide slot antenna for SAR applications. *IEEE Antennas Wirel. Propag. Lett.* **2020**, *19*, 1719–1723. [CrossRef]
15. Lu, J.; Zhang, H.; Wang, W.; Liang, X.; Ge, J.; Jin, M.; Wu, W. Broadband dual-polarized waveguide slot filtenna array with low cross polarization and high efficiency. *IEEE Trans. Antennas Propag.* **2018**, *67*, 151–159. [CrossRef]
16. Chen, P.; Hong, W.; Kuai, Z.; Xu, J. A substrate integrated waveguide circular polarized slot radiator and its linear array. *IEEE Antennas Wirel. Propag. Lett.* **2008**, *8*, 120–123. [CrossRef]
17. Montisci, G. Design of circularly polarized waveguide slot linear arrays. *IEEE Trans. Antennas Propag.* **2006**, *54*, 3025–3029. [CrossRef]
18. Dimitrov, K.C.; Lee, Y.; Min, B.W.; Park, J.; Jeong, J.; Kim, H.J. Circularly polarized T-shaped slot waveguide array antenna for satellite communications. *IEEE Antennas Wirel. Propag. Lett.* **2019**, *19*, 317–321. [CrossRef]
19. Kazemi, R.; Yang, S.; Suleiman, S.H.; Fathy, A.E. Design procedure for compact dual-circularly polarized slotted substrate integrated waveguide antenna arrays. *IEEE Trans. Antennas Propag.* **2019**, *67*, 3839–3852. [CrossRef]
20. Al Sharkawy, M.; Kishk, A.A. Wideband beam-scanning circularly polarized inclined slots using ridge gap waveguide. *IEEE Antennas Wirel. Propag. Lett.* **2014**, *13*, 1187–1190. [CrossRef]
21. Xu, J.; Wang, M.; Huang, H.; Wu, W. Circularly polarized patch array fed by slotted waveguide. *IEEE Antennas Wirel. Propag. Lett.* **2014**, *14*, 8–11. [CrossRef]
22. Ferrando-Rocher, M.; Herranz-Herruzo, J.I.; Valero-Nogueira, A.; Rodrigo, V.M. Circularly polarized slotted waveguide array with improved axial ratio performance. *IEEE Trans. Antennas Propag.* **2016**, *64*, 4144–4148. [CrossRef]
23. Tomura, T.; Saito, Y.; Hirokawa, J. 8× 2-element 60-GHz-band circularly polarized post-wall waveguide slot array antenna loaded with dipoles. *IEEE Access* **2020**, *8*, 85950–85957. [CrossRef]
24. Cao, J.; Wang, H.; Mou, S.; Soothar, P.; Zhou, J. An air cavity-fed circularly polarized magneto-electric dipole antenna array with gap waveguide technology for mm-wave applications. *IEEE Trans. Antennas Propag.* **2019**, *67*, 6211–6216. [CrossRef]
25. Hoseini, M.; Forooraghi, K.; Abdolali, A. A Waveguide Slot-Fed Patch Antenna Array. *IEEE Antennas Wirel. Propag. Lett.* **2023**, *22*, 2027–2031. [CrossRef]
26. Dogan, D.; Top, C.B. Circularly polarized Ka-band waveguide slot array with low sidelobes. In Proceedings of the 2012 6th European Conference on Antennas and Propagation (EUCAP), Prague, Czech Republic, 26–30 March 2012; pp. 1105–1109.
27. Zarifi, D.; Farahbakhsh, A.; Zaman, A.U. A Dual-Circularly Polarized Gap Waveguide-Based Linear Array Antenna for 60 GHz-Band. In Proceedings of the 2023 17th European Conference on Antennas and Propagation (EuCAP), Florence, Italy, 26–31 March 2023; pp. 1–3.
28. Chen, Q.; Wu, T.; Wan, Y.; Li, L.; Jin, M. A circularly polarized waveguide slot antenna for random array application. In Proceedings of the 2019 International Conference on Microwave and Millimeter Wave Technology (ICMMT), Guangzhou, China, 19–22 May 2019; pp. 1–3.
29. Wu, X.; Yang, F.; Xu, F.; Zhou, J. Circularly polarized waveguide antenna with dual pairs of radiation slots at Ka-band. *IEEE Antennas Wirel. Propag. Lett.* **2017**, *16*, 2947–2950. [CrossRef]
30. Li, Y.Q.; Zhang, H.; Fu, Y.Q.; Yuan, N.C. RCS reduction of ridged waveguide slot antenna array using EBG radar absorbing material. *IEEE Antennas Wirel. Propag. Lett.* **2008**, *7*, 473–476.
31. Qiu, L.; Xiao, K.; Chai, S.L.; Qi, H.Y.; Mao, J.J. A double-layer shaped-beam traveling-wave slot array based on SIW. *IEEE Trans. Antennas Propag.* **2016**, *64*, 4639–4647. [CrossRef]
32. Yin, W.; Chen, A.; Liang, X.; Shi, L.; Guan, F.; Liu, X.; Zi, J. Frequency scanning single-ridge serpentine dual-slot-waveguide array antenna. *IEEE Access* **2020**, *8*, 77245–77254. [CrossRef]
33. Yuan, W.; Liang, X.; Zhang, L.; Geng, J.; Zhu, W.; Jin, R. Rectangular grating waveguide slot array antenna for SATCOM applications. *IEEE Trans. Antennas Propag.* **2019**, *67*, 3869–3880. [CrossRef]
34. Moradian, M.; Tayarani, M.; Khalaj-Amirhosseini, M. Planar slotted array antenna fed by single wiggly-ridge waveguide. *IEEE Antennas Wirel. Propag. Lett.* **2011**, *10*, 764–767. [CrossRef]
35. Jasik, H.; Mao, Y.; Lin, C. *The Handbook of Antenna Engineering*; National Defense Industry Press: Beijing, China, 1966; pp. 243–249.
36. Jiang, Z. *Antenna Princples*; Beihang University Press: Beijing, China, 1993; pp. 195–199.

sensors

MDPI

Article

Nonlinear Frequency Offset Beam Design for FDA-MIMO Radar

Yanjie Xu, Chunyang Wang, Guimei Zheng * and Ming Tan

College of Air and Missile Defense, Air Force Engineering University, Xi'an 710051, China
* Correspondence: zheng-gm@163.com

Abstract: The beam pattern of frequency diversity array (FDA) radar has a range–angle two-dimensional degree of freedom, which makes it possible to distinguish different targets from the same angle and brings a new approach to anti-jamming of radars. However, the beam pattern of conventional linearly frequency-biased FDA radar is range–angle-coupled and time-varying. The method of adding nonlinear frequency bias among the array elements of the FDA array has been shown to eliminate this coupling property while still allowing for better beam performance of the emitted beam. In this paper, we obtain a decoupled and time-invariant beam direction map using the FDA-multi-input–multi-output (FDA-MIMO) radar scheme and then obtain a sharp pencil-shaped main sphere beam pattern with range–angle dependence using a linear frequency offset scheme weighted by a Chebyshev window. Finally, the anti-interference performance of the proposed method is verified in an anti-interference experiment.

Keywords: nonlinear frequency offset; FDA-MIMO radar; Chebyshev window based frequency offset; array structure design

1. Introduction

Since the first introduction of FDA radars in [1], the important property that their beam patterns are range-dimension dependent and can vary with time has been discovered, which has attracted extensive attention from scholars home and abroad. When the time and frequency offsets are fixed, the emission direction pattern of the frequency-controlled array has a range–angle-dependent property, which is more flexible compared with the conventional phased array and has a very high potential in practical applications [2–6]. However, the conventional fixed-frequency offsets FDA emission direction diagram is "S" shaped, with range–angle coupling characteristics, and is time-varying, which is unfavorable for the practical application of radar. With the exploration of the nature of the FDA directional pattern characteristics, many beam design methods for FDA radars have emerged. The range–angle coupling characteristics of FDA radar is caused by introducing the frequency offset among the array elements, so the frequency offset also needs to be designed and optimized to solve the coupling problem. The first problem to be solved is the time-varying problem. The literature [7] studied the time-varying problem of FDA radar and designed a time-invariant beam pattern, and the literature [8,9] combined FDA radar with MIMO technology to form a time-invariant beam pattern as well. On the basis of the existing research, the literature [10] explored the fundamental reason for the time-varying nature of the beam pattern from the nature of FDA radar and proposed the concept of "time figure", which is of guidance for FDA radar research. After a period of research, the FDA-MIMO radar regime was proved to be one of the available methods to solve the beam patterns' coupling problems [11,12]. After solving the time-varying problem, the beam performance is the next issue to be considered. On the basis of ensuring the time invariance, many beam design schemes have been proposed to provide a more superior performance. A logarithmic frequency offset scheme is proposed in the literature [13], which solves the periodic problem in space and forms a point beam, but with poor performance. The literature [14] uses a particle swarm optimization algorithm to determine the frequency offsets and array element spacing to form a point-like transmit beam map, which also has a lower side flap in the range dimension, but the calculation is too complex. As studies proceeded, the design of FDA radar frequency offsets using a window function weighting scheme became popular. However, window function weighting schemes based on fixed windows have been studied more often, while window function weighting schemes based on flexible windows are still less mentioned [14–22]. Compared with the fixed window function, the flexible window function can have more degrees of freedom in the power proportion of the main and side flaps, and, therefore, a better performance of the beam direction map can be obtained [23]. Overall, the present research on frequency-controlled array transmit beam formation can basically solve the problems of coupling and periodicity, but the present beam design scheme still has the problems of insufficient beam main flap focus, low beam resolution, and high side flap, which is more unfavorable for the specific application of radar. Therefore, it is necessary to design a point-type transmitting beam with narrow main flap and low side flap. For linear frequency bias, the beam direction map is not point-like, which is not favorable for radar detection, and because of this, a frequency bias design is needed to improve the beam direction map performance. In this paper, the time-invariant FDA beam direction diagram is obtained on the basis of the FDA-MIMO radar system, and the beam is further optimized by using the Chebyshev window to weight the linear frequency offsets on the basis of the existing nonlinear frequency offsets schemes, obtaining the results of narrowing the main flap and reducing the side flaps of the transmit beam. To better test its application capability, it is combined with the Minimum Variance Distortion-free Response (MVDR) adaptive beamforming algorithm to test its performance in jamming immunity. The simulation results show that this scheme has a more obvious advantage in the beam pattern performance compared with several window function weighted frequency offset schemes that have been proposed.

2. System Model

From the literature [10], we know that an effective measure to eliminate the time parameter in the transmit–receive diagram is to use a series of mixers and matched filters at the receiver. Analogously to this case, we can also use a combination of FDA-MIMO radar and multiple matched filters at the receiver side to produce beam maps independent of the time parameters. Consider a narrow-band FDA-MIMO radar system with M transmitting and N receiving arrays, and design both transmitting and receiving arrays as conventional linear arrays. Suppose the range between transmitting arrays is d_T, the range between receiving arrays is d_R, and the signal carrier frequency is f_0; such a system configuration is shown in Figure 1.

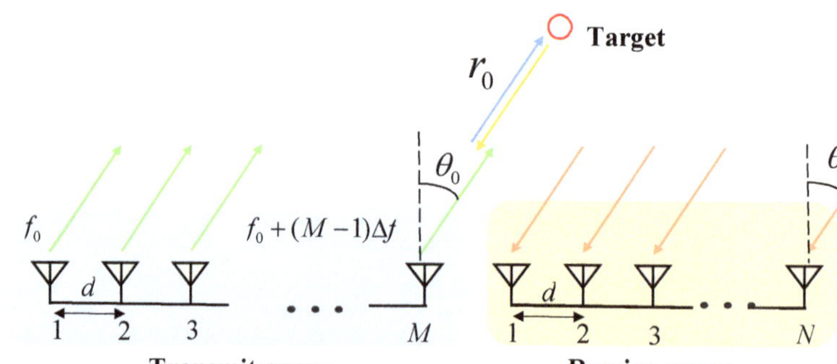

Figure 1. FDA-MIMO Radar System.

As shown in the figure, the frequency offset of the mth element is $f(m)$, and the emission frequency of the mth array element can be expressed as

$$f_m = f_0 + f(m) \tag{1}$$

Suppose each array element emits the same waveform, then the signal emitted by the mth array element at time t can be expressed as Equation (2)

$$s_m(t) = w_m(t)e^{j2\pi f_m t} \tag{2}$$

where $w_m(t)$ is the orthogonal signal envelope emitted by the mth array element, which satisfies

$$\int w_{m1}{}^*(t) \cdot w_{m2}(t-\tau)dt = 0, m1 \neq m2. \tag{3}$$

where $w_{m1}{}^*(t)$ is the conjugate of the $w_{m1}(t)$. With the FDA-MIMO regime radar, the time delay at the target can be expressed as

$$\tau_{m,n} = 2r_0/c - md_T \sin\theta_0/c - nd_R \sin\theta_0/c \tag{4}$$

Then, the received signal related to the mth transmitting array element and the nth receiving array element can be expressed as

$$s_{m,n}(t-\tau_{m,n}) = w_n(t-\tau_{m,n})e^{j\Phi_m(t-\tau_{m,n})} \cdot e^{j2\pi(f_0+f(m))(t-\tau_{m,n})} \tag{5}$$

where $n = 1, 2, \ldots, N$, Φ_m is the phase modulation corresponding to the mth transmitting array element in the FDA-MIMO radar, and the time index can be expressed as

$$t' = t - \frac{2r}{c} \tag{6}$$

Then, under the far-field condition, Equation (2) can be rewritten as

$$s_{m,n}(t') = w_m(t')e^{j\Phi_m(t')}e^{j\varphi_n(t')} \tag{7}$$

In order to eliminate the effect of the time parameter and produce a time-invariant and decoupled transmit beam pattern, we need to use a multi-matching filter approach to process the signal at the receiver side of the FDA-MIMO radar, as shown in Figure 2.

As shown in the figure, the received signal needs to be mixed with the radio frequency of the mixer and later mixed with $w_n(t')$ in the digital signal processor, so that the relative outputs of the mth transmitting array element and the nth receiving array element can be expressed as

$$s_{m,n}^{\text{Output}}(t') = \xi_s e^{j\Phi_m(t')}e^{j\varphi_n(t')} \tag{8}$$

where $\varphi_n(t')$ can be expressed as

$$\begin{aligned}\varphi_n(t') &= 2\pi \int_0^{t'+((md_T\sin\theta/c)+(nd_R\sin\theta/c))}(f_0+f(m))dx \\ &= 2\pi\left((f_0+f(m))t' + f_0\frac{md_T\sin\theta}{c} + f_0\frac{nd_R\sin\theta}{c}\right)\end{aligned} \tag{9}$$

If the radar is aimed at the target at (r_t, θ_t), the array factor AF can be expressed as

$$\begin{aligned}AF = \sum_{m=1}^{M}\sum_{n=1}^{N} s_{m,n}^{\text{output}}(t') &= \xi_s e^{-j2\pi f_0((2r/c)+md_T(\sin\theta-\sin\theta_t)/c)} \\ \cdot \sum_{m=1}^{M} e^{j(4\pi/c)f(m)(r_t-r)} &\sum_{n=1}^{N} e^{j(2\pi f_0 nd_R(\sin\theta-\sin\theta_t)/c)}\end{aligned} \tag{10}$$

The transmit–receive normalized beam pattern can be expressed as

$$B = B_T \cdot B_R = \left| \sum_{m=1}^{M} e^{j(4\pi/c)f(m)(r_t - r)} e^{j(2\pi f_0 m d_T(\sin\theta - \sin\theta_t)/c)} \right|^2 \cdot \left| \sum_{n=1}^{N} e^{j(2\pi f_0 n d_R(\sin\theta - \sin\theta_t)/c)} \right|^2 \quad (11)$$

It is worth noting that the transmit–receive beam pattern in the above formula can be equivalent to the multiplication of the transmit beam pattern and the receive beam pattern at the receiver, which can be represented by B_T and B_R, respectively.

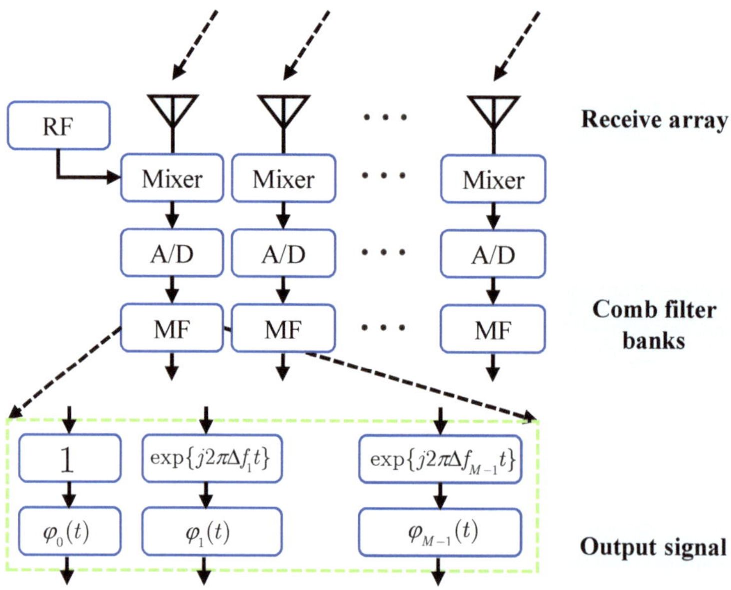

Figure 2. FDA-MIMO and multi-matching filter system.

Equation (11) shows that the FDA-MIMO radar can rely on its own characteristics to produce a time-invariant range-dependent beam pattern after the signal processing process, and that the beam pattern can be maximized at the target point with a maximum value of M^2N^2, regardless of the frequency offset change. Therefore, when designing the beam direction map of the FDA-MIMO radar, the range dependence of the FDA-MIMO radar should be activated by adding the corresponding variables of phase and target spatial position at the transmitter side.

3. Chebyshev Window Weighted Linear Frequency Offsets

For the window function weighted frequency offsets scheme, the parameter selection of the window function is an important element affecting the beam performance. Most of the window functions of variable windows are controlled by two parameters, so it is necessary to use the control variable method to find the best combination of parameters. Ordinarily, when choosing the best position for the performance of a parameter, it is not possible to make both the main flap width and the side flap height optimal. For example, when we set the parameter of the peak side flap of the Chebyshev window to a low level, we can ensure that the side flap energy is low, but the energy will be concentrated in the main flap, resulting in a wide main flap, which we do not expect. When we are designing a beam, we usually give priority to minimizing the width of the main flap if the flap level does not seriously affect the detection. On this basis, the Chebyshev window is utilized for frequency offset design in this paper. The advantage of the Chebyshev window function is that the main flap width of the Chebyshev window is minimum for a given flap height, and all flaps have the same amplitude, so that the beam with minimum main flap width can be obtained by controlling the flaps within an acceptable range when choosing parameters. For nonlinear frequency offset, the proposed frequency offset methods are: log frequency offset, Hamming window based frequency offset (Ham-FDA), and sinusoidal-weighted frequency offset. Our proposed method is a weighted linear frequency offset based on the Chebyshev window, which outperforms the conventional method in both the main flap width and the side flap height. The Log-FDA frequency offset can be expressed as

$$\Delta f_m^{\log} = \delta \cdot \log(m+1) \tag{12}$$

where δ is a constant, and Δf_m represents the frequency offset of the first transmitting array element. Similarly, the frequency offset of Ham-FDA can be expressed as

$$\Delta f_m^{\mathrm{Ham}} = B\left\{0.54 - 0.46\cos\left[\frac{2\pi(m-(M+1)/2)}{M}\right]\right\} \tag{13}$$

where B is the bandwidth and requires that the number of array elements M should be odd.

The frequency shift of the Hamming window weighted linear frequency offsets (HL-FDA) scheme proposed in the literature [17] can be expressed as

$$\Delta f_m^{HL} = m\Delta f\left\{0.54 - 0.46\cos\left[\frac{2\pi(m-(M+1)/2)}{M}\right]\right\} \tag{14}$$

The frequency offset of the proposed algorithm in this paper can be expressed as

$$\Delta f_m^{proposed} = m\Delta f w_c(m) \tag{15}$$

where $w_c(m)$ is the time domain expression for the Chebyshev window function, which can be expressed as [24]

$$w_c(m) = \frac{\sum_{m=-N}^{S} W_c^0(n)\cos\left(\frac{2\pi}{N}mn\right)}{\sum_{m=-N}^{S} W_c^0(n)} \tag{16}$$

where $W_c^0(n)$ is the discrete spectral expression of the Chebyshev window, and S should be defined as $S = (M-1)/2$. Similarly, there are $s = (m-1)/2$. As with the Hamming window weighting scheme, it is also required here that the number of array elements M should be odd. The algorithm proposed in this paper can be seen in the form of a Chebyshev window combined with a linear array of arrays, where the output signal associated with the mth transmitting array element and the nth receiving array element after processing at the receiver side of the FDA-MIMO system, as shown in Equation (17). The array factor AF can be expressed as Equation (18).

$$\begin{aligned} s_{m,n}^{Output}(t') &\approx rect\left(\tfrac{t'}{T_p}\right)\xi_s \exp\left\{j2\pi\left[-f_0\tfrac{2r}{c} - m\Delta f w_c(m)\right)\tfrac{2r}{c} \right. \\ &\left. +f_0\tfrac{md_T\sin\theta}{c} + f_0\tfrac{nd_R\sin\theta}{c}\right]\right\} \\ &= rect\left(\tfrac{t'}{T_p}\right)\xi_s \exp\left\{j2\pi\left[-f_0\tfrac{2r}{c} - m\Delta f\tfrac{\sum_{m=-N}^{N} W_c^0(n)\cos\left(\frac{2\pi}{N}mn\right)}{\sum_{m=-N}^{N} W_c^0(n)}\right)\tfrac{2r}{c}\right. \\ &\left. +f_0\tfrac{nd_T\sin\theta}{c} + f_0\tfrac{md_R\sin\theta}{c}\right]\right\} \end{aligned} \tag{17}$$

$$\begin{aligned} AF &= \sum_{m=0}^{M-1}\sum_{n=0}^{2N_1} s_{m,n}^{Output}(t') \\ &= rect\left(\tfrac{t'}{T_p}\right)\xi_s \sum_{m=0}^{M-1}\sum_{n=0}^{2N_1} e^{j2\pi\left[-f_0\frac{2r}{c} - m\Delta f w_c(m)\right)\frac{2r}{c} + f_0\frac{md_T\sin\theta}{c} + f_0\frac{nd_R\sin\theta}{c}\right]} \end{aligned} \tag{18}$$

The vector of orientation **A** can be expressed as Equation (19), where **a** and **b** are the transmitting and receiving guide vectors, respectively, which can be expressed as Equations (20) and (21).

$$\mathbf{A} = \xi_s\, \mathbf{a}(r,\theta) \otimes \mathbf{b}(\theta) \tag{19}$$

$$\mathbf{a}(r,\theta) = \begin{bmatrix} 1 \\ e^{j2\pi[f_0 \frac{d_T \sin\theta}{c} - m(\Delta f w_c(m))\frac{2r}{c}]} \\ \vdots \\ e^{j2\pi[f_0 \frac{2Md_T sin\theta}{c} - 2M(\Delta f w_c(m))\frac{2r}{c}]} \end{bmatrix} \tag{20}$$

$$\mathbf{b}(\theta) = \begin{bmatrix} 1 \\ e^{j2\pi f_0 d_R \sin\theta/c} \\ \vdots \\ e^{j2\pi f_0 d_R (N-1)\sin\theta/c} \end{bmatrix} \tag{21}$$

Then, the emission direction diagram $B_{T,R}$ can be expressed as

$$\begin{aligned} B_{T,R} &= \left|\mathbf{w}_D{}^{\mathrm{H}} \cdot \mathbf{A}\right|^2 \\ &= \left|\sum_{n=0}^{2M} \exp\{j2\pi(f_0 d_T m \frac{\sin\theta - \sin\theta_0}{c} - 2m\Delta f w_c(m))\}\right|^2 \\ &\times \left|\sum_{n=0}^{N-1} \exp\{j2\pi f_0 d_R n \frac{\sin\theta - \sin\theta_0}{c}\}\right|^2 \end{aligned} \tag{22}$$

4. MVDR Algorithm Model

In order to better test the practical application capability of the proposed frequency offsets scheme, it is combined with the adaptive beamforming algorithm to verify its beam output performance under the conventional MVDR algorithm. Adaptive beamforming is an important method to enhance the signal output signal-to-noise ratio in the array radar signal processing process. By attaching a weighting factor between each array element, it causes the array output power to be maximum in the desired direction and minimum in the jamming direction, by which the output signal-to-noise ratio can converge quickly at the jamming location, thus forming a zero null.

In the FDA-MIMO system, the desired signal, the dummy target, and the noise are statistically independent of each other, and the location of the desired target in the setup space is (r_0, θ_0), The interfering signal is i_j, j is the number of jammings, jamming position can be set to (r_j, θ_j), and noise $n(t)$ is Gaussian white noise. The signal at the kth snap count can be expressed as Equation (23), and the output signal in the FDA-MIMO system can be expressed as Equation (24).

$$\boldsymbol{x}(t) = \boldsymbol{a}(r_0,\theta_0)s(t) + \sum_{j=1}^{J} \boldsymbol{a}(r_j,\theta_j)i_j(t) + \boldsymbol{n}(t) \tag{23}$$

$$\boldsymbol{y}(t) = \boldsymbol{w}^H[\boldsymbol{a}(r_0,\theta_0) \otimes \boldsymbol{b}(\theta_0)]s(t) + \boldsymbol{w}^H \sum_{j=1}^{J} [\boldsymbol{a}_0(r_j,\theta_j) \otimes \boldsymbol{b}_n(\theta_j)]i_j(t) + \boldsymbol{n}(t) \tag{24}$$

To keep the energy of the noise minimum while satisfying the constraints, the optimization function of MVDR can be expressed as

$$\begin{aligned} &\min_\omega \mathbf{w}^H\mathbf{R}\mathbf{w} \\ &\text{s.t.} \begin{cases} \mathbf{w}^H\mathbf{a}(r_0,\theta_0) = 1 \\ \mathbf{w}^H\mathbf{a}(r_j,\theta_j) = 0 \end{cases} \end{aligned} \tag{25}$$

where R is the jamming-plus-noise covariance matrix. The equation can be solved by the Lagrange multiplier method, the optimal weights of the array are obtained as Equation (26), and the output signal-to-jamming-noise ratio (SINR) can be expressed as Equation (27)

$$w_{MVDR} = \frac{\mathbf{R}^{-1}\mathbf{a}(r_0,\theta_0)}{\mathbf{a}(r_0,\theta_0)^H\mathbf{R}^{-1}\mathbf{a}(r_0,\theta_0)} \tag{26}$$

$$R_{SINR} = \frac{\left|w^H\mathbf{a}(r_0,\theta_0)\right|^2}{w^H\mathbf{R}^{-1}w - \left|w^H\mathbf{a}(r_0,\theta_0)\right|^2} \tag{27}$$

Finally, incorporating the guidance vector into the above equation, we obtain

$$R_{SINR} = \frac{\left|w^H\left(1, e^{j2\pi[f_0\frac{d_T\sin\theta}{c} - m(\Delta f w_c(m))\frac{2r}{c}]}, \ldots, e^{j2\pi[f_0\frac{2Md_T\sin\theta}{c} - 2M(\Delta f w_c(m))\frac{2r}{c}]}\right)\right|^2}{w^H\mathbf{R}^{-1}w - \left|w^H\left(1, e^{j2\pi[f_0\frac{d_T\sin\theta}{c} - m(\Delta f w_c(m))\frac{2r}{c}]}, \ldots, e^{j2\pi[f_0\frac{2Md_T\sin\theta}{c} - 2M(\Delta f w_c(m))\frac{2r}{c}]}\right)\right|^2} \tag{28}$$

5. Simulation and Analysis

5.1. Simulation of Frequency Offset Scheme

In our given FDA-MIMO system, the frequency offset affects only the equivalent transmit beammap and does not affect the receive beammap. Therefore, the simulation compares only the equivalent transmit beamplot. In this section, we simulate and compare the performance of Log-FDA, Ham-FDA, HL-FDA, and our proposed FDA frequency offset scheme. The maximum frequency offsets of these methods are set to be approximately equal, and the other relevant parameters are set as follows. The carrier frequency f_0 is 10 GHz, the array element index is $M = N = 15$, the array spacing of the transmitting and receiving arrays is set to $dt = dr = 0.015$ m, bandwidth is set to $B = 27.5$ kHz, and the target position is set to $(100\text{ km}, 20°)$. Figure 3 compares the array element order versus position for each method.

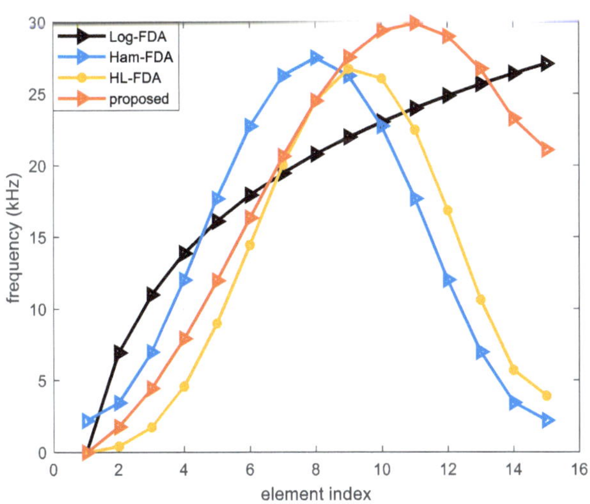

Figure 3. Relationship between array element index and position.

After designing the frequency offset on the FDA-MIMO system, we analyzed the equivalent transmit beamplots of four frequency offset schemes: Log-FDA, Ham-FDA, HL-FDA, and our proposed scheme, as shown in Figure 4.

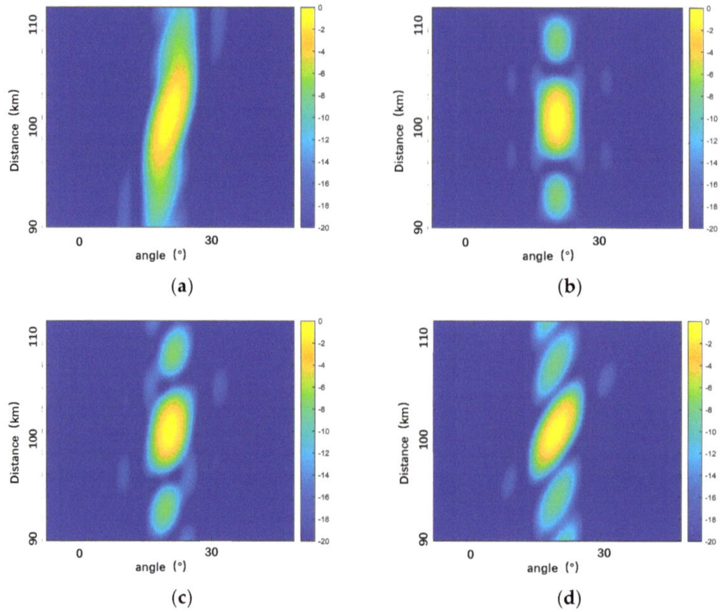

Figure 4. Equivalent transmitting antenna beams of the four schemes: (**a**) Log-FDA; (**b**) Ham-FDA; (**c**) HL-FDA; and (**d**) Proposed.

As shown in Figure 4, several frequency offsets schemes have been proposed to solve the time-varying and coupling problems of the FDA directional map, but the beam performance varies. The Log-FDA scheme can solve the range–angle coupling problem, but the main flap of its formed beam is too wide and does not focus well, and the Ham-FDA and HL-FDA schemes can focus the emitted energy at a point in space. The Ham-FDA and HL-FDA schemes can produce a point-like beam, but the main flap of these two schemes is still very wide, and their side flaps are high in height, which also affects the radar detection performance to some extent. Our proposed method can produce a more focused point-like beam with a lower side flap than the other methods. The 3D view of the beam provides a more intuitive view of the overall effect of the beam, which is shown in Figure 5.

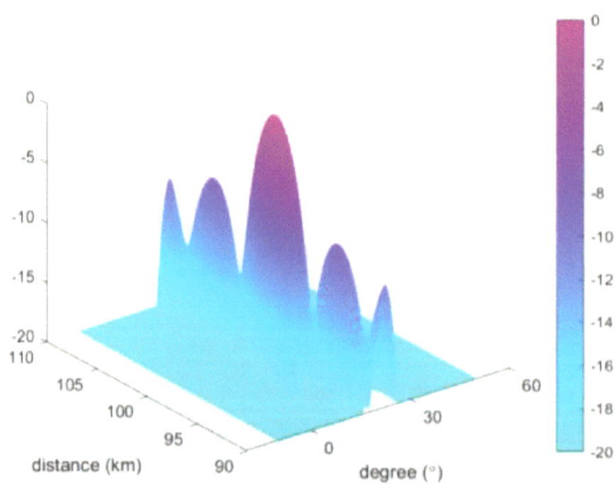

Figure 5. The 3D beam pattern.

As shown in Figure 5, it can be seen that our designed beam produces a lower side flap, which can help suppress the clutter in the side flap direction. It should be noted that, although our method produces a lower side flap, the depression energy between the main flap and the side flap becomes higher, which is the inevitable result of weighting the frequency offsets with a window function. However, this does not affect the detection because the side flap energy is still low compared with the main flap, and the jamming signal inside the side flap can be easily found during the detection process by aligning the main flap to the detection target. The range-dimensional slicing of the beam map can better reflect the performance of the beam, and the comparison of each method is shown in Figure 6.

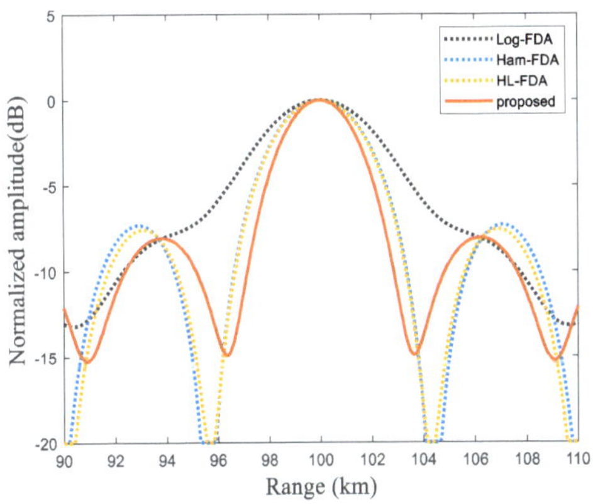

Figure 6. Range-Dimensional Slicing Performance Comparison.

As shown in Figure 6, the proposed scheme has greater advantages at both the main flap and the side flap because the Chebyshev window function is used to evenly disperse the energy in the side flap to the airspace far away from the main flap. Compared with several proposed methods, our proposed scheme has better beam-focusing performance with a narrower main flap and good side flap height, and the half-power beamwidth diagram can be a good comparison of the performance gap of the scheme, as shown in Figure 7.

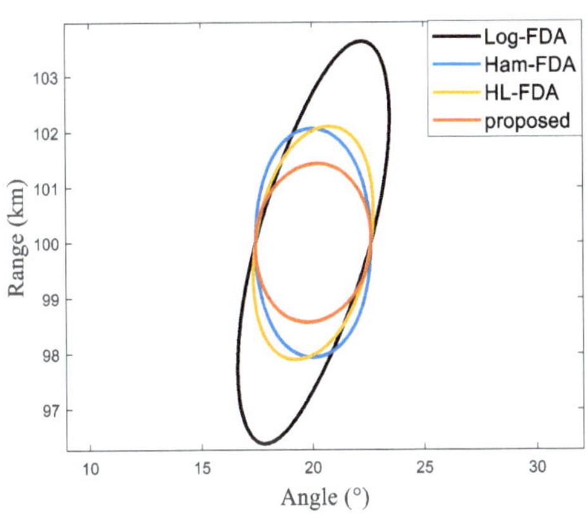

Figure 7. Half-power beamwidth diagram.

As represented in Figure 7, the half-power beamwidth comparison results show the superior performance of our proposed scheme in both angular and range dimensions.

5.2. Analysis of the Anti-Jamming Characteristics of the Proposed Scheme

In this section, we verify the anti-jamming capability of the proposed frequency offsets scheme in the presence of jamming in space and verify its beam performance under MVDR conditions, respectively. In the experiment, a range-dimensional jamming and an angle-dimensional jamming are set up, respectively. It can be seen that Jamming 1 is at the same angle as the target but at a different range; it is located on the side flap of the range dimension, and its beam gain is lower than the gain at the target point. Jamming 2 is at the same range from the target but at a different angle; it is located in the angular dimension of the side flap, and its gain is also lower than that at the target point. Meanwhile, we can see from the range slice and angle slice that after the adaptive beam formation, the beam of FDA-MIMO radar will automatically form a null trap at the jamming to suppress the jamming, but for the linear frequency offsets FDA-MIMO, it still has more energy distribution in the angle dimension, which indicates its weak anti-jamming capability in the angle dimension. The details are shown in Figure 8.

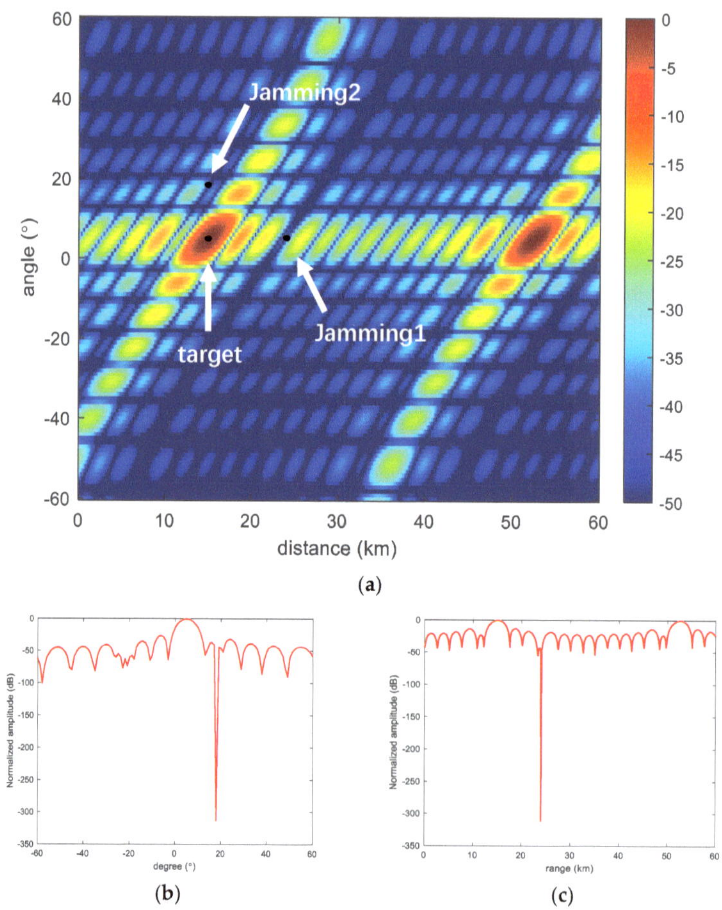

Figure 8. Linear frequency offset FDA-MIMO. (**a**) FDA-MIMO; (**b**) Range-dimensional slicing; and (**c**) Angle-dimensional slicing.

Figure 9 shows the beamforming performance of Log-FDA under MVDR conditions. It can be seen that the main flap of the Log-FDA beam becomes very wide and the position of the peak of the main flap is shifted, which indicates that Log-FDA does not adapt well to the environment where jamming is present.

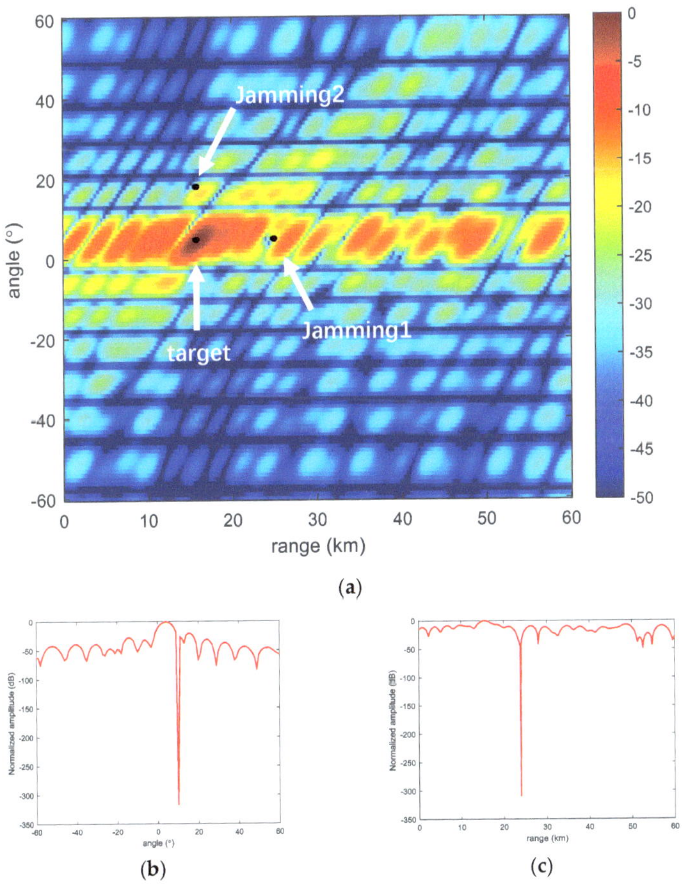

Figure 9. Log-FDA performance. (**a**) Log-FDA; (**b**) Range-dimensional slicing; and (**c**) Angle-dimensional slicing.

Figures 10 and 11 show the beamforming performance of HL-FDA under MVDR conditions. It can be seen that both HL-FDA and the proposed scheme can adapt well to the MVDR algorithm and can form zero traps at jamming locations in the presence of jamming in space. By observing the slice plots, it can be seen that the proposed scheme has a higher side flap in the range dimension but a narrower main flap width, which can better suppress jamming close to the main flap, which is one of the challenges faced by radar detection. Figure 12 shows the comparison of the output SINR in the presence and non-existence of MVDR conditions. It can be seen from the figure that the method in this paper has certain advantages over the traditional linear frequency bias method, which indicates that it is more applicable to the MVDR algorithm.

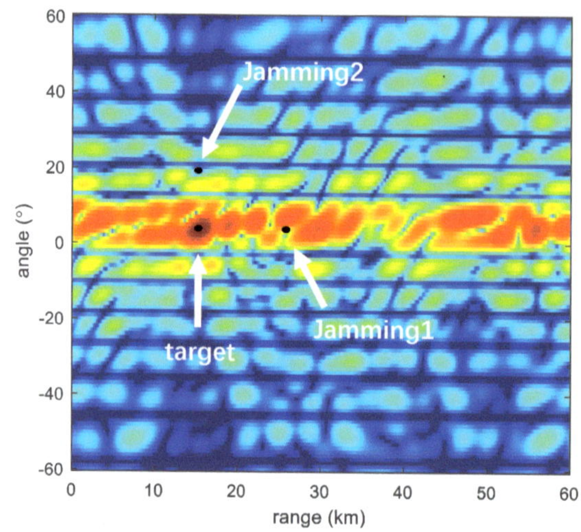

(**a**)

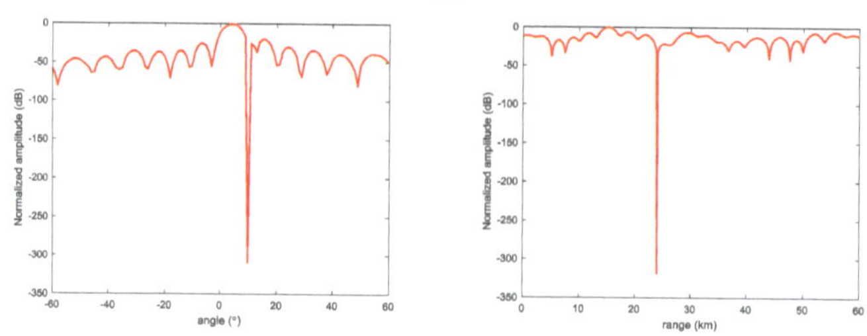

(**b**) (**c**)

Figure 10. HL-FDA performance. (**a**) HL-FDA; (**b**) Range-dimensional slicing; and (**c**) Angle-dimensional slicing.

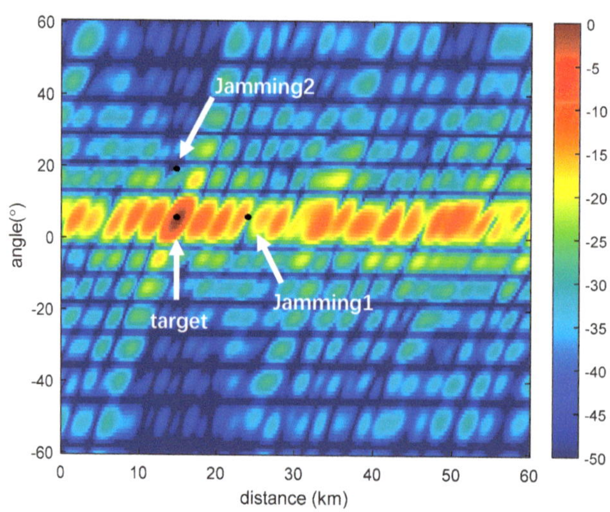

(a)

Figure 11. *Cont.*

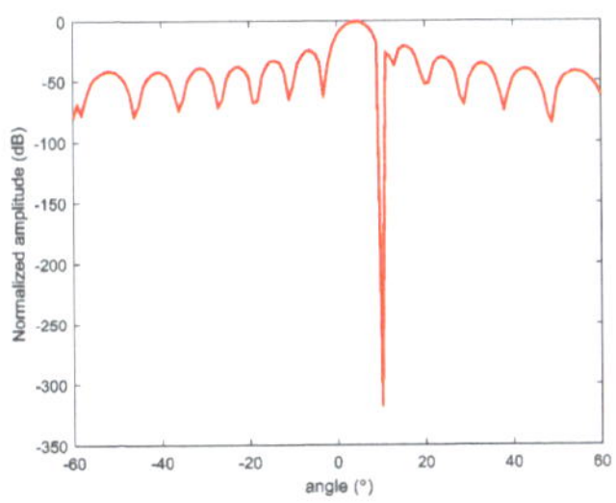

(**b**)

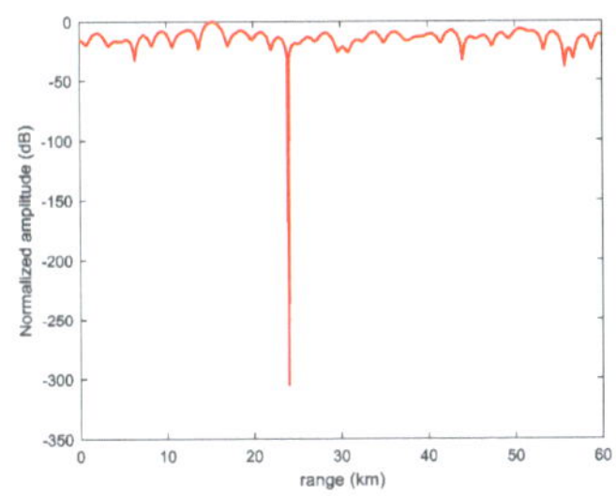

(**c**)

Figure 11. Proposed scheme performance. (**a**) The proposed scheme; (**b**) Range-dimensional slicing; and (**c**) Angle-dimensional slicing.

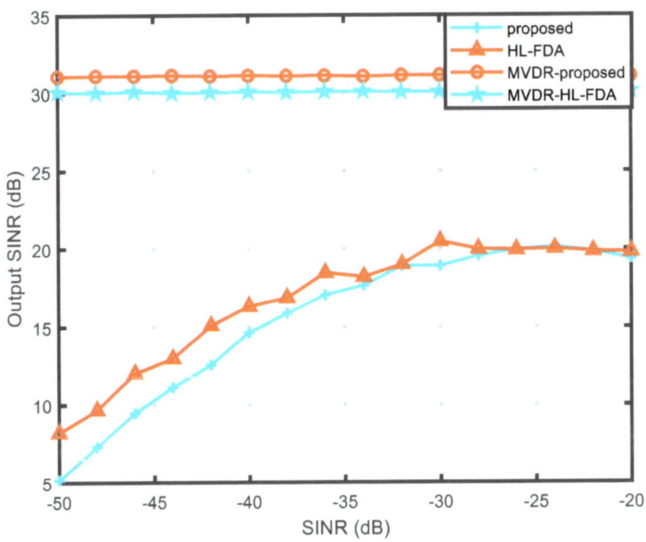

Figure 12. Comparison of output SINR.

6. Experimental Results and Discussion

In this paper, the coupling problem of FDA radar is solved using the FDA-MIMO system, a new nonlinear frequency offset scheme is proposed, and its anti-interference performance is verified relative to several other frequency offset schemes using the MVDR algorithm. Considering that most of the current studies focus on the optimization using fixed window function weighting, this paper utilizes the Chebyshev window function to optimize the FDA frequency offset for the first time. Compared with the currently proposed HL-FDA algorithm and Ham-FDA algorithm, the algorithm proposed in this paper has better performance with the correct selection of parameters and better anti-interference performance when there is interference in space. However, since this method requires the selection of parameters, it may require more computation time in practical applications.

Author Contributions: Methodology, G.Z. and M.T.; Writing—original draft, Y.X.; Supervision, C.W. All authors have read and agreed to the published version of the manuscript.

Funding: This work is supported by the National Natural Science Foundation of China (62201580); by Natural Science Foundation of Shaanxi (2021JM-222); by China Postdoctoral Science Foundation (2019M6622257).

Institutional Review Board Statement: Not applicable.

Informed Consent Statement: Not applicable.

Data Availability Statement: All data used in this paper can be obtained by contacting the authors of this study.

Conflicts of Interest: The authors declare no conflict of interest.

References

1. Antonik, P.; Wicks, M.C.; Griffiths, H.D.; Baker, C.J. Frequency diverse array radars. In Proceedings of the 2006 IEEE Conference on Radar, Verona, NY, USA, 24–27 April 2006; IEEE: Piscataway, NJ, USA, 2006; pp. 215–217.
2. Wang, W.-Q. Overview of frequency diverse array in radar and navigation applications. *IET Radar Sonar Navig.* **2016**, *6*, 1001–1012. [CrossRef]
3. So, H.C.; Amin, M.G.; Blunt, S.; Gini, F.; Wang, W.-Q. Introduction to the special issue on time-frequency modulated array signal processing. *IEEE J. Sel. Top. Signal Process.* **2020**, *2*, 225–227.
4. Xu, J.; Liao, G.; Zhu, S.; Huang, L.; So, H.C. Joint range and angle estimation using MIMO radar with frequency diverse array. *IEEE Trans. Signal Process.* **2015**, *13*, 3396–3410. [CrossRef]
5. Xu, J.; Zhu, S.; Liao, G.; Zhang, Y. An Overview of Frequency Diverse Array Radar Technology. *J. Radars* **2018**, *7*, 167–182.
6. Chen, H.; Wang, W.-Q.; Cui, Y.-S.; Liao, Y. Statistical analysis for time modulation-based frequency diverse array beam-pattern. *IEEE Access* **2019**, *7*, 84142–84154. [CrossRef]
7. Cheng, Q.; Zhu, J.; Xie, T.; Luo, J.; Xu, Z. Time-invariant angle-range dependent directional modulation based on time-modulated frequency diverse arrays. *IEEE Access* **2017**, *5*, 26279–26290. [CrossRef]
8. Xu, J.; Kang, J.; Liao, G.; So, H.C. Mainlobe deceptive jammer suppression with FDA-MIMO radar. In Proceedings of the 2018 IEEE 10th Sensor Array and Multichannel Signal Processing Workshop (SAM), Sheffield, UK, 8–11 July 2018; IEEE: Piscataway, NJ, USA; pp. 504–508.
9. Lan, L.; Liao, G.; Xu, J.; Zhang, Y.; Fioranelli, F. Suppression approach to main-beam deceptive jamming in FDA-MIMO radar using nonhomogeneous sample detection. *IEEE Access* **2018**, *6*, 34582–34597. [CrossRef]
10. Tan, M.; Wang, C.; Li, Z. Correction analysis of frequency diverse array radar about time. *IEEE Trans. Antennas Propag.* **2021**, *69*, 834–847. [CrossRef]
11. Xu, Y.; Shi, X.; Li, W.; Xu, J.; Huang, L. Low-sidelobe range-angle beamforming with FDA using multiple parameter optimization. *IEEE Trans. Aerosp. Electron. Syst.* **2019**, *55*, 2214–2225. [CrossRef]
12. Khan, W.; Qureshi, I.M.; Saeed, S. Frequency diverse array radar with logarithmically increasing frequency offset. *IEEE Antennas Wirel. Propag. Lett.* **2015**, *14*, 499–502. [CrossRef]
13. Xu, Y.; Xu, J. Corrections to range-angle-dependent beamforming of pulsed-frequency diverse array. *IEEE Trans. Antennas Propag.* **2018**, *66*, 6466–6468. [CrossRef]
14. Lan, L.; Liao, G.; Xu, J.; Wen, J. Range-angle pencil-beamforming for non-uniformly distributed array radar. *Multidimens. Syst. Signal Process.* **2018**, *29*, 867–886. [CrossRef]
15. Basit, A.; Qureshi, I.M.; Khan, W.; Rehman, S.U.; Khan, M.M. Beam Pattern Synthesis for an FDA Radar with Hamming Window-Based Nonuniform Frequency Offset. *IEEE Antennas Wirel. Propag. Lett.* **2017**, *16*, 2283–2286. [CrossRef]
16. Zhou, C.; Wang, C.; Gong, J.; Tan, M.; Zhao, Y.; Liu, M. FDA-MIMO Beampattern Synthesis with an Analytical Method. *Int. J. Aerosp. Eng.* **2021**, *2021*, 9051131. [CrossRef]
17. Tan, M.; Wang, C.; Yuan, H.; Bai, J.; An, L. FDA-MIMO Beampattern Synthesis with Hamming Window Weighted Linear Frequency Increments. *Int. J. Aerosp. Eng.* **2020**, *2020*, 4851909. [CrossRef]
18. Basit, A.; Wang, W.; Nusenu, S. Adaptive transmit array sidelobe control using FDA-MIMO for tracking in joint radar-communications. *Digit. Signal Process.* **2020**, *97*, 102619. [CrossRef]
19. Lan, L.; Liao, G.; Xu, J.; Zhu, S.; Zhang, Y. Range-angle-dependent beamforming for FDA-MIMO radar using oblique projection. *Sci. China Inf. Sci.* **2022**, *65*, 152305. [CrossRef]
20. Lan, L.; Liao, G.; Xu, J.; Zhang, Y.; Liao, B. Transceive Beamforming With Accurate Nulling in FDA-MIMO Radar for Imaging. *IEEE Trans. Geosci. Remote Sens.* **2020**, *58*, 4145–4159. [CrossRef]
21. Wen, F.; Gui, G.; Gacanin, H.; Sari, H. Compressive sampling framework for 2D-DOA and polarization estimation in mmWave polarized massive MIMO systems. *IEEE Trans. Wirel. Commun.* **2022**. [CrossRef]
22. Wen, F.; Shi, J.; Gui, G. 3D Positioning method for anonymous UAV based on bistatic polarized MIMO radar. *IEEE Internet Things J.* **2023**, *10*, 815–827. [CrossRef]
23. Liao, Y.; Tang, H.; Chen, X.; Wang, W.-Q. Frequency Diverse Array Beampattern Synthesis With Taylor Windowed Frequency Offsets. *IEEE Antennas Wirel. Propag. Lett.* **2020**, *19*, 1901–1905. [CrossRef]
24. Sun, Z.; Huang, J.; Yang, J.; Zang, T.; He, Z. A High Accuracy Analysis Method of Harmonic/Interharmonic in Power System Based on Chebyshev Window. *Autom. Electr. Power Syst.* **2015**, *39*, 117–123.

sensors

MDPI

Article

A New Conformal Map for Polynomial Chaos Applied to Direction-of-Arrival Estimation via UCA Root-MUSIC

Seppe Van Brandt *, Jo Verhaevert, Tanja Van Hecke and Hendrik Rogier

IDLab, Department of Information Technology, Faculty of Engineering and Architecture, Ghent University-imec, 9052 Gent, Belgium; jo.verhaevert@ugent.be (J.V.); tanja.vanhecke@ugent.be (T.V.H.); hendrik.rogier@ugent.be (H.R.)

* Correspondence: seppe.vanbrandt@ugent.be

Abstract: The effects of random array deformations on Direction-of-Arrival (DOA) estimation with root-Multiple Signal Classification for uniform circular arrays (UCA root-MUSIC) are characterized by a conformally mapped generalized Polynomial Chaos (gPC) algorithm. The studied random deformations of the array are elliptical and are described by different Beta distributions. To successfully capture the erratic deviations in DOA estimates that occur at larger deformations, specifically at the edges of the distributions, a novel conformal map is introduced, based on the hyperbolic tangent function. The application of this new map is compared to regular gPC and Monte Carlo sampling as a reference. A significant increase in convergence rate is observed. The numerical experiments show that the UCA root-MUSIC algorithm is robust to the considered array deformations, since the resulting errors on the DOA estimates are limited to only 2 to 3 degrees in most cases.

Keywords: conformal map; polynomial chaos; root-MUSIC; direction-of-arrival; uniform circular array; deformation; error propagation

1. Introduction

Direction-of-arrival (DOA) estimation is a fundamental enabler of current and next-generation wireless communication systems. With the arrival of 5G and the development of 6G, it is of great importance to understand how DOA techniques perform under imperfect conditions, especially as the effects of hardware impairments become more important when the operating frequency increases [1]. Simulation times can be large and Monte Carlo (MC) analysis is often too time-consuming due to the excessive number of realizations that must be evaluated. Therefore, a fitting stochastic framework for the characterization of uncertainty propagation within these systems is Generalized Polynomial Chaos (gPC) [2,3]. As a versatile technique, gPC has been applied extensively to study the effects of randomness on antenna performance and radio wave propagation [4–11]. With the appropriate approach, gPC can even compete with MC when a high number of random variables are used [12–17]. In the context of localization, gPC has also been used—for example, in [18,19], where the effects of element displacements on DOA estimation were investigated; in [20], where the effects of random element gain and phase variations were studied, and in [21], where the uncertainty in multiple angle-of-arrival measurements was translated into an uncertainty in position estimation.

Being based on polynomials, however, gPC has its own drawbacks, one of the most important ones being slow convergence in the presence of function singularities [22,23]. Conformal maps can alleviate these problems, as demonstrated on Maxwell's source problem in [24]. Conformal maps were used previously in [25] to compute more accurate quadrature rules for integrands with similar non-polynomial behavior, inspired by [26,27].

In this work, we extend the techniques described in [24] by introducing a novel map that compensates for (apparent) singularities on the real axis. We illustrate the effectiveness of this new conformal map by characterizing the effects of elliptical array deformations on root-Multiple Signal Classification for uniform circular arrays (UCA root-MUSIC) [28], as it is a well-established and efficient DOA estimation algorithm. The UCA topology is a relatively simple array geometry that enables DOA estimation of the azimuth angle over a full 360° field of view, besides some more limited capability to provide estimates of the elevation angle. However, the dedicated UCA-root-MUSIC algorithm heavily relies on the inherent circular symmetry of the antenna array. Therefore, this algorithm and array topology are the ideal candidates to study the effects of random deformations on DOA estimation techniques. To be concise and keep the focus on the presence of singularities, we do not include white noise, in contrast to [18–20].

In the next section, key concepts are introduced and the novel conformal map is revealed in the appropriate mathematical framework. In Sections 3 and 4, the computational results are presented and discussed, and in the final section, Section 5, a conclusion is given.

2. Methods

2.1. gPC Approximation

Generalized Polynomial Chaos (gPC) [2,3] approximates a function f of a random variable x by an expansion in a well-chosen orthogonal polynomial basis $\{\Phi_k\}$. The orthogonal polynomials in question are associated with the weight function w, representing the probability density function of the random variable being used, typically according the Askey scheme [2]. If f is a function that is (computationally) expensive to evaluate, a gPC approximation of f can provide a computationally cheap substitute.

Assuming that w is a weight function with support $[-1, 1]$, the gPC orthogonal projection of degree N is defined as

$$P_N f(x) = \sum_{k=0}^{N} t_k \Phi_k(x), \quad t_k = C_k^{-1} \cdot \int_{-1}^{1} f(x)\Phi_k(x)w(x)\mathrm{d}x, \tag{1}$$

with $C_k = \int_{-1}^{1} \Phi_k^2(x)w(x)\mathrm{d}x$. Once the expansion coefficients t_k are known, the first two moments of f can be calculated and estimated by

$$\begin{aligned} \mu &= \int_{-1}^{1} f(x)w(x)\mathrm{d}x = t_0 \\ \mu_2 &= \int_{-1}^{1} f^2(x)w(x)\mathrm{d}x \approx \sum_{k=0}^{N} t_k^2 \cdot C_k, \end{aligned} \tag{2}$$

while the standard deviation on f can be estimated by

$$\sigma = \sqrt{\mu_2 - \mu^2} \approx \left(\sum_{k=1}^{N} t_k^2 \cdot C_k \right)^{\frac{1}{2}}. \tag{3}$$

The error in the L^2-norm $\|\cdot\|_{L^2,w}$ can be written in terms of the standard deviation σ and its estimator:

$$\begin{aligned}
\|f - P_N f\|_{L^2,w} &= \left(\int_{-1}^{1} [f(x) - P_N f(x)]^2 w(x) \mathrm{d}x \right)^{\frac{1}{2}} \\
&= \left(\int_{-1}^{1} \left[\sum_{k=0}^{\infty} t_k \Phi_k(x) - \sum_{k=0}^{N} t_k \Phi_k(x) \right]^2 w(x) \mathrm{d}x \right)^{\frac{1}{2}} \\
&= \left(\int_{-1}^{1} \left[\sum_{k=N+1}^{\infty} t_k \Phi_k(x) \right]^2 w(x) \mathrm{d}x \right)^{\frac{1}{2}} = \left(\sum_{k=N+1}^{\infty} t_k^2 \cdot C_k \right)^{\frac{1}{2}} \\
&= \left(\sum_{k=1}^{\infty} t_k^2 \cdot C_k - \sum_{k=1}^{N} t_k^2 \cdot C_k \right)^{\frac{1}{2}} = \left(\sigma^2 - \sum_{k=1}^{N} t_k^2 \cdot C_k \right)^{\frac{1}{2}}.
\end{aligned} \tag{4}$$

The quality of the gPC approximation will depend upon the analyticity of f in the complex plane [3,22,23]. The analyticity of f can be described by a Bernstein ellipse E_ρ, which is defined as the ellipse with foci -1 and 1 and ρ as the sum of its semiminor and semimajor axis; see Figure 1. From Bernstein's theorem for the convergence of the Chebyshev projection P_N^{Ch} [22,23] and the fact that the gPC projection P_N minimizes the L^2-error [3], it follows that, if f is analytically continuable to the open Bernstein ellipse $E_\rho \subseteq \mathbb{C}$, the error on the Nth degree gPC approximation, according to the L^2-norm, is limited to

$$\|f - P_N f\|_{L^2,w} \leq \|f - P_N^{\mathrm{Ch}} f\|_{L^2,w} \leq \|f - P_N^{\mathrm{Ch}} f\|_\infty \cdot \|1\|_{L^2,w} \leq 2\frac{M\rho^{-N}}{\rho - 1}, \tag{5}$$

with $\|\cdot\|_\infty$ the supremum norm and $|f(x)| \leq M$ for $x \in E_\rho$. It is clear from Equation (5) that it is preferable to have a high ρ, i.e., a large Bernstein ellipse, for optimal convergence. Singularities in the complex plane, close to the interval of interest $[-1, 1]$, can substantially lower ρ and will, therefore, be detrimental to the gPC approximation.

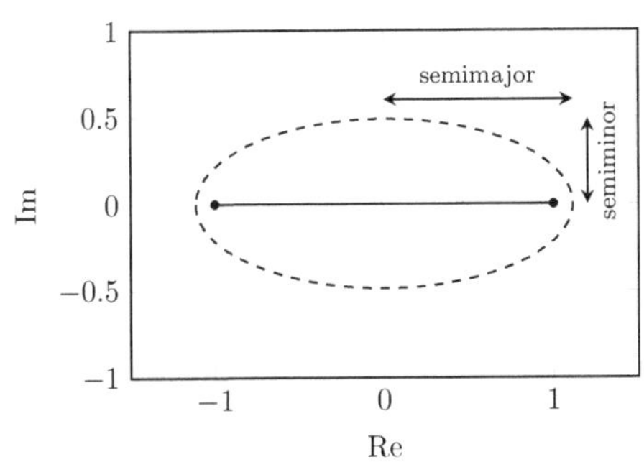

Figure 1. The $E_{1.6}$ Bernstein ellipse. Its size ρ is equal to the sum of its semiminor and semimajor axes—in this case, 1.6. Its foci -1 and 1 are shown by the black dots.

In most practical applications, the integral in Equation (1) is not analytically computable. In 1D cases, the coefficients t_k are generally approximated numerically using a Gauss quadrature rule [3]:

$$t_k \approx C_k^{-1} \cdot \sum_{i=1}^{N+1} f(x_i) \Phi_k(x_i) w_i, \tag{6}$$

with (x_i, w_i) being the $N+1$ quadrature nodes and weights of polynomial accuracy $2N+1$ associated with w. According to [3], Section 3.6, the aliasing error on the polynomial expansion of f is given by

$$A_N f(x) = \sum_{j=N+1}^{\infty} t_j \sum_{k=0}^{N} C_k^{-1} \left[\sum_{i=1}^{N+1} \Phi_j(x_i)\Phi_k(x_i)w_i \right] \Phi_k(x). \tag{7}$$

In essence, the aliasing error follows from the fact that the lower-order polynomials $\{\Phi_k\}$ cannot be distinguished from the higher-order polynomials $\{\Phi_j\}$ on a finite grid. As seen in Equation (7), it can be interpreted as the error that is introduced by using the lower-order discrete expansion of the higher-order polynomials instead of the higher-order polynomials themselves. In [3], it is mentioned that the aliasing error induced by using Formula (6) is usually of the same order as the projection error in Equation (5). Hence, the aliasing error will also benefit from a higher ρ.

2.2. Conformally Mapped gPC

In [24], a framework was established to incorporate conformal maps into the gPC algorithm for enlarging the Bernstein ellipse, based on earlier research from [25]. Consider a map g that is conformal in an open region $\Omega \subseteq \mathbb{C}$ with subdomain $[-1,1] \subseteq \mathbb{R}$, with $g([-1,1]) = [-1,1]$ and $g(\pm 1) = \pm 1$. This map can be used to define a new variable $\tilde{x}$:

$$\begin{cases} x = g(\tilde{x}) \\ \tilde{x} = g^{-1}(x). \end{cases} \tag{8}$$

If x is a random variable, distributed according to the weight function $w(x)$ with support $[-1,1]$, the variable after transformation $\tilde{x}$ will have its own weight function $\tilde{w}(\tilde{x})$ with support $[-1,1]$ [24]:

$$\begin{aligned} \tilde{w}(\tilde{x}) &= [w \circ g](\tilde{x}) \left| \frac{dx}{d\tilde{x}} \right| \\ &= [w \circ g](\tilde{x}) |g'(\tilde{x})|. \end{aligned}$$

The symbol $\circ$ is used to denote function composition, i.e., $[f_1 \circ f_2](x) = f_1(f_2(x))$. By choosing $g' > 0$ in Ω, the above equation becomes

$$\tilde{w}(\tilde{x}) = [w \circ g](\tilde{x}) g'(\tilde{x}). \tag{9}$$

Assuming that the orthogonal polynomials $\{\Phi_k\}$ associated with w are known, the orthogonal polynomials $\{\tilde{\Phi}_k\}$ associated with $\tilde{w}$ can be constructed by the Modified Chebyshev Algorithm [29]. Next, the quadrature points $\tilde{x}_i$ and associated weights $\tilde{w}_i$ of $\tilde{w}$ can be calculated using the Golub–Welsch Algorithm [30]. The Modified Chebyshev Algorithm is based upon a set of integrals called the modified moments, defined by

$$m_k = \int_{-1}^{1} \Phi_k(\tilde{x}) \tilde{w}(\tilde{x}) d\tilde{x}. \tag{10}$$

Only for select combinations of weight function and conformal map can these integrals be calculated analytically. In other cases, one has to make use of quadrature rules (or other computational methods) to evaluate these integrals numerically. As the quadrature nodes and weights belonging to $\tilde{w}$ are not yet known at this point in the algorithm, this integral has to be reformulated as an integral with weight function w, whose quadrature nodes and weights are known. One means of achieving this is by rewriting the integral as

$$m_k = \int_{-1}^{1} \Phi_k(\tilde{x})\tilde{w}(\tilde{x})\mathrm{d}\tilde{x}$$
$$= \int_{-1}^{1} \Phi_k(x)\frac{\tilde{w}(x)}{w(x)}w(x)\mathrm{d}x \tag{11}$$
$$\approx \sum_{i=1}^{N_m} \Phi_k(x_i)\frac{\tilde{w}(x_i)}{w(x_i)} \cdot w_i.$$

with (x_i, w_i) the quadrature nodes and weights of polynomial accuracy $2N_m - 1$ associated with w. Within this research, the value of N_m was set to 50, a value for which sufficient accuracy was reached.

Instead of a polynomial expansion of f, in the conformally mapped gPC algorithm introduced by [24], an expansion of $f \circ g$ is performed, using the newly constructed $\tilde{\Phi}_k$ polynomials:

$$P_N[f \circ g](\tilde{x}) = \sum_{k=0}^{N} \tilde{t}_k\tilde{\Phi}_k(\tilde{x}). \tag{12}$$

By substituting the inverse map $\tilde{x} = g^{-1}(x)$ at both sides of Equation (12), a mapped approximation of f is obtained:

$$\tilde{P}_N f(x) = \sum_{k=0}^{N} \tilde{t}_k\left[\tilde{\Phi}_k \circ g^{-1}\right](x). \tag{13}$$

A schematic overview of this algorithm is shown in Scheme 1.

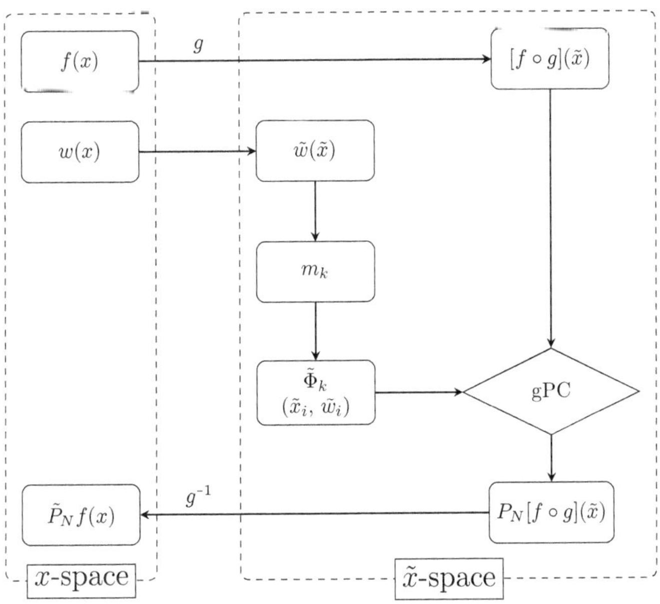

Scheme 1. An overview of the conformally mapped gPC algorithm. A new random variable $\tilde{x}$ is defined by applying a conformal map $x = g(\tilde{x})$ on the random variable x. An adapted quadrature rule $(\tilde{x}_i, \tilde{w}_i)$ is constructed for the new weight function $\tilde{w}(\tilde{x})$ via the modified moments m_k, along with its own orthogonal polynomials $\{\tilde{\Phi}_k\}$. Generalized Polynomial Chaos (gPC) is applied to the composite function $[f \circ g]$ and a mapped gPC approximation of f is found after performing the inverse conformal map $\tilde{x} = g^{-1}(x)$.

Analogously to classic gPC, the coefficients $\tilde{t}_k$ can be approximated by means of discrete projection, yielding

$$\begin{aligned}\tilde{t}_k &= \tilde{C}_k^{-1} \cdot \int_{-1}^{1} [f \circ g](\tilde{x}) \tilde{\Phi}_k(\tilde{x}) \tilde{w}(\tilde{x}) \mathrm{d}\tilde{x} \\ &\approx \tilde{C}_k^{-1} \cdot \sum_{i=1}^{N+1} [f \circ g](\tilde{x}_i) \tilde{\Phi}_k(\tilde{x}_i) \tilde{w}_i,\end{aligned} \tag{14}$$

with $\tilde{C}_k = \int_{-1}^{1} \tilde{\Phi}_k^2(\tilde{x}) \tilde{w}(\tilde{x}) \mathrm{d}\tilde{x}$ and $(\tilde{x}_i, \tilde{w}_i)$ the quadrature nodes and weights of polynomial accuracy $2N+1$ associated with $\tilde{w}$.

An upper bound similar to the one in Equation (5) can be derived for the conformally mapped gPC expansion [3,22–24]:

$$\|f - \tilde{P}_N f\|_{L^2,w} = \|f \circ g - P_N[f \circ g]\|_{L^2,\tilde{w}} \le 2\frac{\tilde{M}\tilde{\rho}^{-N}}{\tilde{\rho} - 1}. \tag{15}$$

In this case, $\tilde{\rho}$ is the size of the open Bernstein ellipse $E_{\tilde{\rho}}$, in which $f \circ g$ is analytically continuable. It is apparent from Equations (5) and (15) that the conformal map g needs to be chosen in such a way that $\tilde{\rho}$ exceeds ρ, thus achieving a faster convergence. In other words, $f \circ g$ needs to be analytically continuable in a larger open Bernstein ellipse than f.

To the best of the authors' knowledge, only one class of conformal map has been applied in the context of polynomial-based methods. This set of maps shifts singularities directly above or below the $[-1, 1]$ interval, away from the origin, in order to achieve a larger Bernstein ellipse. The Ellipse-to-Strip map [25], the Sausage map [25], the Kosloff Tal-Ezer (KTE) map [27] and the Ellipse-to-Slit map [31] all fall into this category. In Figure 2, examples are shown of these established maps.

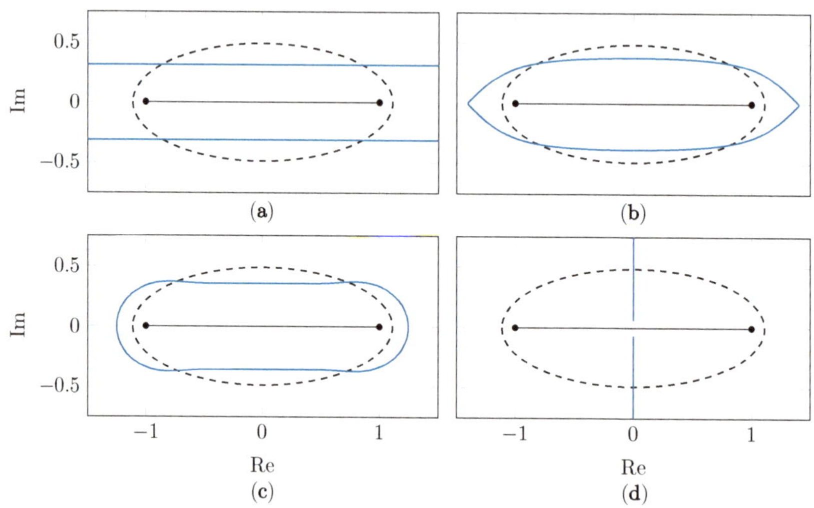

Figure 2. The $E_{1.6}$ Bernstein ellipse (dashed black line) and its image (solid blue line) under the Ellipse-to-Strip map (**a**), the KTE map (**b**), the Sausage map (**c**) and the Ellipse-to-Slit map with slits along the imaginary axis (**d**).

However, it is possible to encounter function singularities in other locations of the complex plane, such as on the real axis, close to the $[-1, 1]$ interval. In these situations, there is a need for a new conformal map, which is proposed further in Section 2.3.

2.3. Tanh Map

Assume that f has a singularity on the real axis at location p with $|p| > 1$. Due to this singularity, the size of the Bernstein ellipse E_ρ associated with f (in the x-space) is limited to $\rho = |p| + \sqrt{p^2 - 1}$. Ideally, one would use a map that gives rise to a Bernstein ellipse $E_{\tilde{\rho}}$, associated with $f \circ g$ in the $\tilde{x}$-space, that is larger than E_ρ. It is clear that a suitable map for this purpose should shift the singularity away from the $[-1, 1]$ interval. We propose the *tanh* map for this purpose:

$$x = g(\tilde{x}; \kappa) = \frac{\tanh(\kappa \tilde{x})}{\tanh(\kappa)}. \tag{16}$$

An illustration of this map for a real x and $\tilde{x}$, and for different values of κ, is shown in Figure 3. The parameter κ is added to make the map adaptable to different values of the singularity's location p.

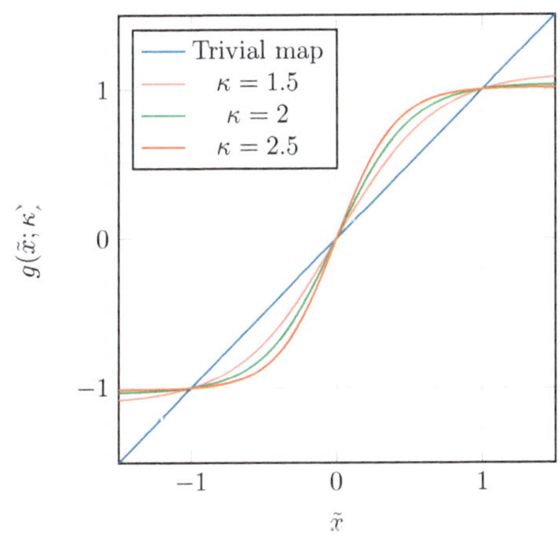

Figure 3. The *tanh* map for real $\tilde{x}$ at different values of κ, together with the trivial map $g(\tilde{x}) = \tilde{x}$.

The $\tanh(z)$ function is periodic with period πj, i.e., $\tanh(z) = \tanh(z \pm \pi j)$, and has simple poles at $\frac{\pi}{2} j \pm \pi j l$, with $l \in \mathbb{Z}$. As the map needs to be bijective in order to define the inverse map, the domain of the above map is restricted to the strip around the real axis between its closest poles:

$$|\mathrm{Im}(\tilde{x})| < \frac{\pi}{2\kappa} j. \tag{17}$$

The inverse of Equation (16) can now be defined as

$$\tilde{x} = g^{-1}(x; \kappa) = \frac{1}{2\kappa} \ln\left(\frac{1 + x \tanh(\kappa)}{1 - x \tanh(\kappa)}\right). \tag{18}$$

Using the *tanh* map, the singularity will shift to a position $|\tilde{p}| = |g^{-1}(p; \kappa)| > |p|$. As $g^{-1}(p; \kappa)$ needs to be defined and $g^{-1}(x; \kappa)$ has branch cuts $\left]-\infty, -\frac{1}{\tanh(\kappa)}\right]$ and $\left[\frac{1}{\tanh(\kappa)}, +\infty\right[$, κ is fundamentally limited to

$$\kappa < \kappa_{\max} = \ln\left(\frac{|p| + 1}{|p| - 1}\right). \tag{19}$$

Two Bernstein ellipses in the $\tilde{x}$-space can be defined: one that only takes into account the shifted singularity $\tilde{p}$ with size

$$\tilde{\rho}_{\text{re}} = |\tilde{p}| + \sqrt{\tilde{p}^2 - 1} = \frac{1}{2\kappa}\left|\ln\left(\frac{1 + p\tanh(\kappa)}{1 - p\tanh(\kappa)}\right)\right| + \sqrt{\frac{1}{4\kappa^2}\ln\left(\frac{1 + p\tanh(\kappa)}{1 - p\tanh(\kappa)}\right)^2 - 1}, \quad (20)$$

and one ellipse that only takes into account the singularities $\pm\frac{\pi}{2\kappa}j$ introduced by the *tanh* map, with size

$$\tilde{\rho}_{\text{im}} = \frac{\pi}{2\kappa} + \sqrt{\frac{\pi^2}{4\kappa^2} + 1}. \quad (21)$$

Depending on the values of p and κ, either $E_{\tilde{\rho}_{\text{re}}}$ or $E_{\tilde{\rho}_{\text{im}}}$ will be the largest ellipse. As the presence of singularities is restrictive for the gPC algorithm, $\tilde{\rho}$ is equal to the size of the smallest of both ellipses, i.e., $\tilde{\rho} = \min(\tilde{\rho}_{\text{re}}, \tilde{\rho}_{\text{im}})$. The largest and optimal value of $\tilde{\rho}$, named $\tilde{\rho}_{\text{eq}}$, is reached when both ellipses overlap ($\tilde{\rho}_{\text{re}} = \tilde{\rho}_{\text{im}}$) at a certain κ_{eq}, which can be found by solving Equations (20) and (21). Figure 4 illustrates this procedure for a singularity located at $p = \pm 1.1125$. The value of κ_{eq} as a function of the singularity position $|p|$ is shown in Figure 5, along with the two regions in $(|p|, \kappa)$-space, one where $\tilde{\rho}_{\text{re}} \leq \tilde{\rho}_{\text{im}}$ and one where $\tilde{\rho}_{\text{re}} \geq \tilde{\rho}_{\text{im}}$.

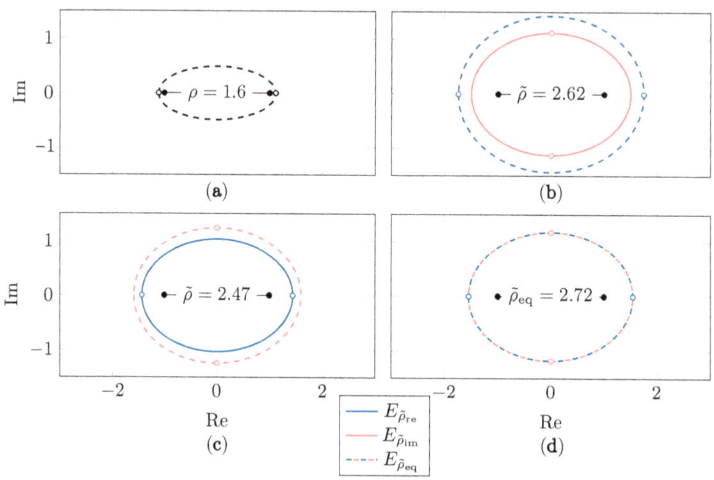

Figure 4. The Bernstein ellipse with size $\rho = 1.6$ (singularity located at $p = \pm 1.1125$) in the x-space (**a**) and the corresponding Bernstein ellipses $E_{\tilde{\rho}_{\text{re}}}$ and $E_{\tilde{\rho}_{\text{im}}}$ when using $\kappa = 1.05\kappa_{\text{eq}}$ (**b**), $\kappa = 0.95\kappa_{\text{eq}}$ (**c**) and $\kappa = \kappa_{\text{eq}}$ (**d**). The singularities are depicted with hollow dots.

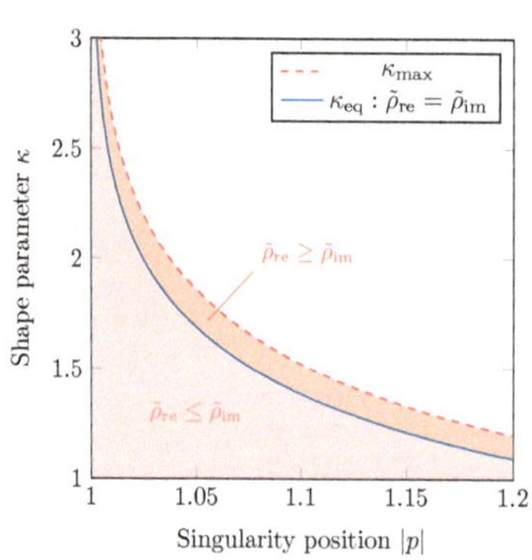

Figure 5. The different regions, denoting the relative sizes of both Bernstein ellipses, in the $(|p|, \kappa)$-space.

Since $E_{\tilde{\rho}_{eq}}$ has the singularities of g, being $\pm\frac{\pi}{2\kappa}j$, on its border, it will map to an infinitely large region $g(E_{\tilde{\rho}_{eq}};\kappa_{eq})$ in the x-space, as can be seen in Figure 6. This is similar to the Ellipse-to-Strip [25] and the Ellipse-to-Slit maps [31]. The condition that f is analytically continuable within this infinite region is rather strict, and if this is not the case, the effective Bernstein ellipse will be smaller than $E_{\tilde{\rho}_{eq}}$. Luckily, the region of analyticity for f will shrink very quickly in comparison to the corresponding Bernstein ellipse $E_{\tilde{\rho}}$. Additionally, when comparing $\tilde{\rho}$ with ρ in Figure 7, one can conclude that, in practice, as long as singularities above and below the $[-1,1]$ interval are reasonably far away, the convergence rate gain does not suffer much when applying the *tanh* map.

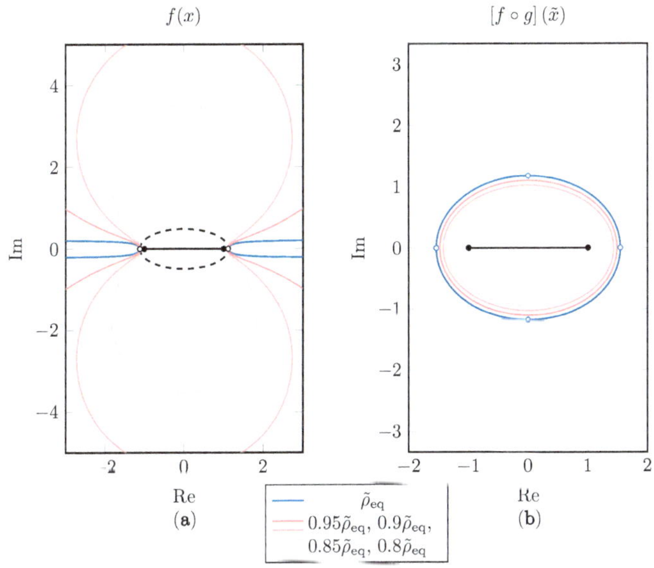

Figure 6. Starting with a Bernstein ellipse $E_{1.6}$ in the x-space, the Bernstein ellipses in the $\tilde{x}$-space with sizes $\tilde{\rho}_{eq} = 2.72$, $0.95\tilde{\rho}_{eq} = 2.59$, $0.9\tilde{\rho}_{eq} = 2.45$, $0.85\tilde{\rho}_{eq} = 2.31$ and $0.8\tilde{\rho}_{eq} = 2.18$ are shown (**b**). The corresponding regions $g(E_{\tilde{\rho}};\kappa_{eq})$ in which analytic continuability of f is assumed are shown in (**a**). The original Bernstein ellipse $E_{1.6}$ is shown with a dashed line as a reference and the singularities are depicted with hollow dots.

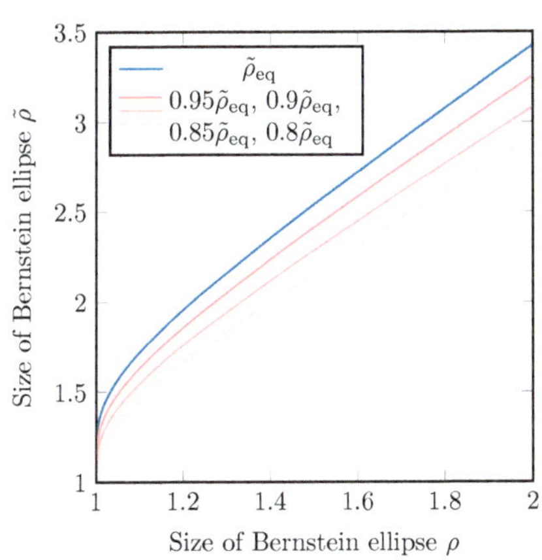

Figure 7. The size of $E_{\tilde{\rho}}$ as a function of the size of E_ρ when using the *tanh* map. $\tilde{\rho}_{eq}$ corresponds to the maximally achievable value of $\tilde{\rho}$.

2.4. Setup

Consider a nine-element uniform circular array (UCA) deployed in the azimuth plane. As the effects of the displacement of individual antenna elements on root-MUSIC have already been studied in earlier research—for example, in [18,19]—to showcase the proposed method, we assume that the array deforms in the shape of an ellipse with constant circumference and constant distance between the elements along the elliptical arc. This deformation is characterized by a single (random) variable: an "extended" eccentricity e defined by

$$e = \begin{cases} \sqrt{1-\frac{b}{a}} & \text{if } a > b \\ -\sqrt{1-\frac{a}{b}} & \text{if } a < b \\ 0 & \text{if } a = b, \end{cases} \tag{22}$$

with $2a$ being the width of the array (along $\phi = 0°$) and $2b$ the height (along $\phi = 90°$). Different array deformations are illustrated in Figure 8.

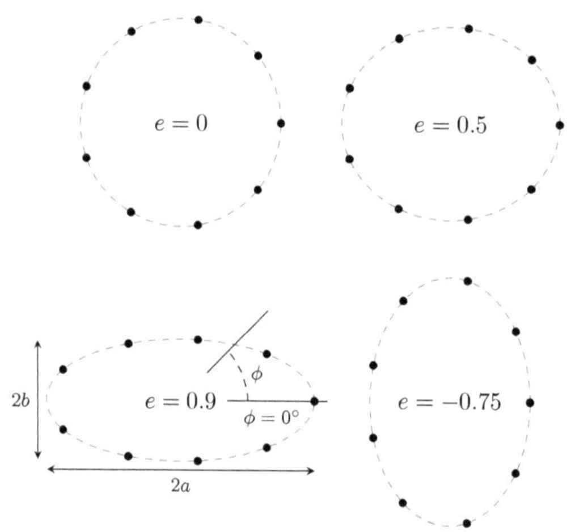

Figure 8. The shape of the deformed UCAs for different values of e.

We limit the eccentricity to the interval $[-0.9, 0.9]$, since values outside of this interval were deemed too unrealistic and the deformation would become too large for the DOA algorithm to provide a DOA estimation. However, it is standard practice [24,25] to rescale random variables to $[-1, 1]$. Therefore, the random variable x is introduced as $e = 0.9 \cdot x$. We assume x to be distributed according to a Beta distribution [32]:

$$\text{Beta}(x; \alpha, \beta) = \frac{(1+x)^{\alpha-1}(1-x)^{\beta-1}}{2^{\alpha+\beta-1}} \cdot \frac{\Gamma(\alpha+\beta)}{\Gamma(\alpha)\Gamma(\beta)}. \tag{23}$$

Figure 9 illustrates three Beta distributions with shape parameters $\alpha = \beta = 1$, $\alpha = \beta = 2$ and $\alpha = \beta = 3$. The orthogonal polynomials $\{\Phi_k\}$ associated with the Beta distribution are the Jacobi polynomials [33].

Since the UCA root-MUSIC algorithm was calibrated with a circular array in mind, as the array deforms, the estimated DOAs will deviate from their correct positions. This error propagation is described by a function $\hat{\phi} = f(x)$, with $\hat{\phi}$ being the DOA estimation corresponding to one specific source.

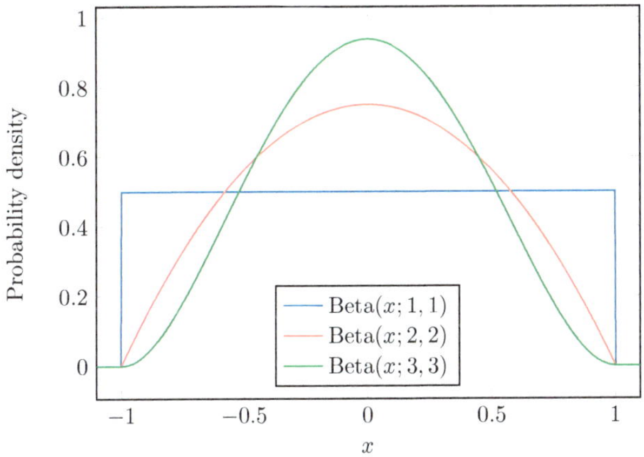

Figure 9. The three different Beta distributions used in this work.

From a practical viewpoint, it is non-intuitive to consider the analytic continuation of $f(x)$ in the complex plane beyond the $[-1, 1]$ interval. However, one can state that f, the relation between deformation and (erroneous) DOA estimation, behaves in a highly erratic manner when x approaches the edges of the interval $[-1, 1]$. This behavior is equivalent to the presence of nearby singularities on the real axis. Therefore, the principles of Sections 2.1–2.3 apply to this case, even though $f(x)$ has no closed form and the analytic continuation is not known.

3. Results

The operating frequency of the considered antenna array is set to 3.5 GHz (wavelength $\lambda \approx 8.57$ cm), being the center of a 5G band [34]. The dipole elements are all of length $\lambda/2$, as is the radius of the UCA. The array is excited by six plane waves along directions $\phi = 50°, 70°, 165°, 220°, 305°$ and $350°$ in the azimuth plane. DOA estimation is performed with the UCA root-MUSIC algorithm [28]. The full-wave NEC2++ simulator [35] is applied to rigorously simulate the complete antenna array, including all mutual coupling effects. For conciseness, we only discuss the behavior of one of the six DOA estimates, being the one corresponding to the source at $\phi = 50°$.

In Figures 10–12, the absolute and relative errors of μ and σ are presented as a function of N for the different Beta distributions. These values are calculated with classic gPC and mapped gPC for varying values of κ. The expansion coefficients are approximated with discrete projection according to Equations (6) and (14). As μ is equal to t_0 and $\tilde{t}_0$, the error shown in subfigures (c) is only the aliasing error introduced by the discrete projection. The error on the estimation of σ in subfigures (d) can, in addition to the aliasing error, be directly linked to the L^2-error according to Equation (4). As there are no analytical solutions to compare the results with, reference values are computed numerically by means of Monte Carlo simulation, using Latin Hypercube Sampling (LHS) with 10^6 samples [36]. These values are displayed in Table 1. The implemented UCA root-MUSIC algorithm has a precision of around 10^{-6} degrees, which is why, in some graphs, convergence halts at a relative error of around 2×10^{-8} and 10^{-6} for μ and σ, respectively.

Table 1. The reference values of μ and σ. Computed with MC using LHS and 10^6 samples.

	μ_{ref}	σ_{ref}
Beta$(x; 1, 1)$	49.510102	2.527614
Beta$(x; 2, 2)$	50.046675	1.347420
Beta$(x; 3, 3)$	50.163825	0.956752

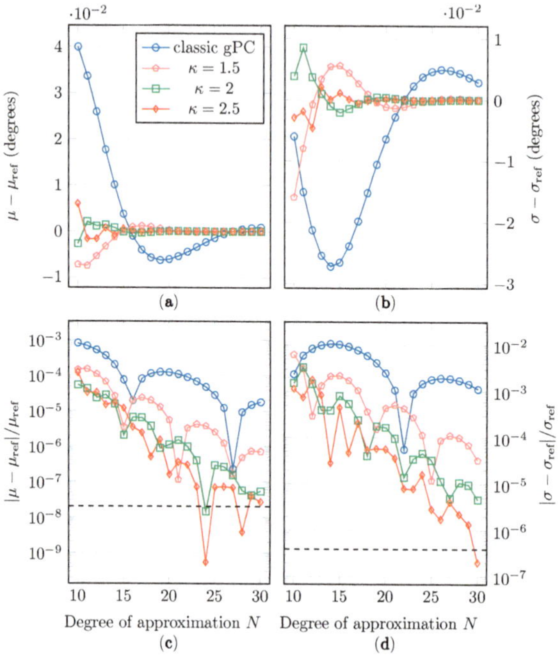

Figure 10. The absolute and relative errors on μ (**a**,**c**) and σ (**b**,**d**) with regard to their reference value when applying the Beta$(x;1,1)$ distribution. The precision floor is shown by a dashed line.

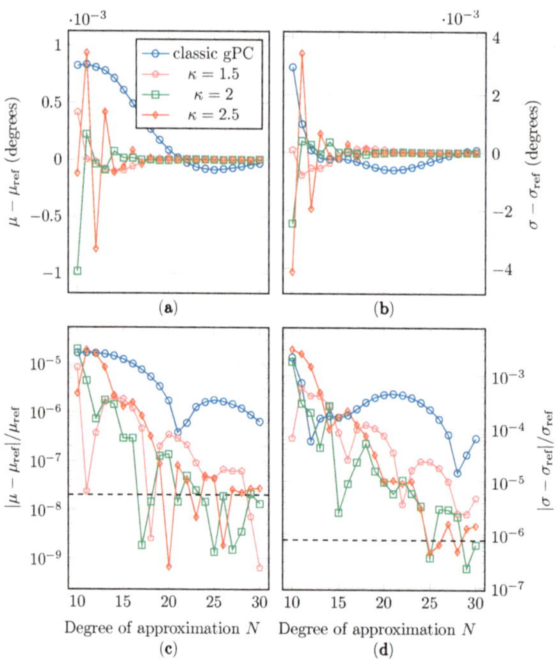

Figure 11. The absolute and relative errors on μ (**a**,**c**) and σ (**b**,**d**) with regard to their reference value when applying the Beta$(x;2,2)$ distribution. The precision floor is shown by a dashed line.

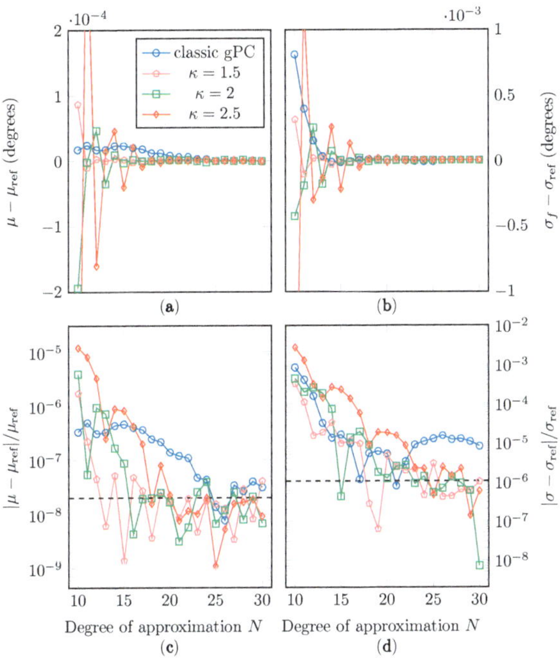

Figure 12. The absolute and relative errors on μ (**a**,**c**) and σ (**b**,**d**) with regard to their reference value when applying the Beta$(x;3,3)$ distribution. The precision floor is shown by a dashed line.

In Figure 13, a comparison between the resulting classic and mapped gPC approximations of f, using the Beta$(x;2,2)$ distribution and $N = 15$, is shown. In Figure 14, the comparison of the resulting empirical cumulative distribution functions (CDFs) is plotted.

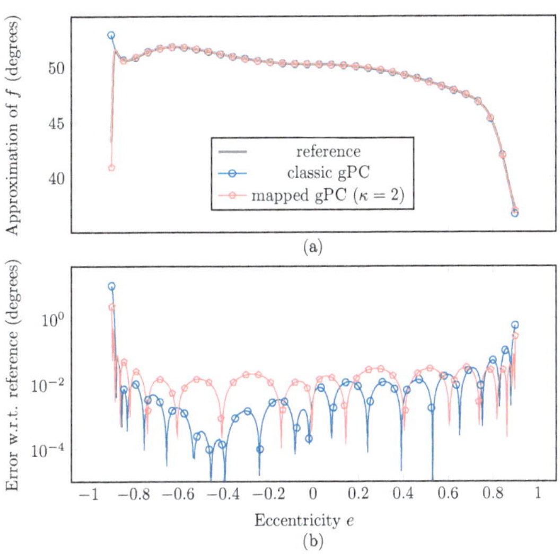

Figure 13. (**a**) The approximation of f using classic and mapped gPC and (**b**) the error on the approximation with regard to the reference curve, with $w(x) = \text{Beta}(x;2,2)$ and $N = 15$. The reference curve was constructed by sampling the full simulation.

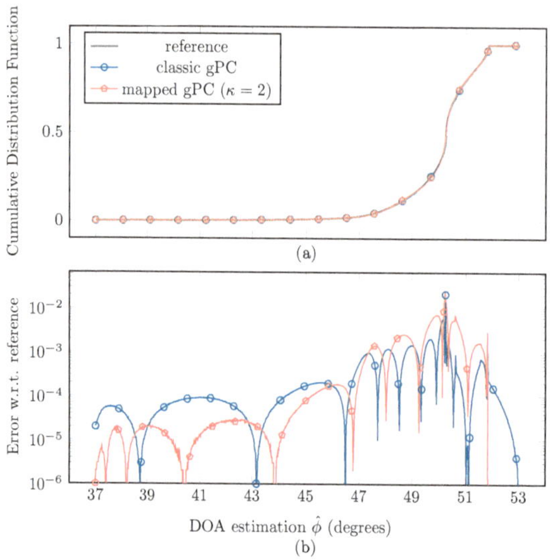

Figure 14. (**a**) The empirical CDFs of $\hat{\phi}$ when sampling the classic and mapped gPC expansion with the full simulation as a reference, each constructed with LHS and 10^6 samples, with $w(x) = \text{Beta}(x; 2, 2)$ and $N = 15$. (**b**) The error of the classic and mapped gPC CDFs in comparison to the reference CDF.

4. Discussion

The discussion is presented in two parts. First, the advantages of the use of the *tanh* map are analyzed, based on the deformed UCA application. Afterwards, the effects of the random array deformations on the UCA root-MUSIC algorithm are discussed.

4.1. Comparison of Classic and Mapped gPC

In Figures 10–12, we see a general improvement when using mapped gPC over classic gPC, as, in all cases, mapped gPC reaches the precision bound the fastest. Comparing the results for the estimation of μ, in subfigures (a) and (c), in which only the aliasing error is present, we can confirm that the aliasing error does indeed benefit from an increase in the size of the Bernstein ellipse. As expected, this increase in convergence rate is also present in the results for the estimation of σ in subfigures (b) and (d), which affirms the principles from Sections 2.2 and 2.3. Note that the error on σ is linked to the L^2-error in Equations (5) and (15) via Equation (4).

One aspect that should be mentioned is that, according to Equations (5) and (15), maximizing the size of the Bernstein ellipse only maximizes the rate of convergence, i.e., the incline of the convergence curves in subfigures (c) and (d). The vertical position of the convergence curves will, however, depend on other factors besides the size of the Bernstein ellipse. Therefore, it is possible that the fastest-converging method is not the one with the smallest error, especially in the cases with $\text{Beta}(x; 2, 2)$ and $\text{Beta}(x; 3, 3)$ as weight functions, where the precision floor is reached relatively quickly.

Two factors influence this phenomenon: first, the $M/\tilde{M}$ parameter in Equations (5) and (15), which is an upper bound of $|f|$ in its supposed region of analyticity, being either E_ρ for classic gPC or $g(E_{\tilde{\rho}}; \kappa)$ for mapped gPC. Although it is difficult to establish closed-form mathematical relations for the value of $M/\tilde{M}$, for the mapped gPC case, one can state that an increase in κ will cause $\tilde{M}$ to either increase or stay the same, as clarified in Equation (24). In other words, an increase in convergence rate due to an increase in κ can be paired with an upward vertical shift in the convergence curve.

$$\kappa_1 < \kappa_2 \Rightarrow g(E_{\tilde{\rho}}; \kappa_1) \subset g(E_{\tilde{\rho}}; \kappa_2) \Rightarrow \tilde{M}_1 \leq \tilde{M}_2 \tag{24}$$

Another factor is the aliasing error, defined by Equation (7), which is a function of the used weight function $w/\tilde{w}$ and the expansion polynomials $\{\Phi_k\}/\{\tilde{\Phi}_k\}$, adding a dependency on w and κ. An exact evaluation of Equation (7) is difficult. Upper bounds to the aliasing error are available for, among others, the Legendre and Chebychev polynomial expansions [37]; however, these are not readily applicable to this mapped gPC context. The dependency of both these factors on κ and w explains why we see different performance for the different κ values as the weight function changes, even though f and its singularities, and therefore also κ_{eq}, stay the same. Unfortunately, it is difficult to establish closed-form mathematical relations for these dependencies.

Figure 13 shows that a better fit is achieved at the extremities of the interval when using mapped instead of classic gPC, resulting in a lower supremum error and L_2-error. As this erratic behavior of f in these regions has a large influence on the statistical moments of the function, a better fit at the edges will have a significant impact on the accuracy of the estimation of μ and σ. Another benefit of the mapped approach is a better approximation of the CDF at the far left and far right sides, as seen in Figure 14.

4.2. Consequences for the UCA Root-MUSIC Algorithm

In the situations studied in this paper, the introduction of a random deformation of the array causes a bias in the DOA estimator of 0.05 to 0.5 degrees and a standard deviation of the DOA estimation of 1 to 2.5 degrees (depending on the distribution shape; see Table 1). All things considered, this makes the UCA root-MUSIC algorithm rather robust against the array deformations studied in this work.

5. Conclusions and Future Work

The non-polynomial behavior of the root-MUSIC DOA estimation as a function of the elliptical deformation of the UCA can be compared to the presence of singularities on the real axis, close to the interval of interest, which have a detrimental effect on the convergence of the classic gPC algorithm. Luckily, using the newly defined *tanh* conformal map, these singularities are moved further away from the domain of the random variable, which makes for a better characterization of the erratic behavior of the DOA estimation and, as a result, causes a considerable increase in the convergence rate of the first- and second-order statistics in comparison to classic gPC.

We conclude from the simulations that the errors induced in the DOA estimation due to the elliptical deformation of the UCA are limited to only a few degrees in most cases. However, when the eccentricity reaches an absolute value of around 0.7, the DOA estimations become very volatile, with much larger errors of up to 10 degrees.

In future work, it can be of interest to develop and research even more specific conformal maps so that each type of function singularity can be dealt with in an efficient manner. As for the DOA estimation in 5G and 6G wireless communication networks, it might be advisable to look at other types of antenna arrays, such as rectangular arrays, which are currently integrated into 5G base stations and handsets [38]. Additionally, it could be interesting to look at other, modified types of MUSIC, such as or spatial/backward/time smoothing MUSIC, which are better equipped to deal with highly correlated signals, as encountered due to multipath propagation in indoor environments [39]. The technique could also be extended to hybrid techniques, such as SpotFi, that rely on a combination of time-of-flight and angle-of-arrival with MUSIC to perform accurate localization with common WiFi infrastructures [40].

Author Contributions: S.V.B. performed the mathematical derivations, wrote the algorithm, performed the simulations and analyzed the data; S.V.B., J.V., T.V.H. and H.R. interpreted the analyzed data; S.V.B. wrote the manuscript and J.V., T.V.H. and H.R. revised it. All authors have read and agreed to the published version of the manuscript.

Funding: This work was supported by the Fonds Wetenschappelijk Onderzoek – Vlaanderen - FWO under Grants n° G0F4918N (EOS ID 30452698) and n° S001521N IoBaLeT: Sustainable Internet of Battery-Less Things.

Institutional Review Board Statement: Not applicable.

Informed Consent Statement: Not applicable.

Data Availability Statement: Not applicable.

Conflicts of Interest: The authors declare no conflicts of interest.

References

1. Abbasi, Q.H.; Jilani, S.F.; Alomainy, A.; Imran, M.A. *Antennas and Propagation for 5G and Beyond*; Institution of Engineering and Technology: Stevenage, UK, 2020.
2. Xiu, D.; Karniadakis, G.E. The Wiener—Askey polynomial chaos for stochastic differential equations. *SIAM J. Sci. Comput.* **2002**, *24*, 619–644. [CrossRef]
3. Xiu, D. *Numerical Methods for Stochastic Computations: A Spectral Method Approach*; Princeton University Press: Princeton, NJ, USA, 2010.
4. Rossi, M.; Dierck, A.; Rogier, H.; Vande Ginste, D. A stochastic framework for the variability analysis of textile antennas. *IEEE Trans. Antennas Propag.* **2014**, *62*, 6510–6514. [CrossRef]
5. Kersaudy, P.; Mostarshedi, S.; Sudret, B.; Picon, O.; Wiart, J. Stochastic Analysis of Scattered Field by Building Facades Using Polynomial Chaos. *IEEE Trans. Antennas Propag.* **2014**, *62*, 6382–6393. [CrossRef]
6. Rossi, M.; Stockman, G.J.; Rogier, H.; Vande Ginste, D. Stochastic Analysis of the Efficiency of a Wireless Power Transfer System Subject to Antenna Variability and Position Uncertainties. *Sensors* **2016**, *16*, 1100. [CrossRef]
7. Du, J.; Roblin, C. Statistical modeling of disturbed antennas based on the polynomial chaos expansion. *IEEE Antennas Wirel. Propag. Lett.* **2016**, *16*, 1843–1846. [CrossRef]
8. Nguyen, B.T.; Samimi, A.; Vergara, S.E.W.; Sarris, C.D.; Simpson, J.J. Analysis of electromagnetic wave propagation in variable magnetized plasma via polynomial chaos expansion. *IEEE Trans. Antennas Propag.* **2018**, *67*, 438–449. [CrossRef]
9. Ding, S.; Pichon, L. Sensitivity Analysis of an Implanted Antenna within Surrounding Biological Environment. *Energies* **2020**, *13*, 996. [CrossRef]
10. Leifsson, L.; Du, X.; Koziel, S. Efficient yield estimation of multiband patch antennas by polynomial chaos-based Kriging. *Int. J. Numer. Model. Electron. Netw. Devices Fields* **2020**, *33*, e2722. [CrossRef]
11. Rogier, H. Generalized Gamma-Laguerre Polynomial Chaos to Model Random Bending of Wearable Antennas. *IEEE Antennas Wirel. Propag. Lett.* **2022**, *21*, 1243–1247. [CrossRef]
12. Babacan, S.D.; Molina, R.; Katsaggelos, A.K. Bayesian compressive sensing using Laplace priors. *IEEE Trans. Image Process.* **2009**, *19*, 53–63. [CrossRef]
13. Zhang, Z.; Batselier, K.; Liu, H.; Daniel, L.; Wong, N. Tensor computation: A new framework for high-dimensional problems in EDA. *IEEE Trans. Comput.-Aided Des. Integr. Circuits Syst.* **2016**, *36*, 521–536. [CrossRef]
14. Salis, C.; Zygiridis, T. Dimensionality reduction of the polynomial chaos technique based on the method of moments. *IEEE Antennas Wirel. Propag. Lett.* **2018**, *17*, 2349–2353. [CrossRef]
15. Kaintura, A.; Dhaene, T.; Spina, D. Review of Polynomial Chaos-Based Methods for Uncertainty Quantification in Modern Integrated Circuits. *Electronics* **2018**, *7*, 30. [CrossRef]
16. Tomy, G.J.K.; Vinoy, K.J. A fast polynomial chaos expansion for uncertainty quantification in stochastic electromagnetic problems. *IEEE Antennas Wirel. Propag. Lett.* **2019**, *18*, 2120–2124. [CrossRef]
17. Lüthen, N.; Marelli, S.; Sudret, B. Sparse polynomial chaos expansions: Literature survey and benchmark. *SIAM/ASA J. Uncertain. Quantif.* **2021**, *9*, 593–649. [CrossRef]
18. Inghelbrecht, V.; Verhaevert, J.; Van Hecke, T.; Rogier, H. The influence of random element displacement on DOA estimates obtained with (Khatri–Rao-) root-MUSIC. *Sensors* **2014**, *14*, 21258–21280. [CrossRef]
19. Inghelbrecht, V.; Verhaevert, J.; Van Hecke, T.; Rogier, H.; Moeneclaey, M.; Bruneel, H. Stochastic framework for evaluating the effect of displaced antenna elements on DOA estimation. *IEEE Antennas Wirel. Propag. Lett.* **2016**, *16*, 262–265. [CrossRef]
20. Van der Vorst, T.; Van Eeckhaute, M.; Benlarbi-Delai, A.; Sarrazin, J.; Horlin, F.; De Doncker, P. Propagation of uncertainty in the MUSIC algorithm using polynomial chaos expansions. In Proceedings of the 11th European Conference on Antennas and Propagation, Paris, France, 19–24 March 2017; pp. 820–822.
21. Van der Vorst, T.; Van Eeckhaute, M.; Benlarbi-Delai, A.; Sarrazin, J.; Quitin, F.; Horlin, F.; De Doncker, P. Application of polynomial chaos expansions for uncertainty estimation in angle-of-arrival based localization. In *Uncertainty Modeling for Engineering Applications*; Springer: Berlin/Heidelberg, Germany, 2019; pp. 103–117.
22. Bernstein, S. *Sur l'ordre de la Meilleure Approximation des Fonctions Continues par des Polynômes de Degré Donné*; Hayez, Imprimeur des Académies Royales: Brussels, Belgium 1912; Volume 4.
23. Trefethen, L.N. *Approximation Theory and Approximation Practice*; Siam: Philadelphia, PA, USA, 2019; Volume 164.

24. Georg, N.; Römer, U. Conformally mapped polynomial chaos expansions for Maxwell's source problem with random input data. *Int. J. Numer. Model. Electron. Netw. Devices Fields* **2020**, *33*, 2776. [CrossRef]
25. Hale, N.; Trefethen, L.N. New quadrature formulas from conformal maps. *SIAM J. Numer. Anal.* **2008**, *46*, 930–948. [CrossRef]
26. Alpert, B.K. Hybrid Gauss-trapezoidal quadrature rules. *SIAM J. Sci. Comput.* **1999**, *20*, 1551–1584. [CrossRef]
27. Kosloff, D.; Tal-Ezer, H. A modified Chebyshev pseudospectral method with an O(N-1) time step restriction. *J. Comput. Phys.* **1993**, *104*, 457–469. [CrossRef]
28. Goossens, R.; Rogier, H.; Werbrouck, S. UCA Root-MUSIC with sparse uniform circular arrays. *IEEE Trans. Signal Process.* **2008**, *56*, 4095–4099. [CrossRef]
29. Gautschi, W. Orthogonal polynomials—Constructive theory and applications. *J. Comput. Appl. Math.* **1985**, *12*, 61–76. [CrossRef]
30. Golub, G.H.; Welsch, J.H. Calculation of Gauss quadrature rules. *Math. Comput.* **1969**, *23*, 221–230. [CrossRef]
31. Hale, N.; Trefethen, L.N. On the Use of Conformal Maps to Speed Up Numerical Computations.Ph.D. Thesis, University of Oxford, Oxford, UK, 2009.
32. Johnson, N.L.; Kotz, S.; Balakrishnan, N. Beta distributions. In *Continuous Univariate Distributions*, 2nd ed.; John Wiley and Sons: New York, NY, USA, 1994; pp. 221–235.
33. Shen, J.; Tang, T.; Wang, L.L. Orthogonal polynomials and related approximation results. In *Spectral Methods*; Springer: Berlin/Heidelberg, Germany, 2011; pp. 47–140.
34. Habibi, M.A.; Nasimi, M.; Han, B.; Schotten, H.D. A comprehensive survey of RAN architectures toward 5G mobile communication system. *IEEE Access* **2019**, *7*, 70371–70421. [CrossRef]
35. Molteno, T.C.A.; Kyriazis, N. *NEC2++: An NEC-2 Compatible Numerical Electromagnetics Code*; University of Otago: Dunedin, New Zealand, 2014.
36. Pebesma, E.J.; Heuvelink, G.B. Latin hypercube sampling of Gaussian random fields. *Technometrics* **1999**, *41*, 303–312. [CrossRef]
37. Hesthaven, J.S.; Gottlieb, S.; Gottlieb, D. *Spectral Methods for Time-Dependent Problems*; Cambridge University Press: Cambridge, UK, 2007; Volume 21.
38. Watanabe, A.O.; Ali, M.; Sayeed, S.Y.B.; Tummala, R.R.; Pulugurtha, M.R. A review of 5G front-end systems package integration. *IEEE Trans. Compon. Packag. Manuf. Technol.* **2020**, *11*, 118–133. [CrossRef]
39. Molodtsov, V.; Kureev, A.; Khorov, E. Experimental Study of Smoothing Modifications of the MUSIC Algorithm for Direction of Arrival Estimation in Indoor Environments. *IEEE Access* **2021**, *9*, 153767–153774. [CrossRef]
40. Kotaru, M.; Joshi, K.; Bharadia, D.; Katti, S. Spotfi: Decimeter level localization using wifi. In Proceedings of the 2015 ACM Conference on Special Interest Group on Data Communication, London, UK, 17–21 August 2015; pp. 269–282.

sensors

MDPI

Article

Reweighted Off-Grid Sparse Spectrum Fitting for DOA Estimation in Sensor Array with Unknown Mutual Coupling

Liangliang Li [1], Xianpeng Wang [1,*], Xiang Lan [1], Gang Xu [2] and Liangtian Wan [3]

[1] State Key Laboratory of Marine Resource Utilization in South China Sea, School of Information and Communication Engineering, Hainan University, Haikou 570228, China; liangliangli0717@126.com (L.L.); xlan@hainanu.edu.cn (X.L.)
[2] State Key Laboratory of Millimeter Waves, Southeast University, Nanjing 210096, China; gangxu@seu.edu.cn
[3] Key Laboratory for Ubiquitous Network and Service Software of Liaoning Province, School of Software, Dalian University of Technology, Dalian 116620, China; wanliangtian@dlut.edu.cn
* Correspondence: wxpeng2016@hainanu.edu.cn

Abstract: In the environment of unknown mutual coupling, many works on direction-of-arrival (DOA) estimation with sensor array are prone to performance degradation or even failure. Moreover, there are few literatures on off-grid direction finding using regularized sparse recovery technology. Therefore, the scenario of off-grid DOA estimation in sensor array with unknown mutual coupling is investigated, and then a reweighted off-grid Sparse Spectrum Fitting (Re-OGSpSF) approach is developed in this article. Inspired by the selection matrix, an undisturbed array output is formed to remove the unknown mutual coupling effect. Subsequently, a refined off-grid SpSF (OGSpSF) recovery model is structured by integrating the off-grid error term obtained from the first-order Taylor approximation of the higher-order term into the underlying on-grid sparse representation model. After that, a novel Re-OGSpSF framework is formulated to recover the sparse vectors, where a weighted matrix is developed by the MUSIC-like spectrum function to enhance the solution's sparsity. Ultimately, off-grid DOA estimation can be realized with the help of the recovered sparse vectors. Thanks to the off-grid representation and reweighted strategy, the proposed method can effectively and efficiently achieve high-precision continuous DOA estimation, making it favorable for real-time direction finding. The simulation results validate the superiority of the proposed method.

Keywords: DOA estimation; sensor array; unknown mutual coupling; off-grid error; Sparse Spectrum Fitting; reweighted sparse recovery

1. Introduction

Parameter estimation has drawn widespread concern and become a research hotspot in the field of array signal processing over the past few decades, especially for direction-of-arrival (DOA) estimation [1,2]. To the best of our knowledge, DOA estimation mainly uses sensor array to sample, analyze, and process spatial signals, achieving azimuth and elevation angles estimation for interested targets. It is one of the research foundations and vital components of parameter estimation, which provides precious angle information, and even prepares for subsequent parameter information like position. It is generally encountered in various real-life applications, such as unmanned aerial vehicles (UAV), vehicle localization, navigation, etc [3,4].

Currently, these efforts towards direction finding can be roughly grouped into two categories, i.e., subspace technologies [5–8] and sparse signal recovery (SSR) attempts [9–13]. The former are represented by the multiple signal classification (MUSIC) algorithm [5] and the estimation of signal parameters via rotational invariance techniques (ESPRIT) algorithm [6], officially opening the era of super-resolution direction finding. Afterwards, subspace estimators [7,8] are successively refined from different aspects, such as accuracy and robustness. These subspace frameworks largely rest with the eigenvalue decomposition (EVD) of the covariance matrix to decouple the subspaces, and then exploit the inherent relationships between the array manifolds and the decoupled subspaces to estimate DOAs. However, they are quite susceptible to the effects of signal-to-noise ratio (SNR), snapshots number, coherent targets, etc. In other words, it is hard for them to obtain the desirable performance under relatively harsh direction finding circumstances, such as unsatisfactory SNR and insufficient snapshots. Motivated by the potential spatial sparsity of the targets, the SSR perspective based on the principle of Compressed Sensing (CS) [14,15] came into being to solve the above problem. Subsequently, a set of sparsity-aware estimators are structured, including convex optimization attempts [9–11] and sparse Bayesian learning (SBL) efforts [12,13]. Simultaneously, plenty of research results have shown that this not only improves the estimation accuracy under the conditions of undesirable SNR and snapshots, but also enhances the robustness to coherent sources [16].

In terms of SBL-based approaches, their estimation accuracy relies heavily on the discretization degree of the investigated spatial domain. The higher the discretization degree (i.e., the denser the grid points), the smaller the grid interval, and the heavier the computational burden, while the lower the discretization degree (i.e., the sparser the grid points), the larger the grid interval, and the greater the mismatches between the desired DOAs and the closest candidate directions. Since it is almost impossible to ensure that all sources fall precisely on the predefined candidate directions, it is hard for the targets to avoid off-grid error caused by such mismatch.

Many attempts have been made in view of off-grid DOA estimation [17–21]. A Sparse Spectrum Fitting with Modeling Uncertainty (SpSFMU) scheme [17] is developed by linearly approximating the off-grid gap in the closest candidate grid points. Compared to its on-grid SpSF framework, it allows continuous DOA estimation, enhancing the estimation accuracy or/and relieving the computational load. Inspired by the linear approximation, a novel off-grid sparse Bayesian inference (OGSBI) approach [18] is presented with the assumption that the off-grid gap is uniformly distributed within the grid interval, where the off-grid gap is ultimately calculated by the expectation maximization (EM) strategy. As shown in [19], a robust block-sparse Bayesian learning framework without noise variance estimation is further derived via employing the sample covariance matrix. Despite these approaches [18,19] being robust to off-grid gap, it is computationally expensive for them to achieve satisfactory estimation accuracy, and their performance remains unacceptable under the condition of very coarse grids. For realizing a satisfactory performance under the coarse grid condition without aggravating the computational load, a robust root off-grid sparse Bayesian learning (Root-OGSBL) method [20] is designed, which dynamically upgrades predefined grid points via solving a polynomial to reduce off-grid error. Inspired by [20], a modified off-grid SBL (Re-OGSBL) approach [21] with a forgotten factor scheme is reported to further refine the dynamic update procedure for grid points.

Although these attempts have some attractive properties, they all either explicitly or implicitly require ideal array manifolds. In reality, there are plenty of array manifold perturbations in complex circumstances, such as unknown mutual coupling [22,23] and gain-phase error [24,25]. It should be noted that for a selected array, the more antennas, the smaller their spacing, and the more likely to cause space electromagnetic fields interaction between sensors. In this way, closely-spaced antennas are greatly vulnerable to unknown mutual coupling effect, destroying the desirable array manifold structure and making these approaches damaged or invalid in complex electromagnetic environments.

Therefore, many efforts have been devoted to DOA estimation in unknown mutual coupling [26–33]. Motivated by the spirit of array compensation, a group of auxiliary antennas are additionally placed on the two boundaries of the initial array to avoid the unknown mutual coupling influence, facilitating the direct utilization of the MUSIC principle [26]. Different from [26], a specific selection matrix [27] is structured to achieve array compensation through choosing sensors at the ends of the original array to be auxiliary ones. This attempt provides acceptable direction finding performance with low computational complexity, although at the cost of array aperture. Subsequently, the parameterized decoupling work [28] is reported to decouple the DOA information from the unknown mutual coupling. Despite preserving the entire array aperture well, it leads to a high computational burden. Except for the attempts achieved by subspace techniques in [26–28], a series of decoupling investigations on unknown mutual coupling interference have been conducted from the perspective of SSR [29–33]. As discussed in [29], an enhanced l_1-SVD (singular value decomposition) approach is introduced via using a selection matrix. Afterwards, a block sparse recovery (BSR) estimator [30] is developed in the data domain by parameterizing the coupled array manifold. Following the idea of parameterized decoupling, a robust BSR framework of array covariance vectors is reported as well [31]. However, it is still subject to the l_1-norm approximation, causing limited recovery performance. Afterwards, weighting techniques have received keen attention for their wide applications in parameter estimation [32,33], data fusion [34,35], error estimation [36,37], etc. Among them, weighting research on DOA estimation, i.e., the reweighted BSR approaches [32,33], are performed to enforce the solution's sparsity for improving the estimation accuracy of the angle parameter. Unfortunately, whether the schemes with off-grid error [17–21] or the frameworks under unknown mutual coupling [26–33], they focus on only one factor that we are interested in, i.e., off-grid error or unknown mutual coupling.

In fact, there are few studies on off-grid DOA estimation under unknown mutual coupling [38–40]. According to [38], a revised Root-OGSBL (Root-SBL) method is reported for off-grid estimation with unknown mutual coupling. Moreover, a novel sparse Bayesian learning with mutual coupling (SBLMC) estimator [39] for MIMO radar is developed by the expectation maximum (EM) principle to iteratively update unknown parameters, such as noise variance, mutual coupling coefficients, and off-grid gap vector. Nevertheless, such estimation superiority is at the expense of computational burden. Subsequently, an off-grid sparse recovery algorithm under unknown mutual coupling (OGSRMC) [40] is presented by iteratively updating DOAs, unknown mutual coupling, and off-grid parameters, which is superior in computational complexity and inferior in estimation accuracy to the SBLMC method.

In this work, a reweighted off-grid SpSF (Re-OGSpSF) scheme is presented for DOA estimation in the environment of unknown mutual coupling. The proposed estimator not only incorporates the off-grid error term into the underlying on-grid sparse recovery model to enhance the robustness, but also utilizes the reweighted strategy to ensure accuracy. Hence, the proposed Re-OGSpSF framework can realize high-precision continuous DOA estimation with low computational burden by adopting a coarse grid interval. Massive experimental results are displayed to validate the above inferences. The main contributions of this paper are listed as follows:

(1) Designing a selection matrix to eliminate the unknown mutual coupling interference;
(2) Formulating an improved off-grid SpSF (OGSpSF) framework for off-grid DOA estimation by joint sparse recovery;
(3) Developing a MUSIC-like weighted matrix to reweight the OGSpSF scheme for strengthening the solution's sparsity.

The remainder of this article is structured as follows: In Section 2, an actual data model affected by unknown mutual coupling is first defined. Then, a novel Re-OGSpSF scheme is explored for direction finding with unknown mutual coupling of the sensor array in Section 3. Subsequently, Section 4 presents the relevant remarks of this work. Afterwards, a series of simulation results are given to validate the superiority of the proposed framework in Section 5. Finally, Section 6 presents the conclusion of this paper.

Additionally, the relevant key notations involved in this work are explained in the following Table 1.

Table 1. Relevant key notations.

Notations	Definitions
$(\cdot)^T$, $(\cdot)^*$, and $(\cdot)^H$	Transpose, conjugate, and conjugate-transpose
$\mathbf{0}_{M\times K}$	$M\times K$ dimensional zero matrix
$\boldsymbol{I}_M$	$M\times M$ dimensional identity matrix
$\text{diag}\{\cdot\}$	Diagonalization operator
$\text{E}\{\cdot\}$	Mathematical expectation operator
$\otimes$ and $\odot$	Kronecker and Khatri–Rao products
$\lvert\cdot\rvert$ and $\det\{\cdot\}$	Absolute value and determinant operators
$\min\{\cdot\}$ and $\max\{\cdot\}$	Return the minimum and maximum value operators
$\|\cdot\|_0$, $\|\cdot\|_1$, and $\|\cdot\|_2$	l_0-norm, l_1-norm, and l_2-norm

2. The Coupled Signal Model

Consider a uniform linear array (ULA) configured with M omnidirectional sensors, each separated by the spacing d. $d \leq \lambda_d/2$, where λ_d refers to the signal wavelength. K narrowband uncorrelated sources $\{s_k\}_{k=1}^K$ are incident on the ULA from distinct directions $\{\theta_k\}_{k=1}^K$, where there is reason to believe that K is known exactly in advance [24,25]. Therefore, the ideal array output can be structured as

$$\boldsymbol{x}(t) = \boldsymbol{A}\boldsymbol{s}(t) + \boldsymbol{n}(t) \tag{1}$$

where $\boldsymbol{x}(t) = [x_1(t), x_2(t), \ldots, x_M(t)]^T \in \mathbb{C}^{M\times 1}$ means the ideal received data. $\boldsymbol{s}(t) = [s_1(t), s_2(t), \ldots, s_K(t)]^T \in \mathbb{C}^{K\times 1}$ reveals the signal vector. $\boldsymbol{n}(t) = [n_1(t), n_2(t), \ldots, n_M(t)]^T \in \mathbb{C}^{M\times 1}$ represents a stochastic Gaussian white noise vector, which follows $\boldsymbol{n}(t) \sim \text{N}(0, \sigma^2\boldsymbol{I}_M)$ with noise power σ^2. $\boldsymbol{A} = [\boldsymbol{a}(\theta_1), \boldsymbol{a}(\theta_2), \ldots, \boldsymbol{a}(\theta_K)] \in \mathbb{C}^{M\times K}$ denotes the array manifold matrix, where $\boldsymbol{a}(\theta_k) = [1, \eta(\theta_k), \ldots, \eta^{M-1}(\theta_k)]^T \in \mathbb{C}^{M\times 1}$ refers to the steering vector with $\eta(\theta_k) = e^{-j2\pi d/\lambda_d \sin\theta_k}$.

For a fixed array, the greater the number of antennas, the closer the sensors, and the greater the possibility of space electromagnetic fields interaction between them. In other words, the sensors are too close to escape such interaction, causing unknown mutual coupling disturbance between closely-spaced sensors. Under such a scenario, the ideal structure of the array manifold is disturbed and then coupled as

$$\hat{\boldsymbol{a}}(\theta_k) = \boldsymbol{G}\boldsymbol{a}(\theta_k) \tag{2}$$

where $\boldsymbol{G} \in \mathbb{C}^{M\times M}$ represents a mutual coupling matrix (MCM). According to [26], it is rational for MCM to depict its intrinsic properties with the complex banded symmetric Toeplitz structure. The mutual coupling coefficients between the sensors are inversely proportional to their spacing, i.e., the greater the antenna distance, the smaller the mutual coupling coefficients. What is more, the magnitude of the mutual coupling coefficient decreases rapidly when the antenna spacing increases [23]. In this way, it is reasonable to assume that the sensors far enough away are immune to the effect of unknown mutual coupling, i.e., the corresponding coefficients in the MCM are zero. Therefore, MCM is typically modeled as a banded symmetric Toeplitz matrix with just a few nonzero coefficients [23,29], i.e.,

$$
\begin{aligned}
G &= \text{Toeplitz}([1, g_1, \ldots, g_H, \mathbf{0}_{1\times(M-H-1)}]) \\
&= \begin{bmatrix}
1 & g_1 & \cdots & g_H & & & & & & \\
g_1 & 1 & g_1 & \cdots & g_H & & & \mathbf{0} & & \\
\vdots & \ddots & \ddots & \ddots & \ddots & \ddots & & & & \\
g_H & \cdots & g_1 & 1 & g_1 & \cdots & g_H & & & \\
 & \ddots & \vdots & \ddots & \ddots & \ddots & \ddots & \ddots & & \\
 & & g_H & \cdots & g_1 & 1 & g_1 & \cdots & g_H & \\
 & & & \ddots & \vdots & \ddots & \ddots & \ddots & \vdots & \\
 & \mathbf{0} & & & g_H & \cdots & g_1 & 1 & g_1 & \\
 & & & & & g_H & \cdots & g_1 & 1 &
\end{bmatrix}_{M\times M}
\end{aligned}
\tag{3}
$$

where $\{g_c\}_{c=1}^{H}$ reveal H unknown non-zero mutual coupling coefficients that satisfy $0 < |g_H| < |g_{H-1}| <, \ldots, < |g_1| < |g_0| = 1$.

Under the condition of unknown mutual coupling, (1) should be rewritten as

$$x(t) = GAs(t) + n(t) \tag{4}$$

By collecting T snapshots, the coupled array output matrix can be formed as

$$X = GAS + N \tag{5}$$

where $X = [x(t_1), x(t_1), \ldots, x(t_T)] \in \mathbb{C}^{M\times T}$ refers to the coupled array output. $\hat{A} = GA = [\hat{a}(\theta_1), \hat{a}(\theta_2), \ldots, \hat{a}(\theta_K)] \in \mathbb{C}^{M\times K}$ reveals the coupled array manifold matrix with $\hat{a}(\theta_k) = Ga(\theta_k)(K = 1, 2, \ldots, K)$. $S = [s(t_1), s(t_2), \ldots, s(t_T)] \in \mathbb{C}^{K\times T}$ is the source matrix. $N = [n(t_1), n(t_2), \ldots, n(t_T)] \in \mathbb{C}^{M\times T}$ indicates the noise matrix.

3. Re-OGSpSF for DOA Estimation in Sensor Array with Unknown Mutual Coupling

In this section, an effective Re-OGSpSF estimator is structured for off-grid DOA estimation with sensor array under the condition of unknown mutual coupling. Under such a framework, an enhanced OGSpSF representation model is constructed by integrating the off-grid error term obtained from the first-order Taylor approximation of the higher-order term into the underlying on-grid sparse recovery one. What is more, a weighted matrix is achieved by the MUSIC-like principle to reweight the OGSpSF scheme for improving the estimation accuracy. Hence, such an off-grid estimator can provide continuous DOA estimation to enforce the estimation accuracy or/and reduce the computational burden when an off-grid case occurs. In this way, it enables effective and efficient high-precision continuous off-grid estimation that would be more suitable for real-time direction finding.

3.1. Eliminating the Effect of Unknown Mutual Coupling

Obviously, the actual steering vector contains unknown mutual coupling coefficients, indicating that numerous DOA estimation approaches, including the SSR manner, will fail to work. In order to study the DOA estimation issue from the SSR perspective, it is quite necessary for the sparse reconstruction model to structure the effective over-complete dictionary. In other words, the unknown mutual coupling interference should be removed first for direction finding. According to (3), the specific selection matrix [29] is designed as

$$J = [\mathbf{0}_{(M-2H)\times H}, I_{(M-2H)}, \mathbf{0}_{(M-2H)\times H}] \tag{6}$$

Then, multiplying the selection matrix J by the coupled steering vector in (2) yields

$$\tilde{a}(\theta_k) = J\hat{a}(\theta_k) = JGa(\theta_k) = \xi_k \vec{a}(\theta_k) \tag{7}$$

where $\xi_k = \sum_{h=-H}^{H} g_{|h|} e^{-j2\pi d/\lambda_d (h+H)\sin\theta_k}$ indicates a constant for each true target. $\vec{a}(\theta_k) = [1, \eta(\theta_k), \ldots, \eta^{\vec{M}-1}(\theta_k)]^T \in \mathbb{C}^{\vec{M}\times 1}$ refers to a new array manifold with $\vec{M} = M - 2H$. It successfully decouples the DOAs parameter from the unknown mutual coupling coefficients, despite not using H antennas on each side of the ULA.

Inspired by (7), (5) can be greatly decoupled as

$$Y = JX = J(GAS + N) = JGAS + JN = \vec{A}\Lambda S + \vec{N} = \vec{A}\vec{S} + \vec{N} \tag{8}$$

where $Y = [y(t_1), y(t_1), \ldots, y(t_T)] \in \mathbb{C}^{\vec{M}\times T}$ reveals the array output free from unknown mutual coupling interference. $\vec{A} = [\vec{a}(\theta_1), \vec{a}(\theta_2), \ldots, \vec{a}(\theta_K)] \in \mathbb{C}^{\vec{M}\times K}$ denotes the decoupled array manifold matrix, like A in (1). $\Lambda = \text{diag}\{\xi_1, \xi_2, \ldots, \xi_K\} \in \mathbb{C}^{K\times K}$ stands for a diagonal matrix that can be integrated with matrix S to generate a revised signal matrix $\vec{S}$. $\vec{N} \in \mathbb{C}^{\vec{M}\times T}$ represents the modified noise matrix.

3.2. SpSF Principle

According to (8), the covariance matrix of Y can be denoted as

$$R = \text{E}\{YY^H\} = \vec{A}R_{\vec{S}}\vec{A}^H + R_{\vec{N}} = \vec{A}\text{diag}\{\rho_1^2, \rho_2^2, \ldots, \rho_K^2\}\vec{A}^H + \sigma^2 I_{\vec{M}} \tag{9}$$

where $R_{\vec{N}} = \sigma^2 I_{\vec{M}}$ denotes the noise covariance matrix. $R_{\vec{S}} = \text{diag}\{\rho_1^2, \rho_2^2, \ldots, \rho_K^2\}$ means the signal covariance matrix, where ρ_k^2 stands for the k-th signal power.

Based on (9), it can be found that the array aperture can be enlarged by performing the covariance vectorization operation. Then, vectoring R yields

$$r_v = \text{vec}(R) = (\vec{A}^* \odot \vec{A})\vec{u} + \vec{n}_v = \vec{A}_v\vec{u} + \vec{n}_v \tag{10}$$

where $\vec{A}_v = \vec{A}^* \odot \vec{A} = [\vec{a}_v(\theta_1), \vec{a}_v(\theta_2), \ldots, \vec{a}_v(\theta_K)] \in \mathbb{C}^{\vec{M}^2\times K}$ is the virtual array manifold matrix, where $\vec{a}_v(\theta_k) = \vec{a}^*(\theta_k) \otimes \vec{a}(\theta_k) \in \mathbb{C}^{\vec{M}^2\times 1}$ denotes the virtual array manifold, unlike the steering vector $\vec{a}(\theta_k)$ in (8). Obviously, such a vectorized signal model effectively increases the degrees of freedom (DOFs) of the virtual array and expands the array aperture. $\vec{u} = [\rho_1^2, \rho_2^2, \ldots, \rho_K^2]^T \in \mathbb{C}^{K\times 1}$ and $\vec{n}_v = \text{vec}(R_{\vec{N}}) = \sigma^2\text{vec}(I_{\vec{M}}) \in \mathbb{C}^{\vec{M}^2\times 1}$.

Inspired by (10), a sparsity-inducing scheme can be structured for DOA estimation. Through discretizing the spatial domain uniformly and densely enough, a set of candidate directions can be achieved, that is, $\bar{\theta} = \{\bar{\theta}_1, \bar{\theta}_2, \ldots, \bar{\theta}_N\} (N \geqslant K)$. In this way, it is reasonable to assume that all source DOAs fall exactly on the candidate directions of set $\bar{\theta}$. Then, an over-complete dictionary can be modeled as

$$\bar{A}_v = \bar{A}^* \odot \bar{A} = [\vec{a}_v(\bar{\theta}_1), \vec{a}_v(\bar{\theta}_2), \ldots, \vec{a}_v(\bar{\theta}_N)] \in \mathbb{C}^{\vec{M}^2\times N} \tag{11}$$

where $\bar{A} = [\vec{a}(\bar{\theta}_1), \vec{a}(\bar{\theta}_2), \ldots, \vec{a}(\bar{\theta}_N)] \in \mathbb{C}^{\vec{M}\times N}$ and $\vec{a}_v(\bar{\theta}_n) = \vec{a}^*(\bar{\theta}_n) \otimes \vec{a}(\bar{\theta}_n) \in \mathbb{C}^{\vec{M}^2\times 1}$.

Combining (11), (10) can be sparsely expressed as

$$r_v = (\bar{A}^* \odot \bar{A})\bar{u} + \vec{n}_v = \bar{A}_v\bar{u} + \vec{n}_v \tag{12}$$

where $\bar{u} = [\bar{\rho}_1^2, \bar{\rho}_2^2, \ldots, \bar{\rho}_N^2]^T \in \mathbb{C}^{N\times 1}$ reveals the K-sparse vector of signal power due to the existence of K sources. K non-zero entries in $\bar{u}$ correspond to the desired DOAs and are equal to that in $\vec{u}$, i.e., $\{\rho_k^2\}_{k=1}^K$. In this way, such a direction-finding problem can be turned into a sparse reconstruction issue, where DOAs can be determined by scanning the locations of K non-zero entries in the sparse vector $\bar{u}$.

To the best of our knowledge, the l_0-norm penalty is considered to be the theoretically optimal choice for measuring sparsity. Unfortunately, as a classical non-convex and non-deterministic polynomial (NP)-hard issue, l_0-norm penalty is mathematically intractable.

That is to say, l_0-norm penalty may not be applicable in actual direction finding. Inspired by [22,33], it is rational for the sparsity-inducing scheme to recover the sparse matrix via using l_1-norm optimization rather than l_0-norm minimization. Through convex approximation, l_1-norm penalty turns the non-convex scenario into a convex one to relieve the computational load. Following this thought, the l_1-norm penalty scheme can be structured as

$$\min\|\bar{\boldsymbol{u}}\|_1 \quad \text{s.t.} \quad \|\boldsymbol{r}_v - \bar{\boldsymbol{A}}_v\bar{\boldsymbol{u}}\|_2^2 \leqslant \varsigma, \quad \bar{\boldsymbol{u}} \geqslant \boldsymbol{0} \tag{13}$$

where ς refers to the regularization parameter, which balances the fitting error and signal sparsity and is crucial to robust sparse recovery. Its detailed explanation can be found in Remark 1 of the Related Remarks in Section 4.

In (9), $\boldsymbol{R}$ is an ideal covariance matrix based on the infinite number of snapshots, which is unavailable in practice. In actual situations, $\boldsymbol{R}$ is usually replaced by the sample covariance matrix under a finite number of snapshots T, i.e.,

$$\bar{\boldsymbol{R}} = \frac{1}{T}\sum_{t=1}^{T} \boldsymbol{Y}(t)\boldsymbol{Y}^H(t) \tag{14}$$

It is easy to find that there is a fitting error between $\boldsymbol{R}$ and $\bar{\boldsymbol{R}}$ caused by the finite snapshots, that is, $\Delta\boldsymbol{R} = \bar{\boldsymbol{R}} - \boldsymbol{R}$. In this way, (10) should be revised as follows:

$$\bar{\boldsymbol{r}}_v = \text{vec}(\bar{\boldsymbol{R}}) = \text{vec}(\boldsymbol{R} + \Delta\boldsymbol{R}) = \boldsymbol{r}_v + \Delta\boldsymbol{r} = \vec{\boldsymbol{A}}_v\vec{\boldsymbol{u}} + \vec{\boldsymbol{n}}_v + \Delta\boldsymbol{r} \tag{15}$$

where $\Delta\boldsymbol{r} = \bar{\boldsymbol{r}}_v - \boldsymbol{r}_v = \text{vec}(\Delta\boldsymbol{R}) = \text{vec}(\bar{\boldsymbol{R}} - \boldsymbol{R})$ refers to the covariance fitting error vector.

Then, (15) can be further sparsely modeled as

$$\bar{\boldsymbol{r}}_v = (\bar{\boldsymbol{A}}^* \odot \bar{\boldsymbol{A}})\bar{\boldsymbol{u}} + \vec{\boldsymbol{n}}_v + \Delta\boldsymbol{r} = \bar{\boldsymbol{A}}_v\bar{\boldsymbol{u}} + \vec{\boldsymbol{n}}_v + \Delta\boldsymbol{r} \tag{16}$$

Inspired by [10,11], $\Delta\boldsymbol{r}$ obeys the distribution as follows:

$$\Delta\boldsymbol{r} \sim \text{AsN}(0_{\vec{M}^2\times 1}, \boldsymbol{W}) \tag{17}$$

where $\boldsymbol{W} = \frac{\boldsymbol{R}^T \otimes \boldsymbol{R}}{T}$. $\text{AsN}(\mu, \varpi)$ indicates the asymptotically normal (AsN) distribution, whose mean and variance are equal to μ and ϖ, respectively. Obviously, $\Delta\boldsymbol{r}$ does not obey the standard normal distribution and ς is hard to calculate at this time. Whereas, with the help of asymptotic characteristic of $\Delta\boldsymbol{r}$, it is easy to determine the parameter ς.

Combining the principle of linear algebra yields

$$\boldsymbol{W}^{-\frac{1}{2}}\Delta\boldsymbol{r} \sim \text{AsN}(0_{\vec{M}^2\times 1}, \boldsymbol{I}_{\vec{M}^2}) \tag{18}$$

As depicted in (18), it can be directly deduced that

$$\left\|\boldsymbol{W}^{-\frac{1}{2}}\Delta\boldsymbol{r}\right\|_2^2 \sim \text{As}\chi^2(\vec{M}^2) \tag{19}$$

where $\text{As}\chi^2(\nu)$ obeys the asymptotically chi-square distribution with ν DOFs.

Let $\bar{\boldsymbol{W}} = \frac{\bar{\boldsymbol{R}}^T \otimes \bar{\boldsymbol{R}}}{T}$ and $\bar{\boldsymbol{n}}_v = \bar{\sigma}^2\text{vec}(\boldsymbol{I}_{\vec{M}})$ be estimates of $\boldsymbol{W}$ and $\vec{\boldsymbol{n}}_v$, respectively. Additionally, $\bar{\sigma}^2$ is the estimated noise power. Combining (16) with (19) yields

$$\min\|\bar{\boldsymbol{u}}\|_1 \quad \text{s.t.} \quad \left\|\bar{\boldsymbol{W}}^{-\frac{1}{2}}(\bar{\boldsymbol{r}}_v - \bar{\boldsymbol{n}}_v - \bar{\boldsymbol{A}}_v\bar{\boldsymbol{u}})\right\|_2 \leqslant \sqrt{\varsigma}, \quad \bar{\boldsymbol{u}} \geqslant \boldsymbol{0} \tag{20}$$

In this way, DOA estimation can be achieved by the spatial spectrum of the recovered sparse vector $\bar{\boldsymbol{u}}$.

3.3. Off-Grid SpSF (OGSpSF) for DOA Estimation

It is clear that the sparse vector $\bar{\boldsymbol{u}}$ in (20) can indeed be recovered for DOA estimation. Unfortunately, it only considers the on-grid case by default. In practice, the spatial domain can hardly be discretized adequately to generate continuous candidate sampling grids, but the target directions are continuous variables. Thus, it is difficult for the predefined grid points to accurately match the actual angles. In other words, there is an inevitable off-grid gap between the desired DOAs and the candidate grid points, degrading the estimation performance to some extent. Inspired by the linear approximation thought in [17], an effective OGSpSF representation model is formed by integrating the off-grid error term derived from the first-order Taylor approximation into the original on-grid sparse recovery one to guarantee the robustness and accuracy.

It is known that the candidate direction $\bar{\theta}_n$ is mismatched with the true DOA θ_K under the off-grid scenario. Furthermore, it is not hard to find that each element of the virtual array manifold $\vec{\boldsymbol{a}}_v(\theta_k)$ in (10) can be represented as $[\vec{\boldsymbol{a}}_v(\theta_k)]_{\vec{m}} = e^{-j2\pi d/\lambda_d \vec{m}\sin\theta_k}$ with integer $\vec{m} \in [1-\vec{M}, \vec{M}-1]$. Resorting to $[\vec{\boldsymbol{a}}_v(\bar{\theta}_n)]_{\vec{m}}$ in (11), $[\vec{\boldsymbol{a}}_v(\theta_k)]_{\vec{m}}$ can be depicted as

$$[\vec{\boldsymbol{a}}_v(\theta_k)]_{\vec{m}} = e^{-j2\pi d/\lambda_d \vec{m}\delta_n} \times [\vec{\boldsymbol{a}}_v(\bar{\theta}_n)]_{\vec{m}} \tag{21}$$

where $[\vec{\boldsymbol{a}}_v(\bar{\theta}_n)]_{\vec{m}} = e^{-j2\pi d/\lambda_d \vec{m}\sin\bar{\theta}_n}$. $\bar{\theta}_n \in \bar{\boldsymbol{\theta}}$ is the sampling grid point closest to θ_k and $\delta_n = \sin\theta_k - \sin\bar{\theta}_n$ indicates the offset parameter. Inspired by the Taylor expansion in algebraic theory, $e^{-j2\pi d/\lambda_d \vec{m}\delta_n}$ can be replaced by its first-order Taylor approximation, i.e., $e^{-j2\pi d/\lambda_d \vec{m}\delta_n} \approx [1 + (-j2\pi d/\lambda_d \vec{m}\delta_n)]$.

Then, (21) can be approximated to

$$\begin{aligned} {}_{\vec{m}} &= e^{-j2\pi d/\lambda_d \vec{m}\delta_n} \times [\vec{\boldsymbol{a}}_v(\bar{\theta}_n)]_{\vec{m}} \\ &\approx [1 + (-j2\pi d/\lambda_d \vec{m}\delta_n)][\vec{\boldsymbol{a}}_v(\bar{\theta}_n)]_{\vec{m}} \\ &\approx [\vec{\boldsymbol{a}}_v(\bar{\theta}_n)]_{\vec{m}} + (-j2\pi d/\lambda_d \vec{m})[\vec{\boldsymbol{a}}_v(\bar{\theta}_n)]_{\vec{m}}\delta_n \end{aligned} \tag{22}$$

According to (22), it can be deduced that the virtual array manifold $\vec{\boldsymbol{a}}_v(\theta_k)$ can be composed of $\vec{\boldsymbol{a}}_v(\bar{\theta}_n)$ and $\vec{\boldsymbol{b}}_v(\bar{\theta}_n)\delta_n$, where $\vec{\boldsymbol{b}}_v(\bar{\theta}_n)$ can be recorded as

$$\vec{\boldsymbol{b}}_v(\bar{\theta}_n) = \boldsymbol{\Gamma}\vec{\boldsymbol{a}}_v(\bar{\theta}_n) \tag{23}$$

where $\boldsymbol{\Gamma} = \mathrm{diag}\{\boldsymbol{\phi}\} \in \mathbb{C}^{\vec{M}^2 \times \vec{M}^2}$ indicates a diagonal matrix with $\boldsymbol{\phi} = [\phi_1, \phi_2, \ldots, \phi_{\vec{M}^2}] = -j2\pi d/\lambda_d[0, 1, \ldots, \vec{M}-1, -1, 0, \ldots, \vec{M}-2, \ldots, 2-\vec{M}, 3-\vec{M}, \ldots, 1, 1-\vec{M}, 2-\vec{M}, \ldots, 0] \in \mathbb{C}^{1\times\vec{M}^2}$. Obviously, its diagonal element corresponding to $[\vec{\boldsymbol{a}}_v(\bar{\theta}_n)]_{\vec{m}}$ is equal to $-j2\pi d/\lambda_d \vec{m}$. By separating the unknown variable δ_n and the known parameter $-j2\pi d/\lambda_d \vec{m}$, $\vec{\boldsymbol{b}}_v(\bar{\theta}_n)$ can finally be known, despite δ_n being hard to determine due to the presence of $\sin\theta_k$.

In this way, another over-complete dictionary $\bar{\boldsymbol{B}}_v$, except for $\bar{\boldsymbol{A}}_v$ in (11), can be further structured to help construct the off-grid sparse recovery model, which takes the following form:

$$\bar{\boldsymbol{B}}_v = \boldsymbol{\Gamma}\bar{\boldsymbol{A}}_v = [\vec{\boldsymbol{b}}_v(\bar{\theta}_1), \vec{\boldsymbol{b}}_v(\bar{\theta}_2), \ldots, \vec{\boldsymbol{b}}_v(\bar{\theta}_N)] \in \mathbb{C}^{\vec{M}^2 \times N} \tag{24}$$

Following (22) and (24), the on-grid sparse representation model in (16) should be refined to an OGSpSF recovery one, i.e.,

$$\vec{\boldsymbol{r}}_v = \mathrm{vec}(\bar{\boldsymbol{R}}) = (\bar{\boldsymbol{A}}_v + \boldsymbol{\Gamma}\bar{\boldsymbol{A}}_v\boldsymbol{\Delta})\bar{\boldsymbol{u}} + \vec{\boldsymbol{n}}_v + \Delta\boldsymbol{r} = \bar{\boldsymbol{A}}_v\bar{\boldsymbol{u}} + \bar{\boldsymbol{B}}_v\bar{\boldsymbol{v}} + \vec{\boldsymbol{n}}_v + \Delta\boldsymbol{r} \tag{25}$$

where $\boldsymbol{\Delta} = \mathrm{diag}\{\delta_1, \delta_2, \ldots, \delta_N\} \in \mathbb{C}^{N\times N}$ denotes a diagonal matrix associated with the unknown DOAs information that can be integrated with $\bar{\boldsymbol{u}}$ to structure a new virtual signal power vector $\bar{\boldsymbol{v}}$, i.e., $\bar{\boldsymbol{v}} = \boldsymbol{\Delta}\bar{\boldsymbol{u}} = [\bar{q}_1^2, \bar{q}_2^2, \ldots, \bar{q}_N^2]^T = [\delta_1\bar{\rho}_1^2, \delta_2\bar{\rho}_2^2, \ldots, \delta_N\bar{\rho}_N^2]^T \in \mathbb{C}^{N\times 1}$.

On the one hand, it is preferable for the estimated spatial spectrum to associate the non-zero signal powers with the nearest sampling grid points. $\bar{\rho}_n^2$ in the sparse vector

$\bar{u}$ means the signal power centered at the sampling grid point $\bar{\theta}_n$. Thus, the following restriction is given as:

$$\gamma_n \leqslant \delta_n \leqslant \beta_n, n = 1, 2, \ldots, N \tag{26}$$

where

$$\gamma_n = \sin\frac{\bar{\theta}_{n-1} + \bar{\theta}_n}{2} - \sin\bar{\theta}_n, n = 2, 3, \ldots, N \tag{27}$$

$$\beta_n = \sin\frac{\bar{\theta}_{n+1} + \bar{\theta}_n}{2} - \sin\bar{\theta}_n, n = 1, 2, \ldots, N-1 \tag{28}$$

where $\gamma_1 = 0$ and $\beta_N = 0$.

On the other hand, as $\bar{u} \geqslant 0$ (i.e., $\bar{\rho}_n^2 \geqslant 0, n = 1, 2, \ldots, N$), the following constraint can be further deduced from (26):

$$\gamma_n\bar{\rho}_n^2 \leqslant \bar{q}_n^2 \leqslant \beta_n\bar{\rho}_n^2, n = 1, 2, \ldots, N \tag{29}$$

Observing (25), when there is no source from the candidate directions set centered on $\bar{\theta}_n$, the corresponding spatial spectrum entries in the sparse vectors $\bar{u}$ and $\bar{v}$ are $\bar{u}_n = 0$ and $\bar{v}_n = 0$, respectively. Hence, $\bar{u}$ takes the same support set (i.e., row sparsity) as $\bar{v}$, which promotes group sparsity between the sparse vectors $\bar{u}$ and $\bar{v}$ [17]. This group sparsity of $[\bar{u}, \bar{v}]$ can be exploited as the objective function, and then an OGSpSF scheme can be built for off-grid DOA estimation, i.e.,

$$\begin{aligned} \min\left\|\boldsymbol{P}^{l_2}\right\|_1 \quad \text{s.t.} \quad \left\|\bar{\boldsymbol{W}}^{-\frac{1}{2}}(\bar{\boldsymbol{r}}_v - \bar{\boldsymbol{n}}_v - \bar{\boldsymbol{A}}_v\bar{\boldsymbol{u}} - \bar{\boldsymbol{B}}_v\bar{\boldsymbol{v}})\right\|_2 \leqslant \sqrt{\varsigma} \\ \gamma_n\bar{\rho}_n^2 \leqslant \bar{q}_n^2 \leqslant \beta_n\bar{\rho}_n^2, \ \bar{\rho}_n^2 \geqslant 0, \ n = 1, 2, \ldots, N \end{aligned} \tag{30}$$

where $\boldsymbol{P}^{l_2} = [\bar{u}, \bar{v}]^{l_2} = [\bar{p}_1, \bar{p}_2, \ldots, \bar{p}_N]^T \in \mathbb{C}^{N\times 1}$, whose nth entry $\bar{p}_n$ equals the l_2-norm of the nth row in $\boldsymbol{P} = [\bar{u}, \bar{v}]$, i.e.,

$$\bar{p}_n = \sqrt{\sum_{\bar{j}=1}^{2}(p_{n,\bar{j}})^2}, \quad n = 1, 2, \ldots, N \tag{31}$$

where $p_{n,\bar{j}}$ is an element in $\boldsymbol{P}$ with the coordinate index $(n, \bar{j})$.

Obviously, the off-grid estimation trouble was finally transformed into a joint sparse recovery problem by exploiting the group sparsity of $\boldsymbol{P} = [\bar{u}, \bar{v}]$. With the help of the recovered sparse vectors $\bar{u}$ and $\bar{v}$ in (30), the problem of off-grid DOAs can be solved.

Specifically, the positions of K peaks, i.e., $n_1, n_2, \ldots, n_K$, can first be determined by plotting the spatial spectrum of the recovered sparse vector $\bar{u}$. Subsequently, the final off-grid DOA estimation can be computed based on the recovered sparse vectors $\bar{u}$ and $\bar{v}$, as follows:

$$\hat{\theta}_k = \arcsin[\sin\bar{\theta}_{n_k} + \bar{v}_{n_k}/\bar{u}_{n_k}], \quad k = 1, 2, \ldots, K \tag{32}$$

where $\bar{u}_{n_k} > 0, k = 1, 2, \ldots, K$ is supposed here.

Such a convex constraint framework reasonably utilizes the group sparsity of $\boldsymbol{P} = [\bar{u}, \bar{v}]$ to facilitate continuous off-grid DOA estimation. Consequently, it is favorable for direction finding under off-grid condition to enhance the robustness and accuracy. What is more, this framework can quickly achieve high-precision continuous DOA estimation by exploiting a coarse grid interval to reduce the computational burden. In this way, such feasible off-grid scheme is applicable to real-time direction finding scenarios.

Unfortunately, the penalty scheme in (30) realizes off-grid DOA estimation by relaxing the l_0-norm constraint to the l_1-norm one, generating an approximation error and compromising the recovery accuracy. In view of the limited recovery performance, the following subsection will reweight the OGSpSF framework in (30) to enhance accuracy.

3.4. Reweighted OGSpSF (Re-OGSpSF) for DOA Estimation in Sensor Array with Unknown Mutual Coupling

As the l_1-norm is just a convex approximation of the l_0-norm, there will inevitably be a difference between these two penalty ways. Different from the impartial l_0-norm optimization, the penalty for larger coefficients outweighs that for smaller coefficients in the l_1-norm constraint scheme, which means that the two sparse vectors $\bar{\boldsymbol{u}}$ and $\bar{\boldsymbol{v}}$ or the sparse vector $\boldsymbol{P}^{l_2}$ in (30) cannot be recovered well. For acquiring better recovery performance in the l_1-norm constraint scheme, a weighted attempt achieved by the MUSIC-like principle [22,33] is carried out to strengthen the solution's sparsity.

First, imposing eigenvalue decomposition on $\bar{\boldsymbol{R}}$ yields

$$\bar{\boldsymbol{R}} = \sum_{\vec{m}=1}^{\vec{M}} \tau_{\vec{m}} \boldsymbol{\alpha}_{\vec{m}} \boldsymbol{\alpha}_{\vec{m}}^H = \boldsymbol{E}_s \boldsymbol{\Lambda}_s \boldsymbol{E}_s^H + \boldsymbol{E}_n \boldsymbol{\Lambda}_n \boldsymbol{E}_n^H \tag{33}$$

where $\tau_{\vec{m}}$ and $\boldsymbol{\alpha}_{\vec{m}}$ stand for the $\vec{m}$th eigenvalue and the corresponding eigenvector of the sample covariance matrix $\bar{\boldsymbol{R}}$, respectively. $\boldsymbol{\Lambda}_s = \text{diag}\{\tau_1, \tau_2, \ldots, \tau_K\} \in \mathbb{C}^{K \times K}$ and $\boldsymbol{\Lambda}_n = \text{diag}\{\tau_{K+1}, \tau_{K+2}, \ldots, \tau_{\vec{M}}\} \in \mathbb{C}^{(\vec{M}-K)\times(\vec{M}-K)}$. The signal subspace $\boldsymbol{E}_s$ and the noise subspace $\boldsymbol{E}_n$ are formed by eigenvectors corresponding to K larger eigenvalues and $\vec{M} - K$ smaller eigenvalues, respectively.

Then, as the decoupled steering vector is orthogonal to its noise subspace, a spatial spectrum function of MUSIC-like is structured as

$$f_{\text{MUSIC}}(\theta) = \arg\min\left\{\det\{\vec{\boldsymbol{a}}^H(\theta)\boldsymbol{E}_n\boldsymbol{E}_n^H\vec{\boldsymbol{a}}(\theta)\}\right\} \tag{34}$$

Since the over-complete dictionary $\bar{\boldsymbol{A}}_v$ in (11) is obtained by sparsely representing $\vec{\boldsymbol{A}}_v$ in (10), it is still orthogonal to its noise subspace, facilitating the weights establishment. Without loss of generality, the over-complete dictionary $\bar{\boldsymbol{A}}_v$ can be partitioned into two sub-matrices, i.e., $\bar{\boldsymbol{A}}_v = [\bar{\boldsymbol{A}}_{v_1}, \bar{\boldsymbol{A}}_{v_2}]$. The former, $\bar{\boldsymbol{A}}_{v_1}$, is assumed to be formed by K array manifolds corresponding to the desired target directions, while the latter, $\bar{\boldsymbol{A}}_{v_2}$, is thought to be made up of the remaining $N - K$ steering vectors. According to (34), the initial weights can be depicted as

$$\hat{z}_n = \det\{\vec{\boldsymbol{a}}^H(\bar{\theta}_n)\boldsymbol{E}_n\boldsymbol{E}_n^H\vec{\boldsymbol{a}}(\bar{\theta}_n)\}, \ n = 1, 2, \ldots, N \tag{35}$$

where the initial weight $\hat{z}_n$ corresponds to the possible target angle $\bar{\theta}_n$. Based on (35), the weights can be further expressed as

$$\bar{z}_n = \hat{z}_n / \max\{\hat{z}_1, \hat{z}_2, \ldots, \hat{z}_N\}, \ n = 1, 2, \ldots, N \tag{36}$$

At last, a robust MUSIC-like weighted matrix achieved by (36) can be established as

$$\boldsymbol{Z} = \text{diag}\{\boldsymbol{z}\} \tag{37}$$

where $\boldsymbol{z} = [\boldsymbol{z}_1, \boldsymbol{z}_2] = [\bar{z}_1, \bar{z}_2, \ldots, \bar{z}_N]$. $\boldsymbol{z}_1$ contains K weights corresponding to the real DOAs, smaller than that in $\boldsymbol{z}_2$ formed by the residual weights. In particular, the weights in $\boldsymbol{z}_1$ satisfy $\boldsymbol{z}_1 \to \boldsymbol{0}$ when the number of snapshots $T \to \infty$. Through applying the weighted matrix to the l_1-norm penalty framework, smaller weights in $\boldsymbol{z}_1$ protect larger coefficients, while larger weights in $\boldsymbol{z}_2$ punish smaller coefficients that are more likely to be zero. Accordingly, they can be punished as fairly as possible, no matter the larger or smaller coefficients in the sparse vectors $\bar{\boldsymbol{u}}$ and $\bar{\boldsymbol{v}}$.

Implanting the weighted matrix $\boldsymbol{Z}$ into the sparsity-inducing framework in (30) yields

$$\begin{gathered} \min\left\|\boldsymbol{Z}\boldsymbol{P}^{l_2}\right\|_1 \quad \text{s.t.} \quad \left\|\bar{\boldsymbol{W}}^{-\frac{1}{2}}(\bar{\boldsymbol{r}}_v - \bar{\boldsymbol{n}}_v - \bar{\boldsymbol{A}}_v\bar{\boldsymbol{u}} - \bar{\boldsymbol{B}}_v\bar{\boldsymbol{v}})\right\|_2 \leqslant \sqrt{\varsigma} \\ \gamma_n\bar{p}_n^2 \leqslant \bar{q}_n^2 \leqslant \beta_n\bar{p}_n^2, \ \bar{p}_n^2 \geqslant 0, \ n = 1, 2, \ldots, N \end{gathered} \tag{38}$$

The reweighted off-grid constraint framework in (38) can be successfully computed by using second order cone (SOC) programming software packages in MATLAB, such as CVX and Sedumi. Similarly, the final off-grid estimation can be realized by calculating (32).

So far, an efficient Re-OGSpSF approach has been developed to solve the off-grid DOA estimation problem in the environment of unknown mutual coupling. The entire procedure of the proposed method is given in Algorithm 1.

Algorithm 1 Re-OGSpSF for DOA estimation in sensor array with unknown mutual coupling

1: **Input:** The coupled received data $\boldsymbol{X}$ in (5);
2: Formulate a decoupled array output $\boldsymbol{Y}$ using (8) by left multiplying the selection matrix $\boldsymbol{J}$ in (6) to eliminate the unknown mutual coupling effect;
3: Compute the sample covariance matrix $\bar{\boldsymbol{R}}$ of $\boldsymbol{Y}$ based on (14);
4: Perform vectorization attempt on $\bar{\boldsymbol{R}}$ to establish a vector data model given in (15);
5: Impose eigenvalue decomposition on $\bar{\boldsymbol{R}}$ by (33) to obtain the noise subspace $\boldsymbol{E}_n$;
6: Structure the over-complete dictionaries $\bar{\boldsymbol{A}}_v$ in (11) and $\bar{\boldsymbol{B}}_v$ in (24) to achieve an enhanced OGSpSF recovery model in (25);
7: Design a MUSIC-like weighted matrix $\boldsymbol{Z}$ using (37) to reinforce the solution's sparsity;
8: Develop an effective Re-OGSpSF framework adopting (38) for off-grid DOA estimation by joint sparse recovery;
9: **Output:** The recovered sparse vectors $\bar{\boldsymbol{u}}$ and $\bar{\boldsymbol{v}}$;
10: Perform a 1-D spectrum search on the vector $\bar{\boldsymbol{u}}$ to determine the indices of its K peaks.
11: Calculate off-grid DOAs based on the recovered sparse vectors $\bar{\boldsymbol{u}}$ and $\bar{\boldsymbol{v}}$ by (32).

4. Related Remarks

Remark 1. *To the best of our knowledge, it is extremely critical for (38) to determine the estimated noise power $\bar{\sigma}^2$ and an appropriate regularization parameter ς. On the one hand, $\bar{\sigma}^2$ can be computed via averaging $\vec{M} - K$ smaller eigenvalues of the sample covariance matrix $\bar{\boldsymbol{R}}$. On the other hand, ς plays an extremely vital role in robust sparse reconstruction, balancing the fitting error and signal sparsity. Inspired by (19), the fitting error in (20), (30), and (38) follows the asymptotic chi-square distribution with $\vec{M}^2$ DOFs. Hence, ς can be determined via the upper bound of the fitting error with a high probability $\vec{\rho}$, i.e.,*

$$\Pr\{\chi^2(\vec{M}^2) \leqslant \varsigma\} = \vec{\rho}, \quad \varsigma = \chi^2_{\vec{\rho}}(\vec{M}^2) \tag{39}$$

where $\Pr\{\cdot\}$ *reveals the probability distribution of the event. In this way, ς can finally be determined by using the function* chi2inv$(\vec{\rho},\vec{M}^2)$ *in MATLAB, where $\vec{\rho} = 0.999$ is enough in this paper.*

Remark 2. *This paper mainly focuses on off-grid DOA estimation under the condition of unknown mutual coupling, and then proposes an enhanced Re-OGSpSF algorithm. For the off-grid error, the off-grid representation is carried out in this article. To be specific, the off-grid error term obtained from the first-order Taylor approximation is integrated into the potential on-grid sparse model to construct the OGSpSF recovery model. To better and more intuitively verify the robustness of the proposed methodology to the off-grid error, a potential reweighted approach using the SpSF principle under the condition of unknown mutual coupling can be obtained by removing the off-grid representation in (21). To distinguish it from the proposed (Re-OGSpSF) framework, it can be defined as the reweighted SpSF (Re-SpSF) method. Specifically, according to the SpSF principle in Section 3.2, DOA estimation can be obtained with the help of the recovered sparse vector $\bar{\boldsymbol{u}}$. However, the approximate penalty for replacing the l_0-norm constraint with the l_1-norm one causes limited reconstruction performance. Thus, the SpSF principle in Section 3.2 is combined with the weighted measure in Section 3.4 to form the Re-SpSF framework, i.e.,*

$$\min\|\boldsymbol{Z}\bar{\boldsymbol{u}}\|_1 \quad \text{s.t.} \quad \left\|\bar{\boldsymbol{W}}^{-\frac{1}{2}}(\bar{\boldsymbol{r}}_v - \bar{\boldsymbol{n}}_v - \bar{\boldsymbol{A}}_v\bar{\boldsymbol{u}})\right\|_2 \leqslant \sqrt{\varsigma}, \quad \bar{\boldsymbol{u}} \geqslant \boldsymbol{0} \tag{40}$$

Due to the off-grid representation, the proposed scheme can provide continuous DOA estimation. From the perspective of estimation accuracy, the proposed Re-OGSpSF method mitigates the interference of off-grid error to a certain extent, enhances the robustness against off-grid error, and improves the estimation accuracy. In terms of angle estimation speed, the proposed Re-OGSpSF method can quickly perform high-precision continuous direction finding by setting a coarse grid interval, which relieves the computational burden and speeds up the DOA estimation. In general, the proposed methodology is effective and efficient in achieving high-precision continuous off-grid DOA estimation, which is more suitable for real-time direction finding in practical applications. This will be demonstrated in the subsequent simulation experiments.

5. Numerical Simulation Results

In this section, extensive simulation experiments are performed to validate the estimation performance of the proposed Re-OGSpSF scheme. In order to show the effectiveness and efficiency, several estimators are simultaneously tested to compare with the proposed approach, including the l_1-SVD algorithm (recorded as SVD method) in [29], the BSR approach in [30], the SRACV method in [31], the reweighted SRACV framework (recorded as ReSRACV method) in [32], the potential Re-SpSF scheme (recorded as Re-SpSF method) in (40), and the Root-SBL estimator in [38]. Furthermore, as a benchmark for performance evaluation, the Cramer–Rao bound (CRB) in [41] is calculated as well. Additionally, the root means square error (RMSE) is computed to measure their accuracy, denoted as

$$\mathrm{RMSE} = \sqrt{\frac{1}{LK}\sum_{l=1}^{L}\sum_{k=1}^{K}(\theta_{l,k} - \theta_k)^2} \tag{41}$$

where $\theta_{l,k}$ reveals the estimated value of the desired DOA θ_k at the lth Monte Carlo running. $L = 200$ stands for the total number of Monte Carlo trials in this article.

In what follows, $M = 10$ sensors are configured to structure a ULA, which are evenly separated by half-wavelength spacing, i.e., $d = \lambda_d/2$. Suppose there are $K = 2$ narrow-band uncorrelated targets from distinct directions incident on the ULA, where DOAs are recorded as $\theta_1 = -7.3^\circ$ and $\theta_2 = 4.2^\circ$, respectively. Moreover, there are three non-zero coefficients in the MCM with $H = 2$, that is, $g_0 = 1$, $g_1 = 0.6864 - j0.4776$, and $g_2 = 0.2069 - j0.1024$. In addition, the discrete grid spacing ϵ of the spatial domain from -90° to 90° is set to 1°.

Figure 1 describes the spatial spectrum of all estimators with SNR $= 0$ dB and $T = 200$. Additionally, their corresponding estimation results are recorded in Table 2. On the one hand, there are two sharp peaks for these approaches in Figure 1, implying that they are able to maintain direction finding under unknown mutual coupling and off-grid conditions. What is more, it is not hard to find that the peaks of the Root-SBL approach are the least sharp and the sidelobe is the highest, while the peaks of the potential Re-SpSF and the proposed methods are the sharpest and the sidelobes are the lowest. On the other hand, Table 2 depicts the proposed method outperforms the Re-SpSF and ReSRACV algorithms, and is the closest to the real DOAs among these estimators. Since the estimated DOAs of Re-SpSF are the same as those of the ReSRACV, Re-SpSF and ReSRACV frameworks show a similar estimation performance. Conversely, the SVD and BSR algorithms are furthest from the desired DOAs and have the worst performance. Moreover, the Root-SBL estimator is overall closer to the true DOAs than the SRACV framework, but not as close as the Re-SpSF and ReSRACV approaches. Hence, the proposed method is superior to other algorithms in terms of resolution and accuracy.

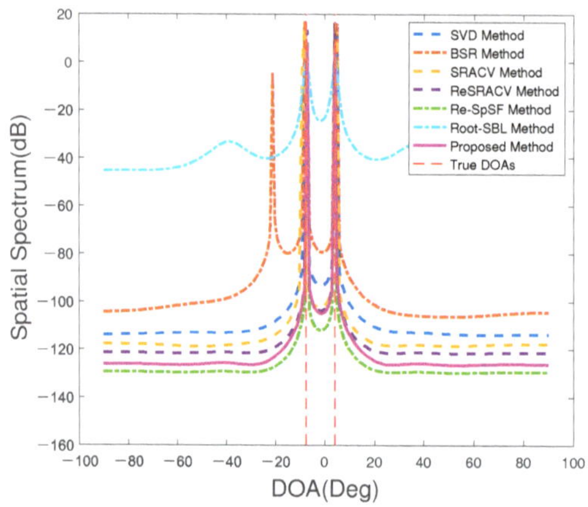

Figure 1. The spatial spectrum of all methods.

Table 2. Comparison of estimation results of all methods.

Methods	DOA 1	DOA 2
SVD Method	−8.0000°	5.0000°
BSR Method	−8.0000°	5.0000°
SRACV Method	−8.0000°	4.0000°
ReSRACV Method	−7.0000°	4.0000°
Re-SpSF Method	−7.0000°	4.0000°
Root-SBL Method	−7.1965°	3.7667°
Proposed Method	−7.3479°	4.5000°

Figure 2 reveals the RMSE versus SNR of all methods with $T = 200$. From Figure 2, with the increase of SNR, the RMSEs of all methods decrease to a certain extent, i.e., these estimators improve the estimation accuracy. Among them, the SVD and BSR methods have the largest RMSEs and the worst performance, where the RMSE of the SVD approach is close to that of the BSR algorithm over the whole interested SNR range. Meanwhile, the RMSE of the SRACV approach is lower than that of them, but higher than that of the ReSRACV and Re-SpSF schemes. That is, the ReSRACV and Re-SpSF estimators are better than the other three approaches in accuracy, mainly due to their reweighted measure. Obviously, the RMSEs for the ReSRACV and Re-SpSF schemes are almost equivalent, revealing that their corresponding estimation performances are very similar. What is more, it can be found that at low SNR (SNR $= -10$ dB), the RMSE of the proposed estimator is slightly greater than that of the ReSRACV algorithm, but less than that of the Root-SBL and Re-SpSF algorithms, especially the Root-SBL algorithm. As SNR improves, the RMSE of the proposed method is the lowest and closest to CRB. Its estimation performance is far better than that of algorithms such as the ReSRACV, Re-SpSF, and Root-SBL schemes. It should be noted that the difference between Re-SpSF and the proposed algorithm is whether an off-grid representation is performed. The Re-SpSF algorithm does not perform off-grid representation, causing a higher RMSE than that of the proposed method. In other words, thanks to the off-grid representation, the proposed method displays good robustness and satisfactory estimation accuracy for the off-grid error. In general, the proposed method has advantages over the other six approaches.

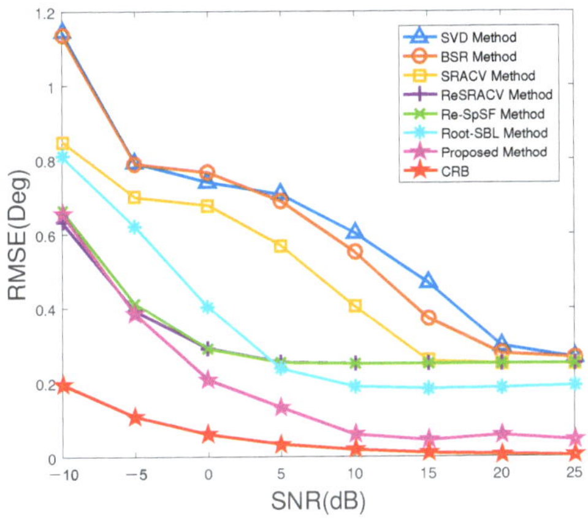

Figure 2. RMSE versus SNR of all methods.

Figure 3 demonstrates the probability of successful detection (PSD) versus SNR of all methods with $T = 200$. If the error between the interested DOAs θ_k and the estimated DOAs $\hat{\theta}_k$ is less than 0.7°, i.e., $|\hat{\theta}_k - \theta_k| < 0.7°$, the signal source can be considered to be successfully detected. As expected, the PSDs of all estimators gradually increase as SNR improves. More importantly, the PSD of the proposed method not only outperforms that of the other six algorithms over the entire selected SNR range, especially for relatively low SNRs, but also reaches 100% faster than that of the other algorithms.

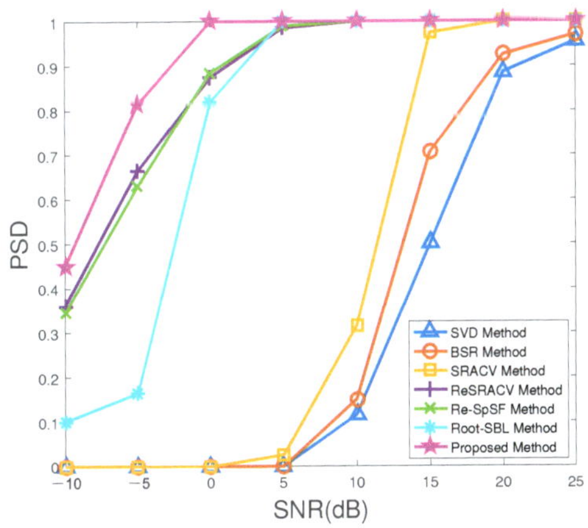

Figure 3. PSD versus SNR of all methods.

Figure 4 illustrates the RMSE versus snapshots of all algorithms with SNR = 5 dB. Additionally, their corresponding RMSEs are recorded in Table 3. Similar to Figure 2, the increase in snapshots number promotes the estimation performance of these approaches to some extent. Simultaneously, the performance of the SVD approach remains similar to that of the BSR framework, being the worst among these estimators. Furthermore, it is obvious that the RMSE of the ReSRACV framework is almost the same as that of the Re-SpSF scheme and much lower than that of the SRACV estimator. This implies that the ReSRACV and Re-SpSF algorithms have the same estimation performance and outperform the SRACV framework. This is mainly attributed to the fact that the ReSRACV framework uses the same reweightwd measure as the Re-SpSF scheme, which reduces the RMSE well. However, it is not difficult to find from Table 3 that the RMSEs of the ReSRACV and Re-SpSF methodologies stop decreasing after reaching a certain level (i.e., RMSE = 0.2500), indicating that their accuracy no longer improves and tends to saturate in such case. This may be because their discrete grid interval (i.e., $\epsilon = 1^{\circ}$) is set too large, resulting in the increase in snapshots not being enough to resist the influence of off-grid error on the estimation accuracy. The RMSE of the Root-SBL estimator is lower than those of the ReSRACV and Re-SpSF schemes, but higher than that of the proposed approach, except for the case of $T = 50$. It means that the Root-SBL methodology is superior to the ReSRACV and Re-SpSF estimators and inferior to the proposed method. Moreover, the RMSE of the proposed algorithm is lower than that of the potential Re-SpSF framework. That is to say, the off-grid representation in the proposed method enhances the robustness to off-grid error well and improves the estimation accuracy. In short, the proposed method takes the lowest RMSE and the best estimation performance, which is closest to CRB. Therefore, the proposed method outperforms the other six estimators.

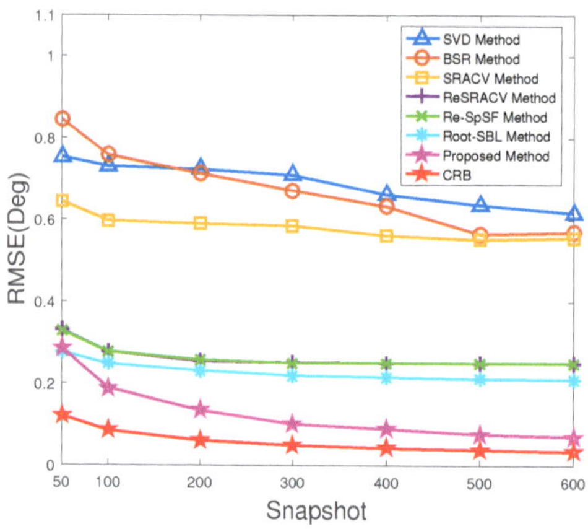

Figure 4. RMSE versus snapshots of all methods.

Table 3. RMSE under different number of snapshots of all methods.

RMSE \ Snapshot / Method	50	100	200	300	400	500	600
SVD Method	0.7544	0.7311	0.7240	0.7103	0.6624	0.6372	0.6177
BSR Method	0.8446	0.7587	0.7134	0.6723	0.6333	0.5665	0.5709
SRACV Method	0.6462	0.5978	0.5905	0.5855	0.5620	0.5530	0.5574
ReSRACV Method	0.3313	0.2790	0.2549	0.2517	0.2500	0.2500	0.2500
Re-SpSF Method	0.3298	0.2783	0.2581	0.2500	0.2500	0.2500	0.2500
Root-SBL Method	0.2773	0.2487	0.2307	0.2189	0.2154	0.2120	0.2107
Proposed Method	0.2878	0.1880	0.1343	0.1007	0.0888	0.0767	0.0703

Figure 5 displays the RMSE versus grid interval of all methods, where SNR $= 0$ dB and $T = 200$. In most of the selected grid intervals, the RMSE of the proposed method is the smallest, closest to 0, and least influenced by the coarse grid intervals. Hence, it shares the best and most stable estimation performance. The main reason is that it not only concerns mutual coupling and off-grid errors to reinforce the robustness, but also exploits the reweighted strategy to improve accuracy. Similarly, the Root-SBL algorithm also considers all factors, unlike the other five regularization methods that focus only on mutual coupling. Clearly, when the grid interval is less than 1°, the RMSE of the Root-SBL estimator is approximately equal to those of the ReSRACV and Re-SpSF approaches, which means that its performance is similar to that of the ReSRACV and Re-SpSF schemes. This is mainly because the influence of off-grid gap is relatively weak in this way, while the reweighted measure in the ReSRACV and Re-SpSF estimators ensures high-precision estimation. As the grid interval increases, the off-grid error dominates among these two errors, and thus the Root-SBL algorithm is significantly better than the ReSRACV and Re-SpSF frameworks at this time. Furthermore, the RMSEs of the SVD, BSR, and SRACV schemes are close, farthest from 0, and most affected by the coarse grid intervals, i.e., their performances are similar, the worst, and the least stable. It is emphasized that the Re-SpSF scheme does not perform off-grid representation compared to the proposed method, so its estimation accuracy is easily affected by off-grid error, especially for coarse grid intervals. Therefore, the proposed method has not only superior estimation accuracy, but also satisfactory robustness to off-grid error.

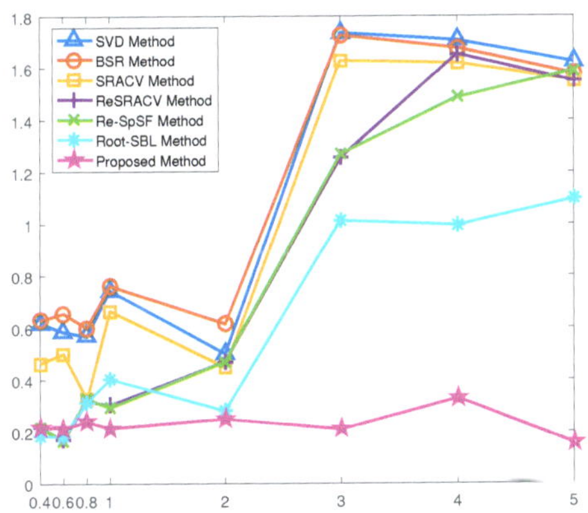

Figure 5. RMSE versus grid interval of all methods.

Figures 6 and 7 indicate the RMSE and PSD versus SNR of the proposed method for different number of antennas, respectively. In Figures 6 and 7, the number of snapshots is set to $T = 200$. According to Figures 6 and 7, as SNR or/and the number of sensors gradually increases, the RMSE of the proposed method decreases overall, while the corresponding PSD gradually enhances to 100%, especially for unsatisfactory SNR. Evidently, the PSD with $M = 11$ sensors is higher than that of other sensor numbers, which can be the first to reach 100%. $M = 9$ antennas has the lowest PSD, which is the slowest to achieve 100%. Therefore, the performance of the proposed method can be well enhanced by improving the SNR and/or the number of antennas. However, it should be pointed out that using a greater number of antennas to improve accuracy will not only increase the estimation cost, but also aggravate the computational burden, which is not conducive to real-time direction finding. In summary, it is a wise choice for practical applications to choose the appropriate number of antennas for balancing the estimation accuracy and computational load.

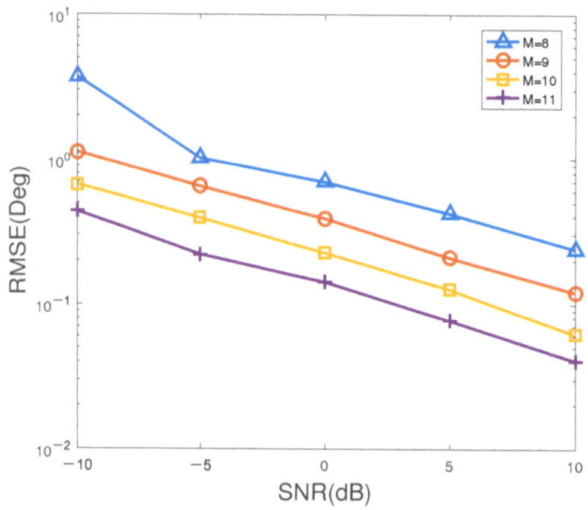

Figure 6. RMSE versus SNR of the proposed method for different number of antennas.

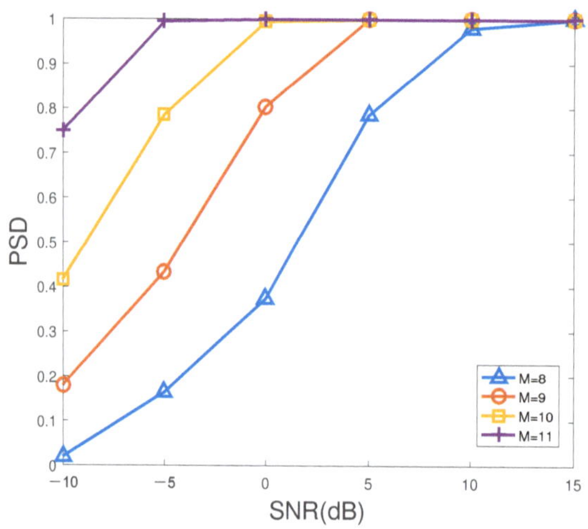

Figure 7. PSD versus SNR of the proposed method for different number of antennas.

Figures 8 and 9 describe the RMSE and PSD versus snapshots of the proposed method for different number of antennas, respectively. In Figures 8 and 9, SNR is fixed at 0 dB. As shown in Figures 8 and 9, the larger the number of snapshots or/and sensors, the smaller the RMSE of the proposed method, and the larger the corresponding PSD, which behaves like the general trend in Figures 6 and 7. On the one hand, it is feasible for direction finding to ensure estimation accuracy by increasing the number of antennas or/and snapshots. On the other hand, it has to be acknowledged that the larger the number of sensors, the higher the estimation cost, and the heavier the computational complexity. Hence, such superior estimation accuracy attributed to multiple antennas is at the expense of computational load. That is to say, a suitable number of sensors is more applicable to the actual direction finding.

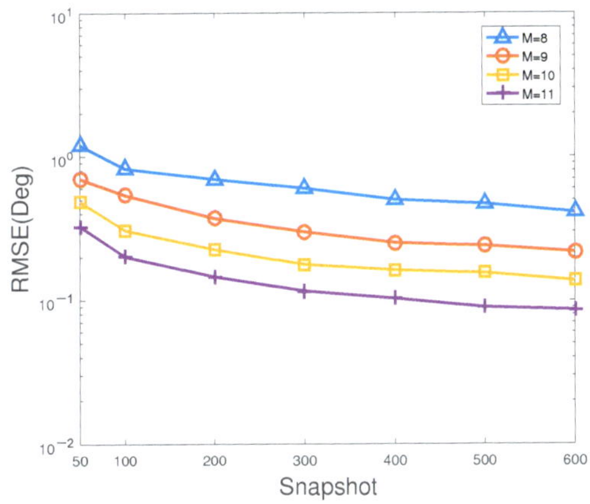

Figure 8. RMSE versus snapshots of the proposed method for different number of antennas.

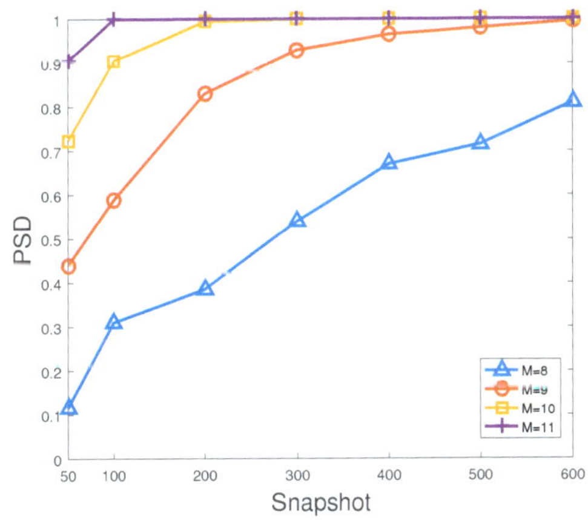

Figure 9. PSD versus snapshots of the proposed method for different number of antennas.

6. Conclusions

In this article, an effective and efficient reweighted sparsity-inducing approach achieved by the OGSpSF framework is presented for off-grid DOA estimation in sensor array with unknown mutual coupling. In the proposed method, the decoupled received data is designed by exploiting a selection matrix to escape the unknown mutual coupling disturbance. Then, an enhanced OGSpSF recovery model is constructed by incorporating the linearly approximated off-grid error term into the potential on-grid sparse model to ensure the robustness. Subsequently, an upgraded Re-OGSpSF framework is explored for off-grid DOA estimation using joint sparse recovery, where a MUSIC-like weighted matrix is further implanted to improve accuracy. Eventually, off-grid DOA estimation can be estimated via the spatial spectrum of the reconstructed sparse vector. Attributed to the off-grid representation and reweighted attempt, the proposed method can efficiently provide high-precision continuous DOA estimation by setting a coarse grid interval, making it more suitable for real-time direction finding. Extensive simulation results show the effectiveness and efficiency of the proposed approach.

Author Contributions: Conceptualization, X.W. and G.X.; methodology, L.L. and X.W.; writing—original draft preparation, L.L.; writing—review and editing, X.W. and L.W.; supervision, X.W. and X.L.; funding acquisition, X.W. All authors have read and agreed to the published version of the manuscript.

Funding: This work was supported by the Important Science and Technology of Hainan Province (No. ZDKJ202010) and the National Natural Science Foundation of China (No. 61961013 and No. 62101088).

Institutional Review Board Statement: Not applicable.

Informed Consent Statement: Not applicable.

Data Availability Statement: Not applicable.

Conflicts of Interest: The authors declare no conflict of interest.

References

1. Shi, J.; Wen, F.; Liu, T. Nested MIMO Radar: Coarrays, Tensor Modeling, and Angle Estimation. *IEEE Trans. Aerosp. Electron. Syst.* **2021**, *57*, 573–585. [CrossRef]
2. Li, J.; Zhang, Q.; Deng, W.; Tang, Y.; Zhang, X.; Wu, Q. Source Direction Finding and Direct Localization Exploiting UAV Array With Unknown Gain-Phase Errors. *IEEE Internet Things J.* **2022**, *9*, 21561–21569. [CrossRef]
3. Wen, F.; Shi, J.; Gui, G.; Gacanin, H.; Dobre, O.A. 3-D Positioning Method for Anonymous UAV Based on Bistatic Polarized MIMO Radar. *IEEE Internet Things J.* **2023**, *10*, 815–827. [CrossRef]
4. Zhang, X.; Wang, F.; Li, H. An Efficient Method for Cooperative Multi-Target Localization in Automotive Radar. *IEEE Signal Process. Lett.* **2022**, *29*, 16–20. [CrossRef]
5. Schmidt, R. Multiple emitter location and signal parameter estimation. *IEEE Trans. Antennas Propagat.* **1986**, *34*, 276–280. [CrossRef]
6. Roy, R.; Kailath, T. Esprit-estimation of signal parameters via rotational invariance techniques. *IEEE Trans. Acoust. Speech Signal Process.* **1989**, *37*, 984–995. [CrossRef]
7. Zhang, X.; Chen, W.; Zheng, W.; Xia, Z.; Wang, Y. Localization of near-field sources: A reduced-dimension music algorithm. *IEEE Commun. Lett.* **2018**, *22*, 1422–1425. [CrossRef]
8. Feng, M.; Yang, Y.; Shu, Q.; Yang, R. An Improved ESPRIT-Based Algorithm for Monostatic FDA-MIMO Radar With Linear or Nonlinear Frequency Increments. *IEEE Commun. Lett.* **2021**, *25*, 2375–2379. [CrossRef]
9. Liu, B.; Gui, G.; Matsushita, S.; Xu, L. Dimension-Reduced Direction-of-Arrival Estimation Based on $l_{2,1}$-Norm Penalty. *IEEE Access* **2018**, *6*, 44433–44444. [CrossRef]
10. Wang, X.; Wang, W.; Li, X.; Liu, J. Real-valued covariance vector sparsity-inducing DOA estimation for monostatic MIMO radar. *Sensors* **2015**, *15*, 28271–28286. [CrossRef]
11. Ottersten, B.; Stoica, P.; Roy, R. Covariance matching estimation techniques for array signal processing applications. *Digit. Signal Process.* **1998**, *8*, 185–210. [CrossRef]
12. Shin, M.; Hong, W.; Lee, K.; Choo, Y. Passive Sonar Target Identification Using Multiple-Measurement Sparse Bayesian Learning. *Sensors* **2022**, *22*, 8511. [CrossRef]
13. Dai, J.; So, H.C. Real-Valued Sparse Bayesian Learning for DOA Estimation With Arbitrary Linear Arrays. *IEEE Trans. Signal Process.* **2021**, *69*, 4977–4990. [CrossRef]
14. Xu, G.; Zhang, B.; Chen, J.; Hong, W. Structured Low-Rank and Sparse Method for ISAR Imaging With 2-D Compressive Sampling. *IEEE Trans. Geosci. Remote Sens.* **2022**, *60*, 1–14. [CrossRef]
15. Xu, G.; Zhang, B.; Yu, H.; Chen, J.; Xing, M.; Hong, W. Sparse Synthetic Aperture Radar Imaging from Compressed Sensing and Machine Learning: Theories, Applications and Trends. *IEEE Geosci. Remote Sens. Mag.* **2022**, *10*, 32–69. [CrossRef]
16. Wang, X.; Wang, W.; Liu, J.; Li, X.;Wang, J. A sparse representation scheme for angle estimation in monostatic mimo radar. *Signal Process.* **2014**, *104*, 258–263. [CrossRef]
17. Zheng, J. Sparse Spectrum Fitting in Array Processing. Ph.D. Thesis, University of Minnesota, Minneapolis, MN, USA, 2013.
18. Yang, Z.; Xie, L.; Zhang, C. Off-grid direction of arrival estimation using sparse bayesian inference. *IEEE Trans. Signal Process.* **2013**, *61*, 38–43. [CrossRef]
19. Zhang, Y.; Ye, Z.; Xu, X.; Hu, N. Off-grid doa estimation using array covariance matrix and block-sparse bayesian learning. *Signal Process.* **2014**, *98*, 197–201. [CrossRef]
20. Dai, J.; Bao, X.; Xu, W.; Chang, C. Root sparse bayesian learning for off-grid doa estimation. *IEEE Signal Process. Lett.* **2016**, *24*, 46–50. [CrossRef]
21. Wen, F.; Huang, D.; Wang, K.; Zhang, L. Doa estimation for monostatic mimo radar using enhanced sparse bayesian learning. *J. Eng.* **2018**, *2018*, 268–273. [CrossRef]
22. Li, L.; Fu, T.; Wang, X.; Huang, M.; Wan, L.; Yang, Y. Doa estimation of strictly noncircular sources in wireless sensor array network via block sparse representation. *IEEE Access* **2020**, *8*, 47500–47508. [CrossRef]

23. Dai, J.; Zhao, D.; Ye, Z.; Zhang, L. DOA estimation and self-calibration algorithm for nonuniform linear array. In Proceedings of the 2010 International Symposium on Intelligent Signal Processing and Communication Systems, Chengdu, China, 6–8 December 2010; pp. 1–4.
24. Li, L.; Fu, T.; Yang, Y.; Huang, M.; Han, Z.; Li, J. Joint angle estimation and array calibration using eigenspace in monostatic mimo radar. *IEEE Access* **2020**, *8*, 60645–60652. [CrossRef]
25. Li, J.; Ding, J.; Jiang, D. Joint direction finding and array calibration method for MIMO radar with unknown gain phase errors. *IET Microw. Antennas Propag.* **2016**, *10*, 1563–1569. [CrossRef]
26. Ye, Z.; Liu, C. On the resiliency of music direction finding against antenna sensor coupling. *IEEE Trans. Antennas Propag.* **2008**, *56*, 371–380. [CrossRef]
27. Zheng, Z.; Zhang, J.; Zhang, J. Joint dod and doa estimation of bistatic mimo radar in the presence of unknown mutual coupling. *Signal Process.* **2012**, *92*, 3039–3048. [CrossRef]
28. Liao, B.; Zhang, Z.; Chan, S. Doa estimation and tracking of ulas with mutual coupling. *IEEE Trans. Aerosp. Electron. Syst.* **2012**, *48*, 891–905. [CrossRef]
29. Dai, J.; Zhao, D.; Ji, X. A sparse representation method for doa estimation with unknown mutual coupling. *IEEE Antennas Wirel. Propag. Lett.* **2012**, *11*, 1210–1213. [CrossRef]
30. Wang, Q.; Dou, T.; Chen, H.; Yan, W.; Liu, W. Effective block sparse representation algorithm for doa estimation with unknown mutual coupling. *IEEE Commun. Lett.* **2017**, *21*, 2622–2625. [CrossRef]
31. Meng, D.; Wang, X.; Huang, M.; Shen, C.; Guo, Y. A sparse representation of array covariance vectors for doa estimation with unknown mutual coupling. In Proceedings of the IEEE 18th International Conference on Communication Technology (ICCT), Chongqing, China, 8–11 October 2018; pp. 1022–1026.
32. Meng, D.; Wang, X.; Huang, M.; Shen, C. Reweighted l_1-norm minimisation for high-resolution doa estimation under unknown mutual coupling. *Electron. Lett.* **2018**, *54*, 1346–1348. [CrossRef]
33. Wang, X.; Yang, L.T.; Meng, D.; Dong, M.; Ota, K.; Wang, H. Multi-uav cooperative localization for marine targets based on weighted subspace fitting in sagin environment. *IEEE Internet Things J.* **2022**, *9*, 5708–5718. [CrossRef]
34. Hu, G.; Gao, S.; Zhong, Y.; Gao, B.; Subic, A. Matrix weighted multisensor data fusion for INS/GNSS/CNS integration. *Proc. Inst. Mech. Eng. G J. Aerosp. Eng.* **2016**, *230*, 1011–1026. [CrossRef]
35. Gao, S.; Zhong, Y.; Li, W. Random weighting method for multi-sensor data fusion. *IEEE Sens. J.* **2011**, *11*, 1955–1961. [CrossRef]
36. Gao, S.; Zhong, Y.; Wei, W.; Gu, C. Windowing-based random weighting fitting of systematic model errors for dynamic vehicle navigation. *Inf. Sci.* **2014**, *282*, 350–362. [CrossRef]
37. Gao, S.; Hu, G.; Zhong, Y. Windowing and random weighting -Based adaptive unscented Kalman filter. *Int J. Adapt Control Signal Process.* **2015**, *29*, 201–223. [CrossRef]
38. Wang, H.; Wang, X.; Huang, M.; Cao, C.; Bi, G. Off-grid doa estimation in mutual coupling via robust sparse bayesian learning. In Proceedings of the 2018 IEEE 23rd International Conference on Digital Signal Processing (DSP), Shanghai, China, 19–21 November 2018; pp. 1–5. [CrossRef]
39. Chen, P.; Cao, Z.; Chen, Z.; Wang, X. Off-grid doa estimation using sparse bayesian learning in mimo radar with unknown mutual coupling. *IEEE Trans. Signal Process.* **2018**, *67*, 208–220.
40. Chen, P.; Cao, Z.; Chen, Z.; Yu, C. Sparse off-grid doa estimation method with unknown mutual coupling effect. *Digit. Signal Process.* **2019**, *90*, 1–9. [CrossRef]
41. Friedlander, B.; Weiss, A. Direction finding in the presence of mutual coupling. *IEEE Trans. Antennas Propag.* **1991**, *39*, 273–284. [CrossRef]

sensors

MDPI

Article

Multi-Target Parameter Estimation of the FMCW-MIMO Radar Based on the Pseudo-Noise Resampling Method

Yao Jiang [1], Xiang Lan [1,*], Jinmei Shi [2], Zhiguang Han [1] and Xianpeng Wang [1]

[1] State Key Laboratory of Marine Resource Utilization in South China Sea, School of Information and Communication Engineering, Hainan University, Haikou 570228, China
[2] College of Information Engineering, Hainan Vocational University of Science and Technology, Haikou 571158, China
* Correspondence: xlan@hainanu.edu.cn

Abstract: Subspace methods are widely used in FMCW-MIMO radars for target parameter estimations. However, the performances of the existing algorithms degrade rapidly in non-ideal situations. For example, a small number of snapshots may result in the distortion of the covariance matrix estimation and a low signal-to-noise ratio (SNR) can lead to subspace leakage problems, which affects the parameter estimation accuracy. In this paper, a joint DOA–range estimation algorithm is proposed to solve the above issues. Firstly, the improved unitary root-MUSIC algorithm is applied to reduce the influence of non-ideal terms in building the covariance matrix. Subsequently, the least squares method is employed to process the data and obtain paired range estimation. However, in a small number of snapshots and low SNR scenarios, even if the impact of non-ideal terms is reduced, there will still be cases where the estimators sometimes deviate from the true target. The estimators that deviate greatly from targets are regarded as outliers. Therefore, threshold detection is applied to determine whether outliers exist. After that, a pseudo-noise resampling (PR) technology is proposed to form a new data observation matrix, which further alleviates the error of the estimators. The proposed method overcomes performance degradation in a small number of snapshots or low SNRs simultaneously. Theoretical analyses and simulation results demonstrate the effectiveness and superiority.

Keywords: target localization; FMCW-MIMO radar; DOA–range estimation; subspace leakage

1. Introduction

As radar is widely used in autonomous driving and military fields, target localization has become one of the most important research directions. Although the traditional ultrasonic radar and laser radar [1–3] can achieve basic target detection, they are vulnerable to environmental effects. In harsh weather scenarios, the detection abilities of the above radars decrease sharply, including (but not limited to) the high costs and low estimation accuracies. To improve the detection performances in such complex communication environments, a FMCW-MIMO radar is proposed as a solution [4]. Compared to traditional radars, FMCW-MIMO radars have great advantages, including high accuracies [5,6], low interception probabilities [7], and strong anti-jamming abilities [8–11]. Meanwhile, they transmit orthogonal waveforms to form virtual array elements and enlarge the array apertures so that the limited array elements can be used to achieve higher degrees of freedom and spatial resolutions [12–14]. Thus, FMCW-MIMO radars have great potential for pedestrian detection or target localization [15–17], especially for multi-target scenarios.

Parameters of the FMCW-MIMO radar include the direction of arrival (DOA), the direction of departure (DOD), range, Doppler frequency (DF), etc. [18–20]. Recently, more focus has been placed on target localization work, which includes direction and range estimation. Fast Fourier transform (FFT) is proposed as a representative technology of the target localization [21]. In [22], a novel FFT algorithm for DOA estimation is proposed with the FMCW-MIMO radar. In the algorithm, the array aperture was expanded with interpolation technology. However, the estimation resolution was still limited by the number of sampling points. To further improve the resolution, a method was proposed in [23], which selects regions of interest in the total samples to reduce redundancy. While this method is FFT-based, the improvement in resolution is unsatisfactory compared to other algorithms. For the two-dimensional estimation problem, such as joint DOA–range estimation, the 2D-FFT [24,25] algorithm has high computational complexity, and the resolution is highly dependent on the number of sampling points.

Based on the disadvantages of the FFT algorithm, subspace-based algorithms have been studied, providing new perspectives for target localization. Parameter estimations based on the FMCW-MIMO system are more about extending the classical subspace-based algorithms into two-dimensional (2D) structures, such as the joint angle-frequency estimation (JAFE) [26,27] and 2D multi-signal classification (2D-MUSIC) algorithm [28]. In the 2D-MUSIC algorithm, the eigenvalue decomposition (EVD) [29] or singular value decomposition (SVD) [30] is employed to distinguish the signal and noise subspaces of the received covariance matrix. Based on the orthogonality between two subspaces, a two-dimensional spectral peak search function is applied to realize the parameter estimation through the joint search in the DOA–range domain [31]. However, the 2D-MUSIC algorithm requires decomposing the huge covariance matrix and selecting a suitable grid step to search for the two-dimensional spectral peaks, leading to great computation complexities and difficulties in real-time processing. To ensure the resolution and reduce the computational complexity simultaneously, Kim et al. attempted to combine the subspace algorithm with FFT [32,33]. First, the target parameters were roughly estimated by FFT. Then the MUSIC algorithm only searched for a small region that was close to the rough result by the FFT. By reducing the search region, the MUSIC algorithm avoids the redundant search computation. In [34,35], the authors reduced the computation complexity in a different way. The two papers respectively used ESPRIT and toot-MUSIC instead of MUSIC to estimate parameters, where the spectral peak search was replaced by the direct formula solution. To estimate the coherent source signals, a double smoothing algorithm was proposed in [36]. The algorithm constructs two double-smoothing matrices, and estimates the angle and delay, respectively, by using the translation invariance structure. However, the performances of the above subspace algorithms degrade rapidly in cases of small snapshot numbers or low SNRs. The reason is the subspace leakage [37–39], which means a part of the real signal subspace is possibly located in the estimated noise subspace.

In low SNRs or a small number of snapshot scenarios, the covariance matrices of sample data are different from theoretical covariance matrices. Diagonal loading [40] and shrinkage-based [41] methods move or scale eigenvalues to improve data covariance, but their signal and noise projection matrices do not change much. In [42], a random matrix theory and technology were introduced to improve the performance, but the performance was satisfactory only when the number of snapshots was considered.

In this paper, we propose a new joint DOA–range estimation algorithm for the FMCW-MIMO radar system to solve the performance degradation problem of the above algorithms under a small number of snapshots and low SNR. First, for angle estimation, we propose an improved algorithm based on the unitary root-MUSIC algorithm [43,44], which mainly optimizes and updates the covariance matrix of two-dimensional data to reduce the impacts of non-ideal items. Then, according to the results of the angle estimation, the least squares method is used to estimate the range information to realize the automatic matching of the angle and range. These angle and range estimators constitute the target estimation set. Finally, for the target estimation set, a threshold detection method based on PR technology is designed to detect and eliminate the abnormal estimation caused by the subspace leakage, which further improves the accuracy and stability of the target localization. The main contributions of the proposed method are summarized as follows:

1. The proposed method achieves a two-dimensional parameter estimation under the FMCW-MIMO radar system, which overcomes the disadvantages of traditional FFT-based and subspace-based algorithms. This algorithm has better performance in regard to a low SNR and a small number of snapshots.
2. We propose a new iterative strategy to optimize the angle estimation and apply the least squares to the signal model of the FMCW-MIMO radar to achieve multi-target localization. Therefore, the range and angle parameters can be estimated in pairs.
3. To further improve the target localization accuracy in case of a small number of snapshots and a low SNR, we introduce a threshold detection and pseudo-noise resampling algorithm in the joint estimation framework. Through the screening of the threshold detection and the improvement of the PR method, the unqualified estimates considered as outliers could be removed and the accuracy of the parameter estimation improved accordingly.

The rest of this article is organized as follows. Section 2 briefly introduces the signal model and parameters of the FMCW-MIMO radar. Section 3 describes the proposed algorithm in detail. In Section 4, numerical examples and simulation results are given and compared with other algorithms. Conclusions are drawn in Section 5.

The symbol definitions are shown in Table 1 to facilitate the derivation of subsequent formulas.

Table 1. Related notation.

Notations	Definitions
uppercase bold italic letter	matrix
lowercase bold italic letter	vector
$\boldsymbol{I}_M$	identity matrix of M order
$E(\cdot)$	mathematical expectation
$diag(\cdot)$	diagonalization of the matrix
$(\cdot)^H$	conjugate transpose of the matrix
$(\cdot)^T$	transpose of the matrix
$(\cdot)^*$	conjugate matrix
$(\cdot)^\dagger$	pseudo-inverse of the matrix
$(\cdot)^\perp$	orthogonal complement of the matrix
$\mathbb{C}^{M\times N}$	complex space of size $M \times N$
$\lvert\cdot\rvert$	modulus operator
$\lVert\cdot\rVert_2$	l_2 norm operation
$\cup$	union

2. Signal Model

As shown in Figure 1, the MIMO radar uses two separate arrays to transmit and receive signals, respectively. The first row is the transmitting array, which contains two transmitting antennas; the receiving array lies on the last row, which consists of four antennas with equal spacings. Each pair of transmitting and receiving antennas can be independently equivalent to a monostatic element, which is marked as the midpoint of the transmit–receive element connection. In this way, a uniform linear array(ULA) composed of eight equivalent virtual elements is obtained, as shown in the middle row.

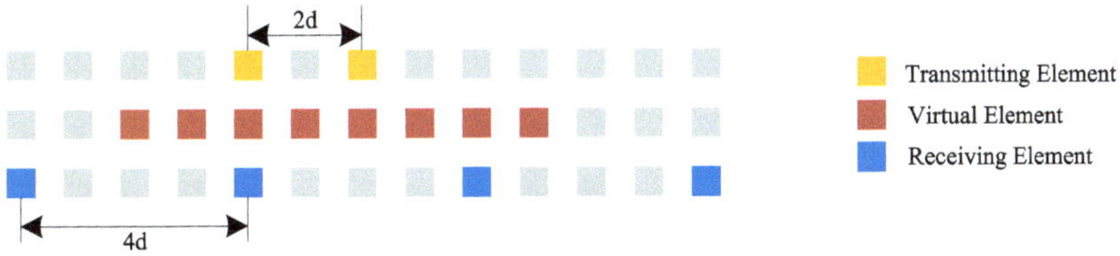

Figure 1. Synthetization of the virtual antenna array.

Figure 2 shows the working principle of the FMCW radar. When transmitting a frequency-modulated continuous signal, the transmitted signal can be specifically expressed as:

$$s(t) = V_T \cos(2\pi f_c t + \pi k t^2 + \phi_0), 0 \le t \le T_c \tag{1}$$

where V_T and f_c denote the amplitude and starting frequency of the chirp, respectively. The slope of the chirp is given by $k = B/T_c$, which is related to the bandwidth B and the scan duration T_c. ϕ_0 represents the initial phase of the signal and τ is the time delay between two chirps (TX and RX).

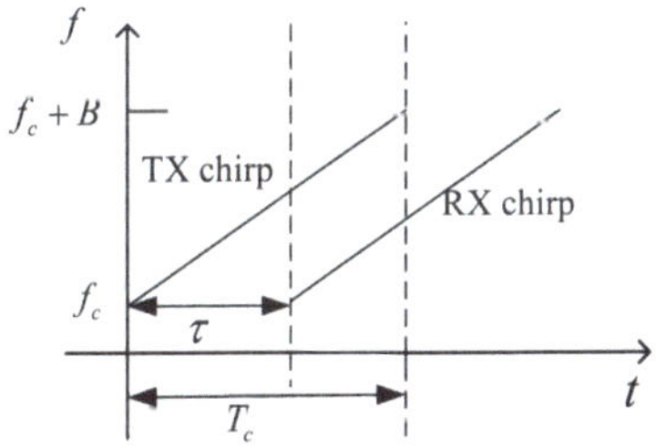

Figure 2. FMCW radar echo diagram.

The target localization of the FMCW MIMO radar is shown in Figure 3. The virtual array constructed by the MIMO part can be viewed as a ULA with spacing d between two adjacent elements. Signals are emitted from the array with the direction of departure θ. When reaching a target, echoes will be generated and impinge on the array from the direction of arrival θ. We assume that there are K far-field targets and M equivalent virtual antennas in the FMCW-MIMO radar system model.

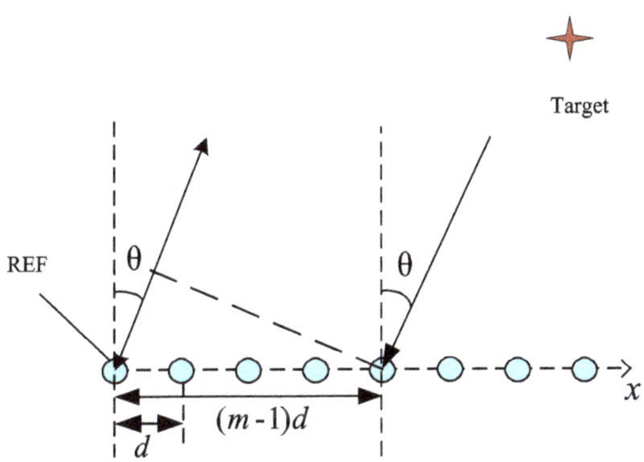

Figure 3. Target localization diagram of FMCW-MIMO radar.

According to the FMCW signal principle, the IF signal is generated after mixing, and the receiving signals are reflected by far-field fixed targets. After applying the Hilbert transform, the IF signal generated by the k-th target is:

$$u(t) = \mu e^{j(2\pi k\tau t + 2\pi f_c \tau)} \tag{2}$$

As receiving elements receive, all IF signals generated by K targets, and the received signals of the m-th element can be expressed as:

$$x_m(t) = \sum_{k=1}^{K} \mu_k e^{j(2\pi k\tau_{m,k} t + 2\pi f_c \tau_{m,k})} + n_m(t) \tag{3}$$

where $\tau_{m,k} = \frac{2r_k}{c} + \frac{2d\times(m-1)\sin\theta_k}{c}$ is the time delay of transmitting and th m-th receiving signals. r_k is the range from the transmitting antenna to the k-th target, $d \times (m-1)$ denotes the range between the transmitting antenna and m-th receiving antenna, $n_m(t)$ is the Gaussian noise collected by the element.

N-point sampling is performed for Equation (3) with sampling interval T_S. The echo signal can finally be changed into:

$$\begin{aligned} x_m[n] = &\sum_{k=1}^{K} \mu_k e^{j2\pi k(\frac{2r_k}{c} + \frac{d(m-1)\sin\theta_k}{c})nT_S} \\ &+ e^{j2\pi f_c(\frac{2r_k}{c} + \frac{d(m-1)\sin\theta_k}{c})} + n_m[n], n = 0, 1, 2, ..., N-1 \end{aligned} \tag{4}$$

Since the echo signals are from far-field, the spacings between the array antennas can be negligible compared to the distance between the transmitting antenna and the target, i.e., $r_k \gg d \times (m-1)\sin\theta_k$, Equation (4) can be reduced as:

$$x_m[n] = \sum_{k=1}^{K} \mu_k e^{j(\frac{4\pi f_c d(m-1)\sin\theta_k}{c} + \frac{4\pi B T_S r_k n}{Tc} + \frac{4\pi f_c r_k}{c})} + n_m[n] \tag{5}$$

The matrix form of the receiving signals can be shown:

$$\boldsymbol{X} = \boldsymbol{AS} + \boldsymbol{N}, \boldsymbol{X} \in \mathbb{C}^{M\times N} \tag{6}$$

Here, the steering vector $\boldsymbol{a}_i = [1, e^{j\frac{2\pi f_c d\sin\theta_i}{c}}, ..., e^{j\frac{2\pi f_c d(M-1)\sin\theta_i}{c}}]^T$, $i = 1, 2, ..., K$ constitutes the matrix $\boldsymbol{A} = [\boldsymbol{a}_1, ..., \boldsymbol{a}_k]$. Moreover, the signal matrix $\boldsymbol{S} = [\boldsymbol{s}_1, ..., \boldsymbol{s}_K]^T$. The noise matrix $\boldsymbol{N}$ is a Gaussian white process with zero mean and covariance $\sigma^2{}_n \boldsymbol{I}_M$.

3. Joint DOA–Range Estimation

This section proposes a novel method of joint DOA–range estimation in the FMCW-MIMO radar. First, for the complex-valued data processing, we introduce unitary transformations to convert complex-valued operations into real-valued operations. Then, in the angle domain, the covariance matrix is iteratively updated with the improved URM algorithm to extract more accurate DOA estimators. According to DOA estimators, we derive the corresponding steering vector and extract the phase. Range estimation is carried out with the least squares method, and the automatic pairing of two-dimensional parameters is directly realized. Finally, our work uses threshold detection for the paired parameter sets and optimizes the data matrix by pseudo-noise resampling. Each step of the proposed algorithm is elaborated in Figure 4.

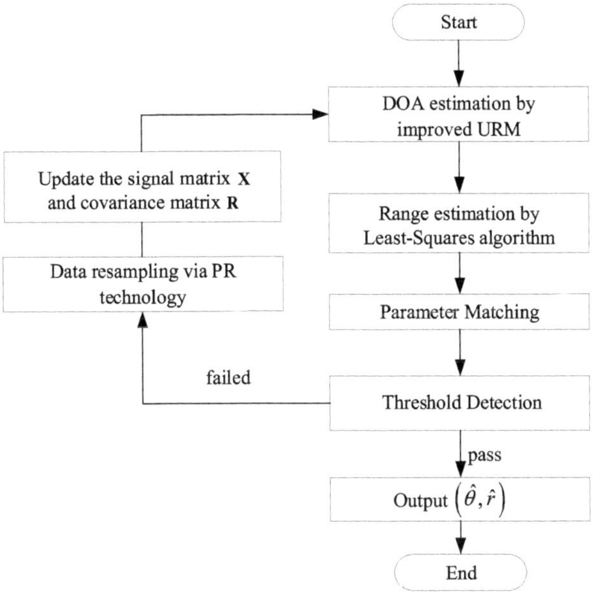

Figure 4. Structure of the proposed algorithm.

3.1. Preliminary Estimation of DOA via URM Algorithm

Assuming that the noise and signals are independent of each other, the covariance matrix of the received data can be denoted:

$$R = E\left(XX^H\right) = AR_sA + \sigma^2 I_M \tag{7}$$

where R_s means the covariance matrix of signals. In practical work, the theoretical value of R_s is hardly obtained and an approach $\hat{R}$ by averaging N snapshots is used instead:

$$\hat{R} = \frac{1}{N}XX^H \tag{8}$$

Next, we decompose the covariance matrix $\hat{R}$ into signal and noise subspaces by the eigenvalue decomposition. For accelerating the decomposition, unitary transformation is used to change the covariance matrix $\hat{R}$ into the real-valued matrix $\hat{C}$:

$$\hat{C} = \frac{1}{2}Q_M^H(\hat{R} + J_M\hat{R}^*J_M)Q_M = Re\left\{Q_M^H\hat{R}Q_M\right\} \tag{9}$$

where J_M is an $M \times M$ exchange matrix with 'ones' on its anti-diagonal and 'zeros'; elsewhere. Q_M is a sparse unitary matrix, defined as:

$$Q_M = \begin{cases} \frac{1}{\sqrt{2}}\begin{bmatrix} I_l & jI_l \\ J_l & -jJ_l \end{bmatrix}, for\ M = 2l \\ \frac{1}{\sqrt{2}}\begin{bmatrix} I_l & 0_l & jI_l \\ 0_l^T & \sqrt{2} & 0_l^T \\ J_l & 0_l & -jJ_l \end{bmatrix}, for\ M = 2l+1 \end{cases} \tag{10}$$

Obviously, Q_M is divided into two cases where the matrix dimension is odd or even. After the eigenvalue decomposition, $\hat{C}$ is expressed as follows:

$$\hat{C} = \hat{E}\hat{\Lambda}\hat{E}^H = \hat{E}_S\hat{\Lambda}_S\hat{E}_S^H + \hat{E}_N\hat{\Lambda}_N\hat{E}_N^H \tag{11}$$

where $\hat{\boldsymbol{E}}_N$ is the noise subspace. $\hat{\Lambda} = diag(\lambda_1, \lambda_2, ..., \lambda_M)$ is the diagonal matrix composed of the eigenvalues λ. $\hat{\boldsymbol{E}}_N$ consists of the eigenvector corresponding to the eigenvalue $\lambda_{K+1}, ..., \lambda_M$. Assuming that the targets are not correlated, the root polynomial is constructed according to the noise subspace [43,44].

$$P_{URM}(z) = \boldsymbol{a}^T(z^{-1})\hat{\boldsymbol{E}}_N\hat{\boldsymbol{E}}_N^H\boldsymbol{a}(z) \tag{12}$$

where $\mathbf{a}(z) = [1, z, ..., z^{M-1}]^T, z = exp(jw)$. After selecting the K roots that are closest to the unit circle, the signal DOA can be calculated according to the following formula:

$$\hat{\theta} = \arcsin(\frac{\arg(z) \times c}{4\pi d f_c}) \tag{13}$$

$\hat{\theta}$ denotes the angle parameters of the targets estimated by the unitary root-MUSIC (URM) algorithm.

3.2. DOA Estimation via Improved URM Algorithm

As mentioned above, the angle information required for multi-target localization can be gained by the URM algorithm. However, due to subspace leakage, the URM algorithm decreases in performance sharply with a low SNR or a small number of snapshots. The reason is the existence of non-ideal terms. If we expand the covariance matrix $\hat{\boldsymbol{R}}$:

$$\begin{aligned} \hat{\boldsymbol{R}} &= \frac{1}{N}\sum_{n=1}^{N}\{\boldsymbol{A}\boldsymbol{S}(n) + \boldsymbol{N}(n)\}\{\boldsymbol{A}\boldsymbol{S}(n) + \boldsymbol{N}(n)\}^H \\ &= \boldsymbol{A}\left\{\frac{1}{N}\sum_{n=1}^{N}\boldsymbol{S}(n)\boldsymbol{S}(n)^H)\right\}\boldsymbol{A}^H + \frac{1}{N}\sum_{n=1}^{N}\boldsymbol{N}(n)\boldsymbol{N}(n)^H \\ &+ \boldsymbol{A}\left\{\frac{1}{N}\sum_{n=1}^{N}\boldsymbol{S}(n)\boldsymbol{N}(n)^H)\right\} + \left\{\frac{1}{N}\sum_{n=1}^{N}\boldsymbol{N}(n)\boldsymbol{S}(n)^H)\right\}\boldsymbol{A}^H \end{aligned} \tag{14}$$

In the preceding formula, $\hat{\boldsymbol{R}}$ was divided into four terms, of which, the first two terms represent the signal covariance matrix and the noise covariance matrix, respectively. The focus is on the latter two items, which are defined as non-ideal terms. Assuming that the noise and signals are uncorrelated, when the number of snapshots is large enough, the non-ideal terms approach zero. However, in practice, the number of snapshots is limited, and the non-ideal terms adversely affect the algorithm, causing subspace estimation to deviate from the real subspace data.

In this section, we propose an improved URM algorithm to enhance the performance. The main idea of improvement is to revise the covariance matrix and reduce the impact of non-ideal terms through optimization iterations. Then, the angle parameters can be extracted from the modified ideal covariance matrix.

Firstly, the URM algorithm is used to estimate the targets and obtain the initial angle set $\hat{\theta}^{(0)} = [\hat{\theta}_1^{(0)}, \hat{\theta}_2^{(0)}, ..., \hat{\theta}_K^{(0)}]$, then the updated array manifold is

$$\hat{\boldsymbol{A}} = [\boldsymbol{a}(\hat{\theta}_1^{(0)}), \boldsymbol{a}(\hat{\theta}_2^{(0)}), ..., \boldsymbol{a}(\hat{\theta}_K^{(0)})] \tag{15}$$

The superscript indicates the number of iterations. Accordingly, the signal $\hat{\boldsymbol{S}}$ can be described as:

$$\hat{\boldsymbol{S}}^{(l)} = \arg\min_{a} ||\boldsymbol{X} - \hat{\boldsymbol{A}}^{(l)}\boldsymbol{S}||_2^2 \tag{16}$$

After the least squares fitting, it can be expressed as:

$$\hat{\boldsymbol{S}}^{(l)} = [(\hat{\boldsymbol{A}}^{(l)})^H\hat{\boldsymbol{A}}^{(l)}]^{((-1)}(\hat{\boldsymbol{A}}^{(l)})^H\boldsymbol{X} \tag{17}$$

Therefore, the noise component can be estimated as:

$$\hat{N}^{(l)} = X - \hat{A}^{(l)}\hat{S}^{(l)} \tag{18}$$

Introducing the real-valued covariance matrix $\hat{C}$ in Equation (11), the non-ideal term T can be found as:

$$\begin{aligned} T &\triangleq \hat{A}^{(l)}\left\{\frac{1}{N}\sum_{n=1}^{N}\hat{S}^{(l)}(n)\hat{N}^{(l)}(n)^{H}\right\} \\ &= \hat{A}^{(l)}\left\{\frac{1}{N}\sum_{n=1}^{N}\left((\hat{A}^{(l)})^{H}\hat{A}^{(l)}\right)^{-1}\left(\hat{A}^{(l)}\right)^{H}X \right. \\ &\left.\times\left(X^{H} - X^{H}\left((\hat{A}^{(l)})^{H}\hat{A}^{(l)}\right)^{-1}\left(\hat{A}^{(l)}\right)^{H}\right)\right\} \\ &= \hat{P}_{A}\left\{\frac{1}{N}\sum_{n=1}^{N}XX^{H}(I_{M} - \hat{P}_{A})\right\} \\ &= \hat{P}_{A}\hat{C}\hat{P}_{A}^{\perp} \end{aligned} \tag{19}$$

where

$$\hat{P}_{A} \triangleq \hat{A}^{(l)}\left((\hat{A}^{(l)})^{H}\hat{A}^{(l)}\right)^{-1}\left(\hat{A}^{(l)}\right)^{H} \tag{20}$$

$$\hat{P}_{A}^{\perp} \triangleq I_{M} - \hat{P}_{A} \tag{21}$$

As shown in the previous formula, the value of the non-ideal term is only related to the real-valued covariance matrix $\hat{C}$ and the array manifold after simplification.

On this basis, the covariance matrix is modified:

$$\hat{C}^{(l)} = \hat{C}^{(l-1)} - \epsilon\left(T + T^{H}\right) \tag{22}$$

The parameters of the modified covariance matrix are estimated again to obtain the angle set $\left\{\hat{\theta}_1^{(l)}, \hat{\theta}_2^{(l)}, \ldots, \hat{\theta}_K^{(l)}\right\}$. When the difference between two consecutive estimation results is less than the preset value, the iteration stops. The improved URM estimation method can be summarized as in Algorithm 1.

Algorithm 1: DOA estimation via the improved URM algorithm.

Input: The received data X

1. The initial covariance matrix $\hat{C}$ is calculated according to (9).

2. The initial angle set $\hat{\theta}^{(0)} = [\hat{\theta}_1^{(0)}, \hat{\theta}_2^{(0)}, \ldots, \hat{\theta}_K^{(0)}]$ are calculated by (13).

Start iteration

3. Update the array manifold $\hat{A}$ according to Equation (15).

4. Calculate the non-ideal terms by Equations (19) and (20).

5. Update the covariance matrix $\hat{C}$ according to (22).

6. Update the new angle set $\left\{\hat{\theta}_1^{(l)}, \hat{\theta}_2^{(l)}, \ldots, \hat{\theta}_K^{(l)}\right\}$ according to the updated covariance matrix $\hat{C}$.

7. $\sum_{k=1}^{K}\|\hat{\theta}_k^{(l-1)} - \hat{\theta}_k^{(l)}\|_2^2$ is less than the preset constant.

Terminate iteration

Output: $\hat{\theta}^{(l)} = [\hat{\theta}_1^{(l)}, \hat{\theta}_2^{(l)}, \ldots, \hat{\theta}_K^{(l)}]$

3.3. Range Estimation

After estimating the angle parameters, the steering vector matrix can be reconstructed as:

$$\tilde{A} = [a(\hat{\theta}_1), a(\hat{\theta}_2), \ldots, a(\hat{\theta}_K)] \tag{23}$$

Combined with Equation (9), we can obtain

$$Y = (\tilde{A}^{\dagger} X)^T = (\tilde{A}^{\dagger} \tilde{A} S + \tilde{A}^{\dagger} N)^T = S^T + (\tilde{A}^{\dagger} N)^T \tag{24}$$

By normalizing the matrix, the real-valued vector matrix can be obtained:

$$Y(:,k) = Im\{\ln\left(Y(:,k)/Y(1,k)\right)\} \tag{25}$$

The 'least squares' is constructed as follows:

$$\min \|Gr_k - Y_k\|_F^2 \tag{26}$$

where G is the least squares matrix:

$$G = \begin{bmatrix} 1 & 1 & \dots & 1 \\ 4\pi BT_s/Tc & 2 \times 4\pi BT_s/Tc & \dots & N \times 4\pi BT_s/Tc \end{bmatrix} \tag{27}$$

where, $\Delta = 4\pi BT_s/Tc$ denotes the phase difference between the adjacent elements and N is the number of snapshots. We define $\hat{r}_k$ as the estimator of the range parameter, which can be expressed as:

$$\hat{r}_k = G_r^{\dagger} Y(:,k) \tag{28}$$

The second element of each column vector of $\hat{r}_k$ is the estimated value of the range parameter. So far, in line with the proposed method, we can estimate the cursory DOA and range parameters of the objectives in pairs.

3.4. *Refined DOA–Range Estimation by Pseudo-Noise Resampling*

In Sections 3.2 and 3.3, we propose a joint DOA–range estimation method based on the FMCW-MIMO radar. However, the abnormal estimators that deviate from real targets still occur in low SNR or a small number of snapshot scenarios. So a joint application of pseudo-noise resampling technology and threshold detection is proposed to mitigate the influence of outliers.

First, we use threshold detection to detect the outliers. The threshold range can be determined by conventional beamforming technology and FFT. Beamforming technology and FFT can quickly estimate the angle and range parameter at the expense of accuracy, to provide us with the approximate range of the targets. Thus, the thresholds for the angle and range parameters are assumed as:

$$\theta_H = [\theta_1^{max} - \theta_1^L, \theta_1^{max} + \theta_1^R] \cup [\theta_2^{max} - \theta_2^L, \theta_2^{max} + \theta_2^R] \cup \dots \cup [\theta_K^{max} - \theta_K^L, \theta_K^{max} + \theta_K^R] \tag{29}$$

$$r_H = [r_1^{max} - r_1^L, r_1^{max} + r_1^R] \cup [r_2^{max} - r_2^L, r_2^{max} + r_2^R] \cup \dots \cup [r_K^{max} - r_K^L, r_K^{max} + r_K^R] \tag{30}$$

where θ_{max} and r_{max} denote the K highest peak of the beamformer and FFT method, respectively, θ_K^L and r_K^L represent the left boundary of the angle parameter and range parameter of each target interval, θ_K^R and r_K^R are the right boundaries.

We denote the rough parameter estimation results with the set $\hat{\theta} = [\hat{\theta}_1, \hat{\theta}_2, \dots, \hat{\theta}_K]$ and $\hat{r} = [\hat{r}_1, \hat{r}_2, \dots, \hat{r}_K]$ in Sections 3.2 and 3.3. If the estimated values satisfy the threshold detection, then the estimated parameters are taken as the final results. Estimators that do not pass the threshold detection are considered outliers, a pseudo-noise approach is presented when an outlier occurs. The idea of this approach is to perturb the raw measurement data matrix X by an artificially generated pseudo-random noise. Here, we define the pseudo-noise matrix $Y \in \mathbb{C}^{M \times N}$ as the pseudo-noise matrix, where:

$$E(Y) = 0, E(YY^H) = \sigma_Y^2 NI, E(YY^T) = 0 \tag{31}$$

Meanwhile, as the variance is related to the variance of the original data matrix, the power of pseudo-random noise σ_Y^2 can be estimated by $\sigma_Y^2 = \rho\sigma^2$, where ρ is a user-defined parameter and σ^2 is given by:

$$\sigma^2 = \frac{1}{M-K} Tr\hat{\Lambda}_N = \frac{1}{M-K} \sum_{i=K+1}^{M} \hat{\lambda}_i \tag{32}$$

$\hat{\lambda}_i$ is a group of eigenvalues arranged in descending order, which is achieved by the decomposition of matrix $\boldsymbol{X}$.

Combing Equations (31) and (32), the resampled data matrix $\boldsymbol{Z}$ is:

$$\boldsymbol{Z} = \boldsymbol{X} + \boldsymbol{Y} \tag{33}$$

We assume that the original matrix X is perturbed for W times and the W groups of resampled data matrix Z are obtained correspondingly. For each resampled data matrix, it is necessary to perform the steps in Sections 3.2 and 3.3. The corresponding estimators construct the W groups of the estimators set $(\hat{\theta}_k, \hat{r}_k), k = 1, 2, ..., K$. Then the threshold detection is applied to the estimator sets, and the W groups of the estimators are divided into two subsets:

$$\begin{aligned} \boldsymbol{B}_1 &= \left\{ (\hat{\theta}_k, \hat{r}_k)^{(1)}, ..., (\hat{\theta}_k, \hat{r}_k)^{(J)} \right\}, \\ \boldsymbol{B}_2 &= \left\{ (\hat{\theta}_k, \hat{r}_k)^{(1)}, ..., (\hat{\theta}_k, \hat{r}_k)^{(W-J)} \right\} \end{aligned} \tag{34}$$

J groups of the estimator sets that pass the detection constitute $\boldsymbol{B}_1$, and $\boldsymbol{B}_2$ is composed of $W - J$ groups of estimator sets rejected by the threshold detection.

If $0 < J < W$, the final target parameter estimator is the average result of the successful J groups of the estimator set:

$$\hat{\theta}_k = \frac{1}{J} \sum_{i=1}^{J} \hat{\theta}_k^{(i)}, \hat{r}_k = \frac{1}{J} \sum_{i=1}^{J} \hat{r}_k^{(i)} \tag{35}$$

If $J = 0$ means that all groups of estimators fail to pass the threshold detection even if the PR technology is performed. Facing the case where all estimators deviate from the true value, we need to select the optimal one among the multiple groups of estimators set in $\boldsymbol{B}_2$ as the final output estimation. Since the angle and range parameters of the target exist in pairs in the set, we can use one of the parameter information pieces as a judgment object. In this paper, the angle parameter was selected as the judgment object.

First, the angle estimators in $\boldsymbol{B}_2$ are divided into K subsets according to different targets:

$$\boldsymbol{B}_{2k} = \left\{ \hat{\theta}_k^{(1)}, .., \hat{\theta}_k^{(W)} \right\}, i = 1, ..., K \tag{36}$$

where $\boldsymbol{B}_{2k}$ represents the set of W group angle estimators for the k-th target.

According to Equation (13), the angle estimation is related to the corresponding polynomial root.

$$|\boldsymbol{D}_k| = \left\{ \left|\hat{z}_k^{(1)}\right|, .., \left|\hat{z}_k^{(W)}\right| \right\}, k = 1, ..., K \tag{37}$$

In the above DOA algorithm, the root polynomial P_{URM} has (M-1) pairs of symmetrically mirrored roots about the unit circle, and the K roots closest to the unit circle are what we need. So, we choose the $\hat{z}$ with the smallest modulus in $\boldsymbol{D}_k$ to ensure that the deviation between the corresponding estimated value $\hat{\theta}_k$ in $\boldsymbol{B}_{2k}$ and the theoretical value is minimum. Then the paired range parameters will be determined. Finally, the corresponding target parameters $(\hat{\theta}_k, \hat{r}_k)$ can be obtained.The above steps can be summarized as Algorithm 2.

Algorithm 2: Refined DOA–range estimation by pseudo-noise resampling.
Input: The received data $\boldsymbol{X}$, initial parameter estimator $(\hat{\theta}_k, \hat{r}_k)$.
1. Set the threshold area and apply the threshold detection to these estimators.
2. If it passes the detection, terminate this algorithm and output the estimator of $\hat{\theta}$ and $\hat{r}$.
3. If not, W groups of the resampled data are obtained by using the pseudo-noise matrix.
(1) For the updated X, use the algorithms in Sections 3.2 and 3.3 to estimate the DOA and range.
(2) Perform the threshold detection again for W group estimators and categorize them as (34).
If $J > 0$, then estimate target parameters via (35).
If $J = 0$, then select target parameters by (36) and (37).
Output: $\{\hat{\theta}_k, \hat{r}_k\}$

4. Simulations

In this section, the proposed algorithm is compared with conventional algorithms. The parameter settings of the FMCW-MIMO radar are shown in Table 2.

In this experiment, three different targets are set at $(\theta_1, r_1) = [-8.78°, 10.73\text{ m}]$, $(\theta_2, r_2) = [2.43°, 32.78\text{ m}]$, and $(\theta_3, r_3) = [12.07°, 23.66\text{ m}]$, respectively. The number of Monte Carlo trials is MC = 500.

Table 2. Parameters of the FMCW-MIMO radar.

Parameter	Value	Parameter	Value
c	3×10^8 m/s	T_s	25 ns
f_s	2.46×10^9 Hz	B	1.2×10^8 Hz
λ	120 mm	M	8
d	60 mm	L	3

4.1. 2D Point Cloud of the Target

First, to verify the localization performance, the proposed algorithm was simulated with SNR = 10, N = 50, where N is the number of snapshots. The results are shown in Figure 1 and the X and Y axes represent the angle and range parameters, respectively.

In Figure 5, the three target points can be distinguished, and the estimated targets are very close to the preset ones, which proves the reliability of our proposed algorithm.

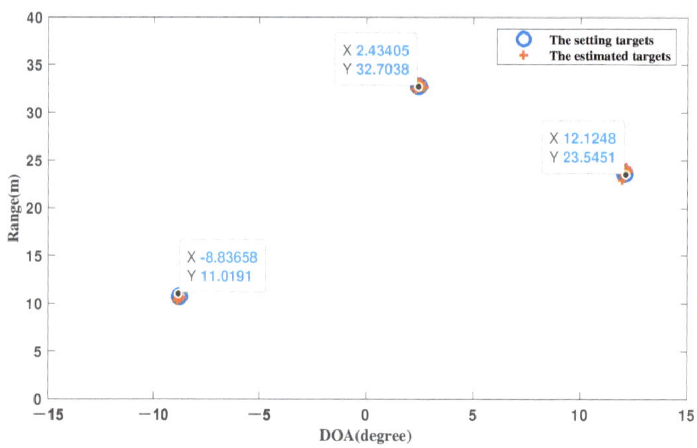

Figure 5. The 2D point cloud of the estimated targets.

4.2. Performance Analysis

In this part, we set multiple SNRs and the number of snapshots to explore the differences between the proposed algorithm and the comparison algorithms in diverse environments. The 2D-MUSIC, 2D-ESPRIT, and DFT-root-MUSIC (DFT-RM) algorithms are used as comparison algorithms, among which 2D-MUSIC and 2D-ESPRIT are two-dimensional joint DOA–range estimation algorithms. Meanwhile, DFT-RM estimates the angle and range with the RM algorithm and DFT algorithm, respectively.

4.2.1. RMSE Performance of DOA

The *RMSE* (root mean square error) is usually used as an important index to evaluate the algorithm's performance. The *RMSE* of the FMCW MIMO radar's DOA can be defined as:

$$RMSE_\theta = \frac{1}{K}\sum_{k=1}^{K}\sqrt{\frac{1}{MC}\sum_{mc=1}^{MC}(\theta_{mc,k}-\theta_k)^2} \quad (38)$$

where K denotes the number of targets, $\theta_{mc,k}$ stands for the angle estimation of the k-th target in the mc-th Monte Carlo experiment.

Figure 6 shows the *RMSEs* of the DOA estimation against SNR when the number of snapshots is fixed at 50. The results show that the proposed method is superior to other algorithms in accuracy and stability. The main reason is that the proposed method reduces the influences of non-ideal terms, which improves the accuracy of the location. The 2D-ESPRIT algorithm loses the array aperture when decomposing the subspace, so the *RMSE* is larger than that of the 2D-RM and proposed algorithm.

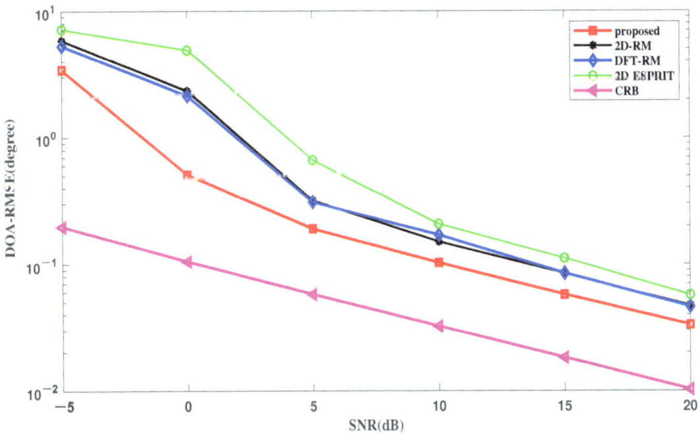

Figure 6. RMSE of the DOA estimation versus SNR.

The trends of the *RMSE* with different snapshot numbers are revealed in Figure 7. In this simulation, the SNR is fixed at 5 dB, and the number of snapshots increases from 10 to 150. When the number of snapshots is less than 70, the performance advantage of the proposed algorithm is obvious, and even if the number of snapshots is greater than 70, it is also one of the algorithms with the smallest errors. When the number of snapshots exceeds 70, the proposed algorithm has a similar performance compared to 2D-RM and DFT-RM. As the number of snapshots increases, the deviation of the sample covariance matrix from the theoretical covariance matrix gradually shrinks. Therefore, the proposed method improves in an unobvious manner and gradually approaches 2D-RM with a large number of snapshots.

The CRB curve of the FMCW-MIMO radar is also given to evaluate the estimation performance of the algorithms. It shows that the proposed algorithm is closer to the CRB curve than the comparison algorithms in terms of the DOA estimation, which means that the proposed algorithm outperforms other algorithms, especially with low SNRs and a small number of snapshots.

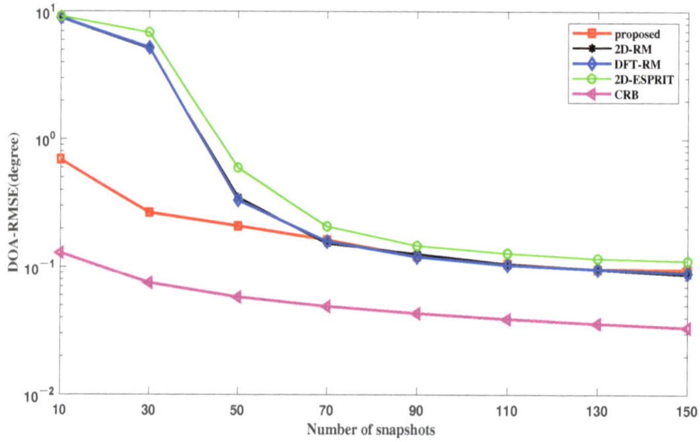

Figure 7. RMSE of the DOA estimation versus snapshots.

4.2.2. RMSE Performance of Range

In this simulation, we evaluated the range parameter estimation performances of different algorithms versus different SNRs and snapshots, as shown in Figures 8 and 9. Similar to the previous section, the *RMSE* formula for the range of the FMCW MIMO radar can be written as:

$$RMSE_r = \frac{1}{K}\sum_{k=1}^{K}\sqrt{\frac{1}{MC}\sum_{mc=1}^{MC}(r_{mc,k} - r_k)^2} \tag{39}$$

Evaluating the range estimation is the same as the DOA estimation performance evaluation, with two different groups of parameter settings: SNR varies from −5 to 20 dB with N = 50, and the number of snapshots changes from 10 to 150 with SNR = 5 dB.

An explanatory point is needed for this simulation; DFT-RM uses the FFT algorithm for the range estimation, which highly depends on the number of snapshots. When the snapshot number is not enough, the range resolution will decrease, resulting in low accuracy and failure of the FFT algorithm.

Therefore, in Figure 8, when the number of snapshots is fixed at 50, the range estimation of the DFT-RM tends to be a straight line. In Figure 9, since the number of snapshots is always set in a low state, the DFT-RM performance is always unstable, resulting in a large estimation error. The performances of other algorithms all improve with the increase of the SNR and the number of snapshots, but the proposed one is closer to the CRB curve intuitively, as seen in Figures 8 and 9. Therefore, the proposed method still has obvious advantages over the other conventional algorithms in the range domain.

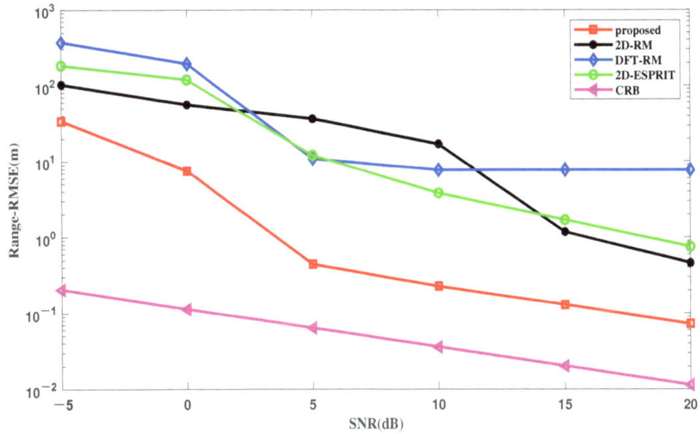

Figure 8. RMSE of the range estimation versus the SNR.

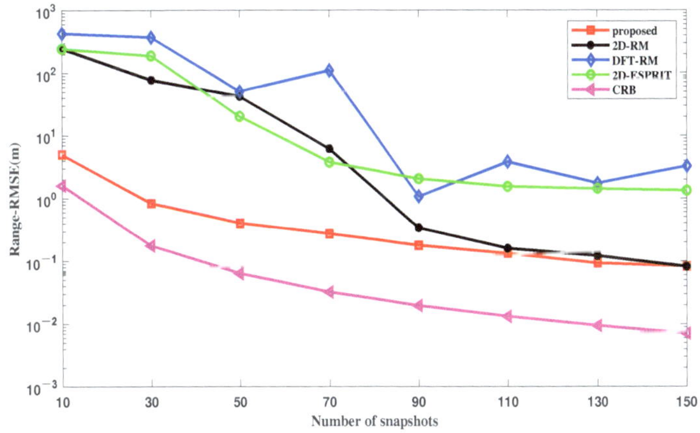

Figure 9. RMSE of the range estimation versus the snapshots.

4.2.3. PSD versus SNR

The probability of successful detection (*PSD*) is another indicator to measure the performance, which can be defined by the following formula:

$$PSD = \frac{H}{D} \times 100 \tag{40}$$

where D represents the number of experiments and H denotes the number of successful tests.

In this section, to demonstrate the trend of each algorithm completely, DOA and range estimates are considered successful in meeting $|\theta_{mc,k} - \theta_k| \leq 0.7^\circ$ and $|r_{mc,k} - r_k| \leq 1.5$ m.

Figures 10 and 11 describe the *PSD* of the range and angle estimation under different SNRs when the number of N = 50. It is clear that the algorithm we proposed is better than other methods under different SNRs, especially in low SNRs. The reason is that the proposed algorithm reduces the negative impacts of non-ideal items. The algorithms, which are highly dependent on the sampling points or snapshots, suffer from non-ideal items and have more failed estimators.

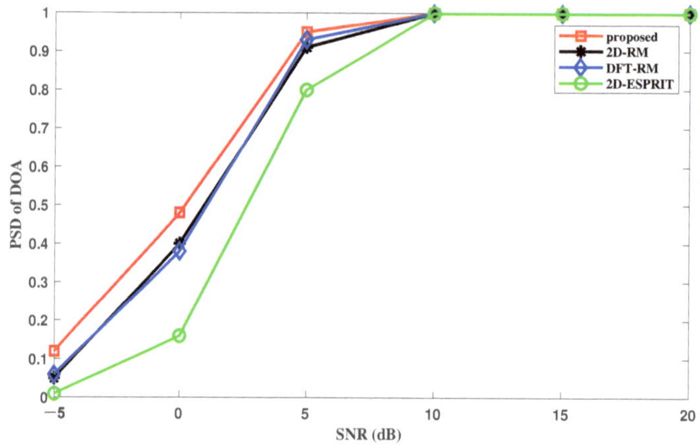

Figure 10. PSD of the DOA estimation versus SNR.

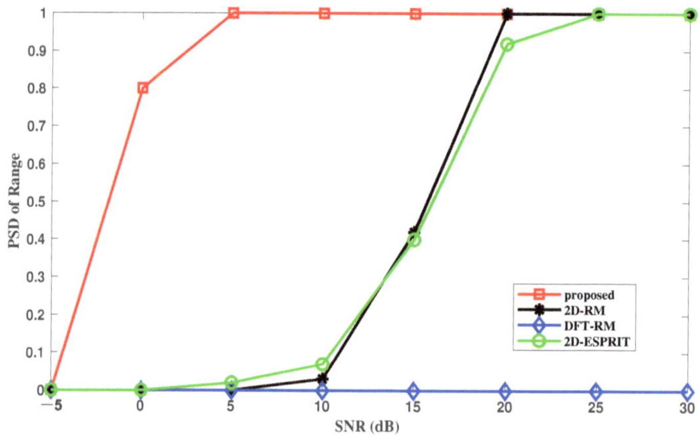

Figure 11. PSD of the range estimation versus SNR.

4.2.4. PSD versus Snapshots

In this section, we investigate the PSD of the proposed and compared algorithms with fixed SNRs and varying numbers of snapshots. In Figure 12, the proposed algorithm has a high success rate when the number of snapshots is only 30, and achieves full detection when the number of snapshots is 50. Compared with the RM, FFT-RM, and ESPRIT algorithms, the proposed method is more stable in the angle domain.

For the range estimation, Figure 13 shows the PSD curves of all algorithms from 10 to 150 snapshots. Compared to other algorithms, the proposed method has a higher success rate in all scenarios. Notice that the DFT-RM method has a low success rate level as this method requires more snapshots than others.

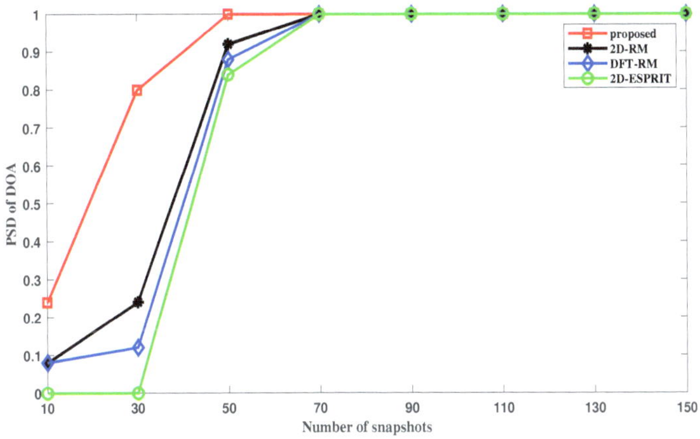

Figure 12. PSD of the DOA estimation versus the snapshots.

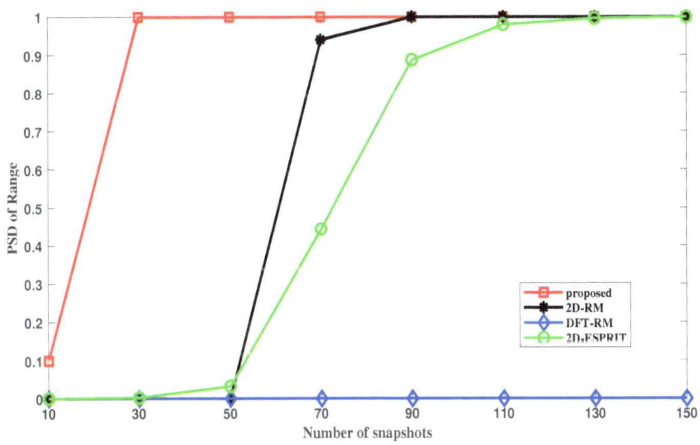

Figure 13. PSD of the range estimation versus the snapshots.

5. Conclusions

In this paper, the algorithm applies iterative updating to reduce the non-ideal items and approach a more accurate covariance matrix. Based on the covariance matrix, a rough angle estimation and range estimation were obtained by U-root-MUSIC and the least squares algorithms, respectively. Threshold detection was then used to refine the rough results, where the outliers were further processed. Simulations have shown that this method improves the estimation accuracy effectively, especially in conditions with a small number of snapshots and low SNRs. In future work, we will optimize the proposed algorithm to make it suitable for more practical application scenarios.

Author Contributions: Conceptualization, Y.J. and X.L.; methodology, Y.J. and X.L.; writing—original draft preparation, Y.J.; writing—review and editing, X.L. and X.W.; supervision, X.L.; formal analysis, Z.H. funding acquisition, X.L. and J.S. All authors have read and agreed to the published version of the manuscript.

Funding: This work was supported by the Natural Science Foundation of Hainan Province (no. 621MS018), the National Natural Science Foundation of China Youth Science Fund Project (no. 62101165), National Natural Science Foundation of China (no. 61861015 and 61961013), Key Research and Development Program of Hainan Province (no. ZDYF2019011), National Key Research and Development Program of China (no. 2019CXTD400), Young Elite Scientists Sponsorship Program by CAST (no. 2018QNRC001), the Scientific Research Setup Fund of Hainan University (no. KYQD(ZR) 1731 and KYQD(ZR)-21134), Hainan Provincial Natural Science Foundation of China, no. 620RC669, and the Key Project of Higher Education and Teaching Reform in Hainan Province, no. Hnjg2022ZD-54.

Institutional Review Board Statement: Not applicable.

Informed Consent Statement: Not applicable.

Data Availability Statement: Not applicable.

Conflicts of Interest: The authors declare no conflict of interest.

References

1. Premachandra, C.; Murakami, M.; Gohara, R.; Ninomiya, T.; Kato, K. Improving landmark detection accuracy for self-localization through baseboard recognition. *Int. J. Mach. Learn. Cybern.* **2017**, *8*, 1815–1826. [CrossRef]
2. Cavanini, L.; Benetazzo, F.; Freddi, A.; Longhi, S.; Monteriu, A. SLAM-based autonomous wheelchair navigation system for AAL scenarios. In Proceedings of the 2014 IEEE/ASME 10th International Conference on Mechatronic and Embedded Systems and Applications (MESA), Senigallia, Italy, 10–12 September 2014; pp. 1–5.
3. Paulet, M.V.; Salceanu, A.; Neacsu, O.M. Ultrasonic radar. In Proceedings of the 2016 International Conference and Exposition on Electrical and Power Engineering (EPE), Iasi, Romania, 20–22 October 2016; pp. 551–554.
4. Bleh, D.; Rösch, M.; Kuri, M.; Dyck, A.; Tessmann, A.; Leuther, A.; Wagner, S.; Weismann-Thaden, B.; Stulz, H.P.; Zink, M.; et al. *W*-Band Time-Domain Multiplexing FMCW MIMO Radar for Far-Field 3-D Imaging. *IEEE Trans. Microw. Theory Tech.* **2017**, *65*, 3474–3484. [CrossRef]
5. Stove, A.G. Linear FMCW radar techniques. In *IEE Proceedings F (Radar and Signal Processing)*; IET: London, UK, 1992; Volume 139, pp. 343–350.
6. Mahafza, B.R. *Radar Systems Analysis and Design Using MATLAB*; Chapman and Hall/CRC: Boca Raton, FL, USA, 2005.
7. Nakagawa, K.; Mitsumoto, M.; Kai, K. Development of Millimeter-wave Radar for Latest Vehicle Systems. In *Advanced Microsystems for Automotive Applications 2005*; Springer: Berlin/Heidelberg, Germany, 2005; pp. 257–268.
8. Cong, J.; Wang, X.; Lan, X.; Huang, M.; Wan, L. Fast target localization method for FMCW MIMO radar via VDSR neural network. *Remote Sens.* **2021**, *13*, 1956. [CrossRef]
9. Richards, M.A. *Fundamentals of Radar Signal Processing*; McGraw-Hill Education: New York, NY, USA, 2014.
10. Griffiths, H.; Cohen, L.; Watts, S.; Mokole, E.; Baker, C.; Wicks, M.; Blunt, S. Radar spectrum engineering and management: Technical and regulatory issues. *Proc. IEEE* **2014**, *103*, 85–102. [CrossRef]
11. Kim, Y.D.; Son, G.J.; Song, C.H.; Kim, H.K. On the deployment and noise filtering of vehicular radar application for detection enhancement in roads and tunnels. *Sensors* **2018**, *18*, 837. [CrossRef]
12. Fishler, E.; Haimovich, A.; Blum, R.; Chizhik, D.; Cimini, L.; Valenzuela, R. MIMO radar: An idea whose time has come. In Proceedings of the 2004 IEEE Radar Conference (IEEE Cat. No. 04CH37509), Philadelphia, PA, USA, 29 April 2004; pp. 71–78.
13. Wan, L.; Sun, L.; Liu, K.; Wang, X.; Lin, Q.; Zhu, T. Autonomous vehicle source enumeration exploiting non-cooperative UAV in software defined internet of vehicles. *IEEE Trans. Intell. Transp. Syst.* **2020**, *22*, 3603–3615. [CrossRef]
14. Wang, X.; Yang, L.T.; Meng, D.; Dong, M.; Ota, K.; Wang, H. Multi-UAV cooperative localization for marine targets based on weighted subspace fitting in SAGIN environment. *IEEE Internet Things J.* **2021**, *9*, 5708–5718. [CrossRef]
15. Wang, X.; Wan, L.; Huang, M.; Shen, C.; Han, Z.; Zhu, T. Low-complexity channel estimation for circular and noncircular signals in virtual MIMO vehicle communication systems. *IEEE Trans. Veh. Technol.* **2020**, *69*, 3916–3928. [CrossRef]
16. Wang, X.; Huang, M.; Wan, L. Joint 2D-DOD and 2D-DOA estimation for coprime EMVS–MIMO radar. *Circuits Syst. Signal Process.* **2021**, *40*, 2950–2966. [CrossRef]
17. Feger, R.; Wagner, C.; Schuster, S.; Scheiblhofer, S.; Jager, H.; Stelzer, A. A 77-GHz FMCW MIMO radar based on an SiGe single-chip transceiver. *IEEE Trans. Microw. Theory Tech.* **2009**, *57*, 1020–1035. [CrossRef]
18. Yang, Y.; Blum, R.S. MIMO radar waveform design based on mutual information and minimum mean-square error estimation. *IEEE Trans. Aerosp. Electron. Syst.* **2007**, *43*, 330–343. [CrossRef]
19. Fuhrmann, D.R.; San Antonio, G. Transmit beamforming for MIMO radar systems using signal cross-correlation. *IEEE Trans. Aerosp. Electron. Syst.* **2008**, *44*, 171–186. [CrossRef]
20. Xu, L.; Li, J. Iterative generalized-likelihood ratio test for MIMO radar. *IEEE Trans. Signal Process.* **2007**, *55*, 2375–2385. [CrossRef]
21. Saponara, S.; Neri, B. Radar sensor signal acquisition and multidimensional FFT processing for surveillance applications in transport systems. *IEEE Trans. Instrum. Meas.* **2017**, *66*, 604–615. [CrossRef]

22. Zhu, W. DOA estimation for broadband signal that use FFT interpolation method. In Proceedings of the 2013 IEEE 4th International Conference on Software Engineering and Service Science, Beijing, China, 23–25 May 2013; pp. 959–962.
23. Amjadian, A.; Mirzakuchaki, S. Reducing FFT Operations for FMCW Radars Range Detection. *Authorea*, 6 June 2022.
24. Wan, R.; Song, Y.; Mu, T.; Wang, Z. Moving target detection using the 2D-FFT algorithm for automotive FMCW radars. In Proceedings of the 2019 International Conference on Communications, Information System and Computer Engineering (CISCE), Haikou, China, 5–7 July 2019; pp. 239–243.
25. Song, M.; Lim, J.; Shin, D.J. The velocity and range detection using the 2D-FFT scheme for automotive radars. In Proceedings of the 2014 4th IEEE International Conference on Network Infrastructure and Digital Content, Beijing, China, 19–21 September 2014; pp. 507–510.
26. Han, X.; Shu, T.; He, J.; Yu, W. Polarization-angle-frequency estimation with linear nested vector sensors. *IEEE Access* **2018**, *6*, 36916–36926. [CrossRef]
27. Lemma, A.N.; Van Der Veen, A.J.; Deprettere, E.F. Analysis of joint angle-frequency estimation using ESPRIT. *IEEE Trans. Signal Process.* **2003**, *51*, 1264–1283. [CrossRef]
28. Xie, R.; Hu, D.; Luo, K.; Jiang, T. Performance analysis of joint range-velocity estimator with 2D-MUSIC in OFDM radar. *IEEE Trans. Signal Process.* **2021**, *69*, 4787–4800. [CrossRef]
29. Si, W.; Wang, Y.; Hou, C.; Wang, H. Real-valued 2D MUSIC algorithm based on modified forward/backward averaging using an arbitrary centrosymmetric polarization sensitive array. *Sensors* **2017**, *17*, 2241. [CrossRef]
30. Orovi, I.; Stanković, S.; Draganić, A. Time-frequency analysis and singular value decomposition applied to the highly multicomponent musical signals. *Acta Acust. United Acust.* **2014**, *100*, 93–101. [CrossRef]
31. Zhang, X.; Xu, L.; Xu, L.; Xu, D. Direction of departure (DOD) and direction of arrival (DOA) estimation in MIMO radar with reduced-dimension MUSIC. *IEEE Commun. Lett.* **2010**, *14*, 1161–1163. [CrossRef]
32. Kim, B.s.; Jin, Y.; Lee, J.; Kim, S. High-Efficiency Super-Resolution FMCW Radar Algorithm Based on FFT Estimation. *Sensors* **2021**, *21*, 4018. [CrossRef] [PubMed]
33. Kim, B.s.; Jin, Y.; Lee, J.; Kim, S. Low-complexity MUSIC-based direction-of-arrival detection algorithm for frequency-modulated continuous-wave vital radar. *Sensors* **2020**, *20*, 4295. [CrossRef] [PubMed]
34. Kim, S.; Oh, D.; Lee, J. Joint DFT-ESPRIT estimation for TOA and DOA in vehicle FMCW radars. *IEEE Antennas Wirel. Propag. Lett.* **2015**, *14*, 1710–1713. [CrossRef]
35. Wang, C.; Zheng, W.; Gong, P.; Li, Z. Joint angle and range estimation in the fda-mimo radar: The reduced-dimension root music algorithm. *Wirel. Pers. Commun.* **2020**, *115*, 2515–2533. [CrossRef]
36. Oh, D.; Ju, Y.; Nam, H.; Lee, J.H. Dual smoothing DOA estimation of two-channel FMCW radar. *IEEE Trans. Aerosp. Electron. Syst.* **2016**, *52*, 904–917. [CrossRef]
37. Guerci, J.; Bergin, J. Principal components, covariance matrix tapers, and the subspace leakage problem. *IEEE Trans. Aerosp. Electron. Syst.* **2002**, *38*, 152–162. [CrossRef]
38. Wallman, J.J.; Barnhill, M.; Emerson, J. Robust characterization of leakage errors. *New J. Phys.* **2016**, *18*, 043021. [CrossRef]
39. Shaghaghi, M.; Vorobyov, S.A. Subspace leakage analysis of sample data covariance matrix. In Proceedings of the 2015 IEEE International Conference on Acoustics, Speech and Signal Processing (ICASSP), South Brisbane, Australia, 19–24 April 2015; pp. 3447–3451.
40. Carlson, B.D. Covariance matrix estimation errors and diagonal loading in adaptive arrays. *IEEE Trans. Aerosp. Electron. Syst.* **1988**, *24*, 397–401. [CrossRef]
41. Chen, Y.; Wiesel, A.; Eldar, Y.C.; Hero, A.O. Shrinkage algorithms for MMSE covariance estimation. *IEEE Trans. Signal Process.* **2010**, *58*, 5016–5029. [CrossRef]
42. Mestre, X.; Lagunas, M.Á. Modified subspace algorithms for DoA estimation with large arrays. *IEEE Trans. Signal Process.* **2008**, *56*, 598–614. [CrossRef]
43. Congfeng, L.; Guisheng, L. Fast algorithm for Root-MUSIC with real-valued egendecomposition. In Proceedings of the 2006 CIE International Conference on Radar, Shanghai, China, 16–19 October 2006; pp. 1–4.
44. Yin, D.; Zhang, F. Uniform linear array MIMO radar unitary root MUSIC Angle estimation. In Proceedings of the 2020 Chinese Automation Congress (CAC), Shanghai, China, 6–8 November 2020; pp. 578–581.

sensors MDPI

Communication

Leveraging Deep Learning for Practical DoA Estimation: Experiments with Real Data Collected via USRP

Hyeonjin Chung, Hyunwoo Park and Sunwoo Kim *

Department of Electronics and Computer Engineering, Hanyang University, Seoul 04763, Korea
* Correspondence: remero@hanyang.ac.kr; Tel.: +82-02-2220-4822

Abstract: This paper presents an experimental validation of deep learning-based direction-of-arrival (DoA) estimation by using realistic data collected via universal software radio peripheral (USRP). Deep neural network (DNN) and convolutional neural network (CNN) structures are designed to estimate the DoA. Two types of data are used for training networks. One is the data synthesized by the signal model, and the other is the data collected by USRP. Here, the signal model considers both mutual coupling and multipath signals. Experimental results show that the estimation performance is most accurate when training DNN and CNN with the collected data. Furthermore, the estimation tends to be poor in the indoor environment, which suffers from the strong non-line-of-sight (NLoS) signals.

Keywords: deep learning; direction-of-arrival estimation; deep neural network; convolutional neural network; universal software radio peripheral

1. Introduction

Direction-of-arrival (DoA) estimation is one of the long-studied research topics in array signal processing. DoA estimation algorithms have been adopted in various applications, such as localization and radar [1]. Traditional DoA estimation algorithms such as MUSIC [2] and ESPRIT [3] are proposed based on the characteristic of the signal model. Although they can ideally achieve high estimation accuracy and resolution, unexpected problems (e.g., multipath effect [4], gain-phase error [5], mutual coupling [6], antenna misalignment [7], etc.) may exist in practice. In this case, the signal model cannot capture the characteristics of a real received signal, thereby causing degradation of the DoA estimation performance.

There have been various studies to deal with problems that may cause model mismatch. One of these studies is coherent DoA estimation, which can estimate multipath signals impinging from different directions [4,8,9]. On the other hand, there are problems induced by hardware impairments such as gain-phase error, mutual coupling, and antenna misalignment. There have been efforts to calibrate these errors without using reference signals, where [5,10,11] deal with gain-phase error, [6,12,13] deal with mutual coupling between antennas in an array, and [7,14] deal with the errors in the steering vector that can be caused by antenna misalignment. However, the performance of the aforementioned works may degrade when several problems simultaneously occur or there are more unexpected problems.

After the introduction of deep learning [15], DoA estimation algorithms based on various types of neural network (NN) have been proposed in [16–20]. One of the benefits of using deep learning is that the user does not have to know the exact signal model if there are sufficient training data. For this reason, it is expected that the deep learning-based DoA estimation does not suffer from model mismatch problems by using data that capture all kinds of errors (e.g., actual measured data). In [16,17], deep neural network (DNN)-based DoA estimation was proposed, where these works report that the DNN-based estimation is more accurate than the traditional DoA estimation. After the proposal of the DNN-based

DoA estimation, convolutional neural network (CNN)-based DoA estimation has been studied in [18,19]. The CNN-based DoA estimation shows better estimation accuracy and resolution compared to the DNN-based DoA estimation. In [20], the DoA estimation based on unsupervised learning was proposed, where unsupervised learning can make data collection easier since data labeling is not required. In recent works, there have been efforts to exploit features of classical DoA estimation, rather than solely depending on NN. Refs. [21–23] respectively employ DNN, CNN, and recurrent neural network (RNN) to estimate the ideal noiseless covariance matrix, which is denoted as a pseudo covariance matrix. Then, the classical DoA estimation such as MUSIC and root-MUSIC estimates the DoA with pseudo covariance matrix. In [24], the residual neural network (ResNet) first estimates the candidates of DoAs. From the candidates, the classical maximum likelihood estimation (MLE) [25] picks the final DoAs. A combination of these two methods enhances the accuracy while achieving lower complexity than only using MLE.

However, the existing works on the deep learning-based DoA estimation lack experimental validation, even though the deep learning-based DoA estimation is expected to be effective in a practical situation where there are many problems that cause a model mismatch. In this paper, we validate deep learning-based DoA estimation with realistic data collected by a universal software radio peripheral (USRP). In the experiment, two types of data—data synthesized by the signal model and data collected by USRP—are used for training networks. The estimation accuracy is then analyzed according to the type of training data.

2. System Model

In this paper, we consider one transmitter, which is equipped with an omni-directional antenna. A receiver is equipped with a uniform linear array (ULA), which has M antenna elements. The spacing between adjacent antennas is set to half-wavelength $\lambda/2$, where λ denotes the wavelength of the transmitted signal.

To generate the data for training DNN and CNN, the signal model should be defined. Here, the generated data are expected to be well-suited for training if the signal model can capture the state of the hardware systems. Among the many kinds of hardware-induced problems, the gain-phase error in our systems is calibrated using the method in [26]. The antenna spacing is designed to be half-wavelength so that there is no antenna misalignment. However, the current systems cannot calibrate mutual coupling and multipath effects. Thus, in this paper, we consider mutual coupling and multipath effects to design the received signal model.

An array manifold vector whose DoA is θ, $\mathbf{a}(\theta)$ can be given as follows:

$$\mathbf{a}(\theta) = \left[1, e^{j\pi\cos\theta}, \ldots, e^{j(M-1)\pi\cos\theta}\right]^T \in \mathbb{C}^{M\times 1}. \tag{1}$$

To capture mutual coupling and multipath effects, we model a received signal $\mathbf{X} \in \mathbb{C}^{M\times D}$ as:

$$\mathbf{X} = \mathbf{C}\sum_{p=0}^{P} \alpha_p \mathbf{a}(\theta_p)\mathbf{s}^T + \mathbf{N} \in \mathbb{C}^{M\times D}. \tag{2}$$

where $\mathbf{C} \in \mathbb{C}^{M\times M}$ denotes the mutual coupling matrix [27]. P denotes the number of non-line-of-sight (NLoS) paths. α_p and θ_p respectively denote the channel gain and the DoA of the p-th signal path. Specifically, α_0 and θ_0 denote the channel gain and the DoA of the line-of-sight (LoS) path. $\mathbf{s} = [s_1, \ldots, s_D]^T \in \mathbb{C}^{D\times 1}$ denotes a signal vector, whose power equals σ_s^2. D is a number of signal snapshots. $\mathbf{N} \in \mathbb{C}^{M\times D}$ is a noise matrix, whose entries all follow $\mathcal{CN}(0, \sigma^2)$. σ^2 denotes the power of the noise. $\mathbf{R_X}$, the covariance matrix of $\mathbf{X}$, can be defined as:

$$\mathbf{R_X} = \mathbb{E}\left[\mathbf{X}\mathbf{X}^H\right] \approx \frac{\mathbf{X}\mathbf{X}^H}{D} \in \mathbb{C}^{M\times M}. \tag{3}$$

3. Deep Learning Network Structure for DoA Estimation

This section introduces two network structures for DoA estimation, which are respectively based on DNN and CNN. A scheme of deep learning-based DoA estimation is depicted in Figure 1. In the presence of multipath signals and mutual couplings, the deep learning network aims to estimate the DoA of the LoS path using the covariance matrix.

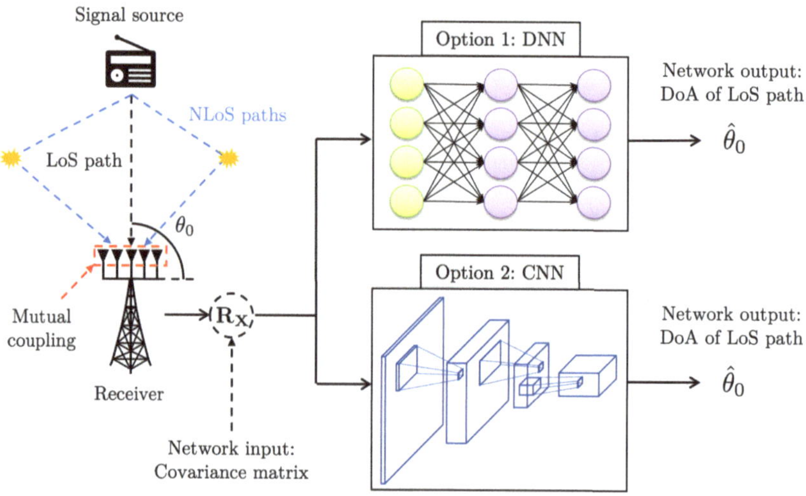

Figure 1. A scheme of deep learning-based DoA estimation. The proposed DNN or CNN structure estimates the DoA of the LoS path in the presence of multipath signals and mutual coupling.

3.1. DoA Estimation via Deep Neural Network

Since the input of the DNN has to be a real vector, the input of the DNN $\chi_{\text{DNN}} \in \mathbb{R}^{2M^2 \times 1}$ is formulated as:

$$\chi_{\text{DNN}} = \text{vec}\left(\left[\text{real}\left(\frac{\mathbf{R_X}}{\|\mathbf{R_X}\|_\text{F}}\right), \text{imag}\left(\frac{\mathbf{R_X}}{\|\mathbf{R_X}\|_\text{F}}\right)\right]\right), \tag{4}$$

where vec$(\cdot)$ denotes the vectorization. real$(\cdot)$ and imag$(\cdot)$ respectively denote the real and imaginary values of the argument. $\|\cdot\|_\text{F}$ denotes the Frobenius norm.

Letting L, $\mathbf{d}^{(l)}$, and $I^{(l)}$ respectively denote the number of dense layers, the output of the l-th layer, and the size of $\mathbf{d}^{(l)}$, $\mathbf{d}^{(l)}(j)$ can be given by:

$$\mathbf{d}^{(l)}(j) = \text{ReLU}\left(\sum_{i=1}^{I^{(l-1)}} \mathbf{U}^{(l)}(j,i)\mathbf{d}^{(l-1)}(j) + \mathbf{v}^{(l)}(j)\right), \quad l = 1, \ldots, L, \tag{5}$$

where $\mathbf{d}^{(0)} = \chi_{\text{DNN}}$. ReLU$(\cdot)$ denotes the ReLU function, which is widely used as the activation function of the neuron [15]. $\mathbf{U}^{(l)}$ and $\mathbf{v}^{(l)}$ denote the weights and the bias of the l-th layers. The loss function of the DNN is given by MSE, $\left(\hat{\theta}_0 - \theta_0\right)^2$, where $\hat{\theta}_0$ denotes the estimated value of the DoA of the LoS path. The weights and the bias that minimize the loss function can be denoted as:

$$\left\{\hat{\mathbf{U}}^{(l)}, \hat{\mathbf{v}}^{(l)}\right\}_{l=1}^{L} = \underset{\left\{\mathbf{U}^{(l)}, \mathbf{v}^{(l)}\right\}_{l=1}^{L}}{\text{argmin}} \left(\hat{\theta}_0 - \theta_0\right)^2, \tag{6}$$

where $\hat{\mathbf{U}}^{(l)}$ and $\hat{\mathbf{v}}^{(l)}$ denote the weights and the bias of the l-th dense layer that minimize the loss function. (6) is implemented by back propagation. A parameter setting for the DNN structure used in this paper is summarized in Table 1.

Table 1. Parameter setting for DNN structure.

Parameter	Value (or Type)
Number of layers (L)	2
Size of layers ($I^{(l)}$ for $l = 1, \ldots, L$)	600, 600
Loss function	MSE
Optimizer	Adam
Activation function	ReLU
Batch size	100

3.2. DoA Estimation via Convolutional Neural Network

Since the input of the CNN can be a three-dimensional matrix, the input of the CNN $\chi_{\text{CNN}} \in \mathbb{R}^{M \times M \times 2}$ is formulated as:

$$\chi_{\text{CNN}} = \left[\text{real}\left(\frac{\mathbf{R_X}}{\|\mathbf{R_X}\|_F}\right); \text{imag}\left(\frac{\mathbf{R_X}}{\|\mathbf{R_X}\|_F}\right)\right], \tag{7}$$

where ; denotes an operator that overlaps matrices with the same dimension.

An output of the k-th convolutional layer, $\mathbf{C}^{(k)}$, can be represented as in [18]:

$$\begin{aligned} \mathbf{C}^{(k)}(:,:,j) &= \text{ReLU}\left(\mathbf{W}_j^{(k)} * \mathbf{C}^{(k-1)} + \mathbf{B}_j^{(k)}\right), \\ &\text{for } j = 1, \ldots, J(k),\ k = 1, \ldots, K, \end{aligned} \tag{8}$$

where $*$ denotes the convolution. $\mathbf{C}^{(k)}(:,:,j)$ denotes the j-th channel of 3D tensor $\mathbf{C}^{(k)}$ and $\mathbf{C}^{(0)} = \chi_{\text{CNN}}$. $J(k)$ and K are the number of kernels and the number of convolutional layers. $\mathbf{W}_j^{(k)} \in \mathbb{R}^{Q_k \times Q_k}$ denotes the j-th kernel in the k-th layer, where Q_k is a dimension of the kernels in the k-th layer. $\mathbf{B}_j^{(k)}$ denotes the bias for the j-th kernel in the k-th convolutional layer.

After undergoing K convolutional layers, all values of $\mathbf{C}^{(K)}$ are summed to yield the output. The loss function of the CNN is given by MSE, $(\hat{\theta}_0 - \theta_0)^2$. The convolution kernel and the bias that minimize the loss function can be given by:

$$\left\{\hat{\mathbf{W}}_j^{(k)}, \hat{\mathbf{B}}_j^{(k)}\right\} = \underset{\left\{\mathbf{W}_j^{(k)}, \mathbf{B}_j^{(k)}\right\}}{\text{argmin}} \ (\hat{\theta}_0 - \theta_0)^2, \text{ for } j = 1, \ldots, J(k),\ k = 1, \ldots, K, \tag{9}$$

where $\hat{\mathbf{W}}_j^{(k)}$ and $\hat{\mathbf{B}}_j^{(k)}$ denote the j-th convolution kernel and the bias of the k-th layer that minimize the loss function. (9) is implemented by back propagation. A parameter setting for CNN structure used in this paper is summarized in Table 2.

Table 2. Parameter setting for CNN structure.

Parameter	Value (or Type)
Number of layers (K)	3
Number of kernels ($J(k)$ for $k = 1, \dots, K$)	$50, 150, 300$
Size of kernels (Q_k for $k = 1, \dots, K$)	$3, 2, 1$
Loss function	MSE
Optimizer	Adam
Activation function	ReLU
Batch size	100

4. Experimental Results and Discussion

4.1. Experimental Setup

In this paper, we use two types of data. One is synthesized data generated based on the signal model in (2). We generated 4,000,000 synthesized data via MATLAB. Here, M, D, and the maximum mutual coupling strength were respectively set to 4, 512, and 0.05. α_0 was fixed to 1. P was randomly set between 0 and 10, and α_p was randomly set between 0 and 0.5. The signal-to-noise ratio (SNR) of synthesized data was also randomly set between 0 dB and 20 dB, where the SNR is defined as $10\log(\sigma_s^2/\sigma^2)$ [dB].

Another type of data is that collected with USRP. Note that this data may differ from the signal model in (2). If so, the estimation is expected to be inaccurate when the network is trained with synthesized data. Figure 2 shows a transmitter and a receiver used for the experiment. The transmitter mainly consists of USRP 2954R and the transmitting antenna. The receiver mainly consists of USRP 2955 and a receiving antenna array. Although USRP 2954R and USRP 2955 support a frequency range of 10 MHz–6 GHz, we set the carrier frequency to 5.8 GHz, which is the center frequency of antennas. For this reason, the spacing between patch antennas in the array is designed to the half-wavelength of 5.8 GHz. USRP 2954R generates a 5.8 GHz cosine wave, and the transmitting antenna emits the wave. Then, USRP 2955 receives the cosine wave via the antenna array at a sampling rate of 1 MHz. By using GNU radio, the covariance matrices of the received signals are collected with USRP 2955.

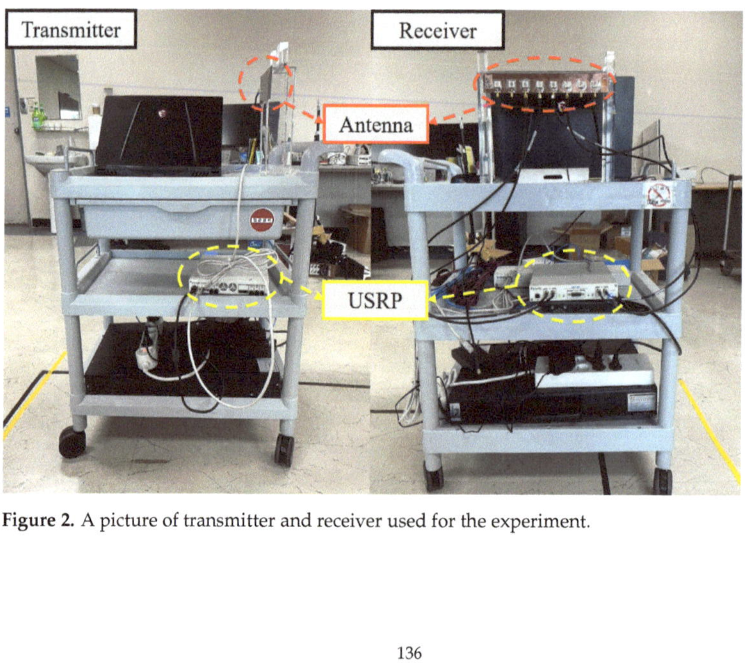

Figure 2. A picture of transmitter and receiver used for the experiment.

As shown in Figure 3, we data in two different environments, the indoor hallway and the outdoor parking lot. In the indoor hallway, the NLoS signals were expected to be stronger than those in the outdoor parking lot. For this reason, the DoA estimation was expected to be inaccurate in the indoor hallway. DoA estimation range is restricted to $[40°, 140°]$ since the radiation pattern of each patch antenna in an array is directional. From $40°$ to $140°$ in $10°$ increments, a total of 17,600 covariance matrices were collected, where half of the data were collected in the indoor hallway while the other half weere collected in the outdoor parking lot. During the experiment, the transmitting power was set to 20 dBm, and the distance between transmitter and receiver was fixed to 6 m. With the collected data, the DoA estimation accuracy is analyzed in the following subsection.

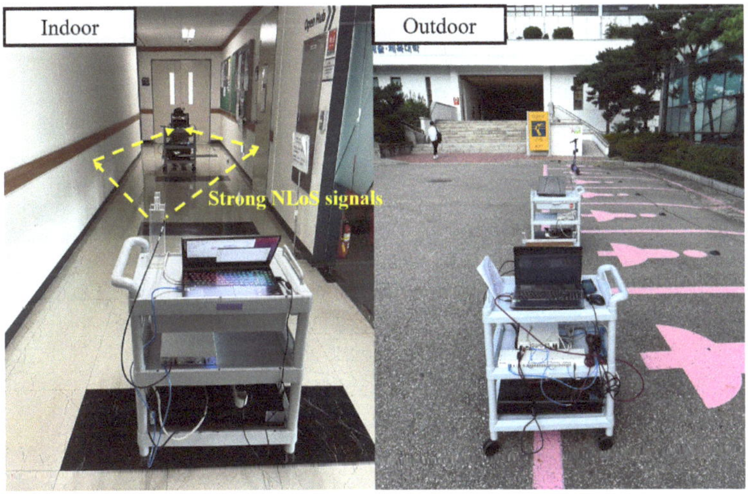

Figure 3. A picture of the indoor hallway and outdoor parking lot. The NLoS signals were expected to be strong in the indoor hallway.

4.2. Peformance Analysis and Discussion

Before analyzing the DoA estimation performance, we checked the similarity between collected data and synthesized data. Since the DoAs of the multipath signals weere not measured, we compared the collected covariance matrices with ideal covariance matrices. Ideal covariance matrices are the covariance matrices calculated without considering multipath signals and other hardware-induced errors. The ideal covariance matrix is defined as:

$$\mathbf{R}_{\text{ideal}}(\theta) = \mathbf{a}(\theta)\mathbf{a}(\theta)^H \in \mathbb{C}^{M\times M},\ \theta \in \Theta, \tag{10}$$

where $\mathbf{R}_{\text{ideal}}(\theta)$ denotes an ideal covariance matrix according to θ. Θ is a set consisting of labeled DoAs of collected data, which equals $\{40°, 50°, 60°, 70°, 80°, 90°, 100°, 110°, 120°, 130°, 140°\}$.

The correlation between collected covariance matrices and ideal covariance matrices is defined as:

$$\rho(\theta) = \mathbb{E}\left[\frac{\langle \mathbf{R}_{\text{col}}(\theta), \mathbf{R}_{\text{ideal}}(\theta)\rangle}{||\mathbf{R}_{\text{col}}(\theta)||_{\text{F}}||\mathbf{R}_{\text{ideal}}(\theta)||_{\text{F}}}\right],\ \theta \in \Theta, \tag{11}$$

where $\rho(\theta)$ denotes the correlation according to DoA. $\mathbf{R}_{\text{col}}(\theta)$ denotes the collected covariance matrix whose DoA label is θ. $\langle \mathbf{A}, \mathbf{B}\rangle$ denotes the correlation between two matrices, which equals $\text{real}(\text{trace}(\mathbf{A}^H\mathbf{B}))$. $\mathbb{E}[\cdot]$ denotes the mean calculated using collected data. $\rho(\theta) \in [0, 1]$, where $\rho(\theta) = 1$ when $\mathbf{R}_{\text{col}}(\theta)$ can be represented as $a\mathbf{R}_{\text{ideal}}(\theta)$. Here, a is a constant. Figure 4 shows the correlation between collected covariance matrices and ideal covariance matrices according to DoA and the experiment environment. As expected, the data collected in the indoor environment has a low correlation since it suffers from strong multipath signals. On the other hand, the data collected in the outdoor environment has a higher correlation since there are fewer objects that can make multipath signals.

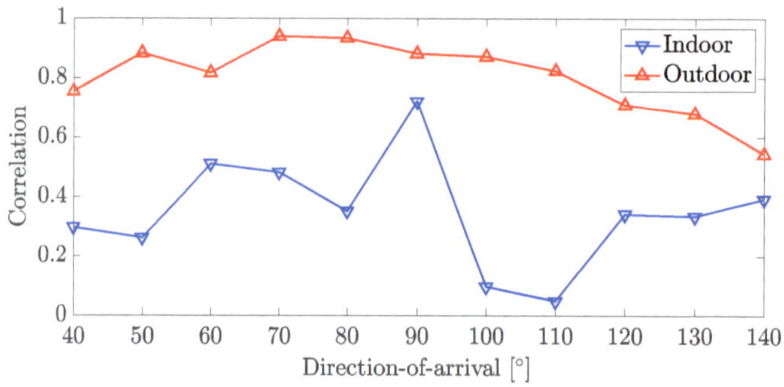

Figure 4. Correlation between collected covariance matrices and ideal covariance matrices according to DoA and experiment environment.

To analyze the DoA estimation performance, we compared five algorithms. One was MUSIC [2]; two were based on DNN and CNN in Section 3.1, and were trained with 4,000,000 synthesized data. The other two algorithms were also based on DNN and CNN in Section 3.1, but they were trained with 75% of the 17,600 collected data points . When using the collected data for training networks, 25% of the 17,600 collected data points were used for testing DoA estimation accuracy. The root mean squared error (RMSE) is defined as $\sqrt{\mathbb{E}\left[(\hat{\theta}_0 - \theta_0)^2\right]}$, where θ_0 and $\hat{\theta}_0$ respectively denote the true DoA and the DoA estimated using the test data.

Figure 5 presents two results for the indoor environment, the RMSE of the DoA estimation algorithms and the histogram of estimation results. Both results are derived based on indoor collected data. As expected, Figure 5a,b show that the estimation accuracy is poor in the indoor environment due to the strong NLoS signals. Furthermore, the results also show that the signal model in (2) fails to capture the indoor propagation characteristics since estimation accuracy decreases when using synthesized data for training. However, the DNN and the CNN trained with collected data show much better performance than others. The RMSE of the DNN and the CNN trained with collected data do not surpass 3.5° in every DoA. To be more specific, Figure 5b shows that estimation results tend to gather around the actual DoA when using collected data. When using synthesized data, however, there is a difference between the mean of estimation results and the actual DoA. Moreover, the variance of the estimation is high.

Figure 6 presents two results for the outdoor environment, the RMSE of the DoA estimation algorithms and the histogram of estimation results. Since the NLoS signals are expected to be much weaker in the outdoor environment, the RMSE of all algorithms are much lower than those in Figure 5a. The RMSE tended to increase when the DoA got far from 90°. We think that this is due to the directivity of the antenna elements. The gain of each antenna element was 5 dBi when the DoA was 90°. However, the gain dropped to 2 dBi when the DoA was 40° or 140°. Overall, the DNN trained with synthesized data were more accurate than MUSIC except in a few DoAs, but the RMSE of the CNN trained with synthesized data was unexpectedly high when the DoA was 40°. Meanwhile, the DNN and the CNN trained with collected data were more accurate than others. To be more specific, when using synthesized data, there was a difference between the mean of estimation results and the actual DoA. However, this difference was smaller than that in Figure 5b. When using collected data, estimation results tended to gather around the actual DoA.

Table 3 shows a total RMSE of all DoA estimation algorithms. In the indoor environment, the algorithms except those using collected data showed poor performance. Among them, the DNN and CNN trained with synthesized data showed slightly better performance than MUSIC. Although the performance of all algorithms improved in the outdoor, the DNN and CNN trained collected data showed much better performance than others. Since the RMSE of the CNN trained with synthesized data soared when the DoA was 40°—its total RMSE was larger than that of MUSIC.

Table 4 shows the training time, computation time, and computational complexity of each algorithm. Here, the computational complexity of CNN is derived using [28]. The training time is proportional to the amount of training data. When training a network with 4,000,000 synthesized data points, it took 3500 and 4400 seconds to train the DNN and CNN. On the other hand, it took 190 and 230 seconds to train the DNN and CNN when using 13,200 collected data points. Although the DNN and CNN-based DoA estimation take a long time for training, their computational complexity is much less than MUSIC once the networks are trained.

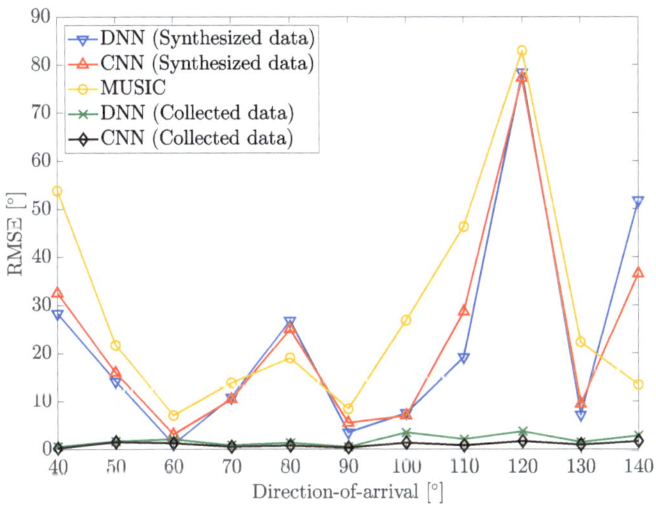

(**a**)

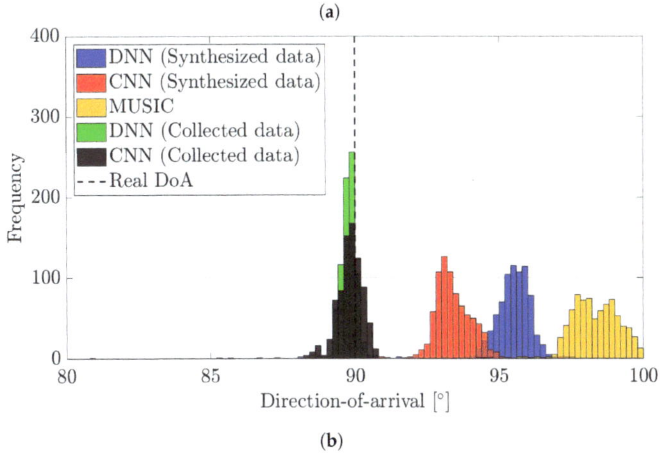

(**b**)

Figure 5. Performance analysis of DoA estimation algorithms in the indoor environment. The first figure shows the RMSE according to DoA, and the second figure is a histogram of estimation results when the actual DoA is 90°. (**a**) RMSE. (**b**) Histogram.

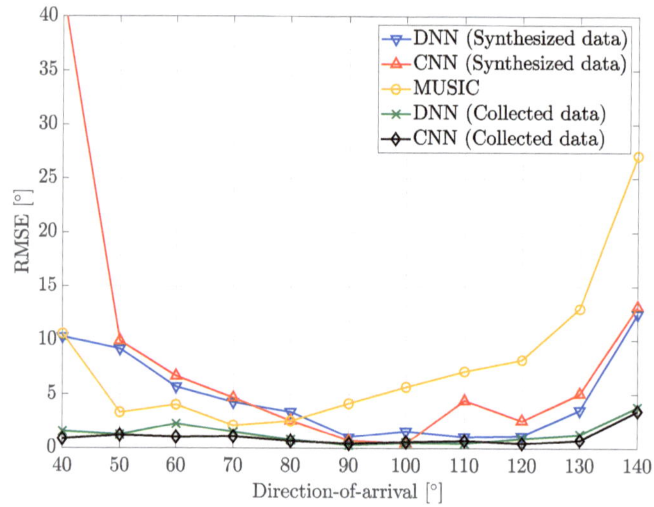

(**a**)

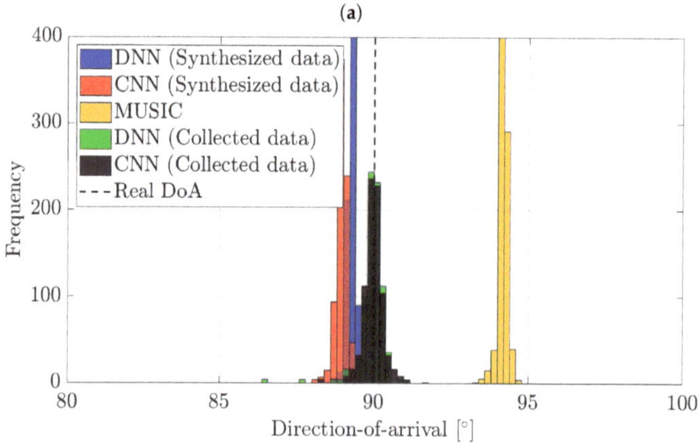

(**b**)

Figure 6. Performance analysis of DoA estimation algorithms in the outdoor environment. The first figure shows the RMSE according to DoA, and the second figure is a histogram of estimation results when the actual DoA is 90°. (**a**) RMSE. (**b**) Histogram.

From all results, we conclude that training with collected data enables accurate DoA estimation. However, collecting sufficient data can be difficult in practice. One of the solutions to this problem is using the synthesized data that well capture the characteristics of the realistic wave. Another solution is to use unsupervised learning such as [20]. Unsupervised learning can make collecting data much easier since data labeling is not required.

Table 3. A total RMSE of DoA estimation algorithms.

	DNN (Synthesized Data)	CNN (Synthesized Data)	MUSIC	DNN (Collected Data)	CNN (Collected Data)
Indoor	31.8°	30.6°	36.2°	2.1°	1.2°
Outdoor	6.2°	14.1°	10.6°	1.6°	1.3°

Table 4. Analysis on training time, computation time, and computational complexity.

	Training Time [s]	Computation Time [μs]	Computational Complexity
DNN (Synthesized data)	3500	33	$\mathcal{O}\left(\left(M^2 + I^{(2)}\right)I^{(1)}\right)$
DNN (Collected data)	190		
CNN (Synthesized data)	4400	35	$\mathcal{O}\left(M^2 \sum_{i=2}^{3} Q_i^2 J(i-1)J(i)\right)$
CNN (Collected data)	230		
MUSIC	-	7161	$\mathcal{O}(M^3)$

5. Conclusions

We present the experimental validation of the deep learning-based DoA estimation using USRP. The DNN and the CNN structures are designed to estimate the DoA of the LoS path with the covariance matrix. In the experiment, two types of data are exploited. One is the data synthesized with the signal model, and the other is the data collected by USRP. The experimental results show that the DoA estimation is most accurate when training DNN and CNN with the collected data. Furthermore, the DoA estimation performance is poor in the indoor environment, which suffers from the strong NLoS signals. However, collecting sufficient data may not be feasible in practice. We expect that this can be resolved by better signal modeling and unsupervised learning.

Author Contributions: Conceptualization, H.C., H.P. and S.K.; methodology, H.C., H.P. and S.K.; software, H.C. and H.P.; validation, H.C., H.P. and S.K.; formal analysis, H.C. and H.P.; investigation, H.C. and H.P.; resources, S.K.; data curation, H.C. and H.P.; writing—original draft preparation, H.C. and S.K.; writing—review and editing, H.C. and S.K.; visualization, H.C. and H.P.; supervision, S.K.; project administration, S.K.; funding acquisition, S.K. All authors have read and agreed to the published version of the manuscript.

Funding: This work was supported by the Future Combat System Network Technology Research Center program of Defense Acquisition Program Administration and Agency for Defense Development (UD190033ED).

Institutional Review Board Statement: Not applicable.

Informed Consent Statement: Not applicable.

Data Availability Statement: Not applicable.

Conflicts of Interest: The authors declare no conflict of interest.

Abbreviations

The following abbreviations are used in this manuscript:

DoA	Direction-of-arrival
USRP	Universal software radio peripheral
DNN	Deep neural network
CNN	Convolution neural network
RNN	Recurrent neural network
ResNet	Residual neural network
MLE	Maximum likelihood estimation
LoS	Line-of-sight
NLoS	Non-line-of-sight
MUSIC	Multiple signal classification
ESPRIT	Estimation of signal parameters via rotational invariance techniques
ULA	Uniform linear array
MSE	Mean squared error
SNR	Signal-to-noise ratio
RMSE	Root mean squared error

References

1. Chen, J.C.; Yao, K.; Hudson, R.E. Source localization and beamforming. *IEEE Signal Process. Mag.* **2002**, *19*, 30–39. [CrossRef]
2. Schmidt, R. Multiple emitter location and signal parameter estimation. *IEEE Trans. Antennas Propag.* **1986**, *34*, 276–280. [CrossRef]
3. Roy, R.; Kailath, T. ESPRIT-estimation of signal parameters via rotational invariance techniques. *IEEE Trans. Acoust. Speech Signal Process.* **1989**, *37*, 984–995. [CrossRef]
4. Du, W.; Kirlin, R. Improved spatial smoothing techniques for DOA estimation of coherent signals. *IEEE Trans. Signal Process.* **1991**, *39*, 1208–1210. [CrossRef]
5. Cao, S.; Ye, Z.; Xu, D.; Xu, X. A Hadamard Product Based Method for DOA Estimation and Gain-Phase Error Calibration. *IEEE Trans. Aerosp. Electron. Syst.* **2013**, *49*, 1224–1233. [CrossRef]
6. Ye, Z.; Dai, J.; Xu, X.; Wu, X. DOA Estimation for Uniform Linear Array with Mutual Coupling. *IEEE Trans. Aerosp. Electron. Syst.* **2009**, *45*, 280–288. [CrossRef]
7. Swindlehurst, A.; Kailath, T. A performance analysis of subspace-based methods in the presence of model errors. I. The MUSIC algorithm. *IEEE Trans. Signal Process.* **1992**, *40*, 1758–1774. [CrossRef]
8. Han, F.M.; Zhang, X.D. An ESPRIT-like algorithm for coherent DOA estimation. *IEEE Antennas Wirel. Propag. Lett.* **2005**, *4*, 443–446. [CrossRef]
9. Cadzow, J.; Kim, Y.S.; Shiue, D.C. General direction-of-arrival estimation: A signal subspace approach. *IEEE Trans. Aerosp. Electron. Syst.* **1989**, *25*, 31–47. [CrossRef]
10. Liu, A.; Liao, G.; Zeng, C.; Yang, Z.; Xu, Q. An Eigenstructure Method for Estimating DOA and Sensor Gain-Phase Errors. *IEEE Trans. Signal Process.* **2011**, *59*, 5944–5956. [CrossRef]
11. Chen, P.; Chen, Z.; Cao, Z.; Wang, X. A New Atomic Norm for DOA Estimation With Gain-Phase Errors. *IEEE Trans. Signal Process.* **2020**, *68*, 4293–4306. [CrossRef]
12. Liao, B.; Zhang, Z.G.; Chan, S.C. DOA Estimation and Tracking of ULAs with Mutual Coupling. *IEEE Trans. Aerosp. Electron. Syst.* **2012**, *48*, 891–905. [CrossRef]
13. Lin, M.; Yang, L. Blind Calibration and DOA Estimation With Uniform Circular Arrays in the Presence of Mutual Coupling. *IEEE Antennas Wirel. Propag. Lett.* **2006**, *5*, 315–318. [CrossRef]
14. Kintz, A.L.; Gupta, I.J. A Modified MUSIC Algorithm for Direction of Arrival Estimation in the Presence of Antenna Array Manifold Mismatch. *IEEE Trans. Antennas Propag.* **2016**, *64*, 4836–4847. [CrossRef]
15. Schmidhuber, J. Deep Learning in neural networks: An Overview. *Neural Netw.* **2015**, *61*, 85–117. [CrossRef]
16. Huang, H.; Yang, J.; Huang, H.; Song, Y.; Gui, G. Deep Learning for Super-Resolution Channel Estimation and DOA Estimation Based Massive MIMO System. *IEEE Trans. Veh. Technol.* **2018**, *67*, 8549–8560. [CrossRef]
17. Liu, Z.; Zhang, C.; Yu, P.S. Direction-of-Arrival Estimation Based on Deep Neural Networks With Robustness to Array Imperfections. *IEEE Trans. Antennas Propag.* **2018**, *66*, 7315–7327. [CrossRef]
18. Wu, L.; Liu, Z.; Huang, Z. Deep Convolution Network for Direction of Arrival Estimation With Sparse Prior. *IEEE Signal Process. Lett.* **2019**, *26*, 1688–1692. [CrossRef]
19. Papageorgiou, G.K.; Sellathurai, M.; Eldar, Y.C. Deep Networks for Direction-of-Arrival Estimation in Low SNR. *IEEE Trans. Signal Process.* **2021**, *69*, 3714–3729. [CrossRef]
20. Yuan, Y.; Wu, S.; Wu, M.; Yuan, N. Unsupervised Learning Strategy for Direction-of-Arrival Estimation Network. *IEEE Signal Process. Lett.* **2021**, *28*, 1450–1454. [CrossRef]
21. Mei, R.; Tian, Y.; Huang, Y.; Wang, Z. 2D-DOA Estimation in Switching UCA Using Deep Learning-Based Covariance Matrix Completion. *Sensors* **2022**, *22*, 3754. [CrossRef] [PubMed]
22. Wu, X.; Yang, X.; Jia, X.; Tian, F. A Gridless DOA Estimation Method Based on Convolutional Neural Network With Toeplitz Prior. *IEEE Signal Process. Lett.* **2022**, *29*, 1247–1251. [CrossRef]
23. Merkofer, J.P.; Revach, G.; Shlezinger, N.; van Sloun, R.J.G. Deep Augmented MUSIC Algorithm for Data-Driven DoA Estimation. In Proceedings of the ICASSP 2022—2022 IEEE International Conference on Acoustics, Speech and Signal Processing (ICASSP), Singapore, 23–27 May 2022; pp. 3598–3602. [CrossRef]
24. de Oliveira, M.L.L.; Bekooij, M.J.G. Deep-MLE: Fusion between a Neural Network and MLE for A Single Snapshot DOA Estimation. In Proceedings of the ICASSP 2022—2022 IEEE International Conference on Acoustics, Speech and Signal Processing (ICASSP), Singapore, 23–27 May 2022; pp. 3673–3677. [CrossRef]
25. Forster, P.; Larzabal, P.; Boyer, E. Threshold performance analysis of maximum likelihood DOA estimation. *IEEE Trans. Signal Process.* **2004**, *52*, 3183–3191. [CrossRef]
26. Ettus Research. gr-doa. 2019. Available online: https://github.com/EttusResearch/gr-doa (accessed on 26 September 2022).
27. Friedlander, B.; Weiss, A. Direction finding in the presence of mutual coupling. *IEEE Trans. Antennasm Propag.* **1991**, *39*, 273–284. [CrossRef]
28. Maji, P.; Mullins, R. On the Reduction of Computational Complexity of Deep Convolutional Neural Networks. *Entropy* **2018**, *20*, 305. [CrossRef] [PubMed]

sensors

MDPI

Article

Coordinated Positioning Method for Shortwave Anti-Multipath Based on Bayesian Estimation

Tao Tang, Linqiang Jiang *, Paihang Zhao and Na-e Zheng

Institute of Information Engineering, PLA Strategic Support Force Information Engineering University, Zhengzhou 450001, China
* Correspondence: 13253566023@163.com

Abstract: Coordinated positioning based on direction of arrival (DOA)–time difference of arrival (TDOA) is a research area of great interest in beyond-visual-range target positioning with shortwave. The DOA estimation accuracy greatly affects the accuracy of coordinated positioning. With existing positioning methods, the elevation angle's estimation accuracy in multipath propagation decreases sharply. Accordingly, the positioning accuracy also decreases. In this paper, the elevation angle is modeled as a random variable, with its probability distribution reflecting the characteristics of multipath propagation. A new coordinated positioning method based on DOA–TDOA and Bayesian estimation with shortwave anti-multipath is proposed. First, a convolutional neural network is used to learn the three-dimensional spatial spectrogram to make an intelligent decision on the number of single and multiple paths, and to obtain a probability distribution of the elevation angle under multiple paths. Second, the elevation angle's estimated value is modified using the elevation angle's probability distribution. The modified elevation angle's estimated value is substituted into a DOA pseudo-linear observation equation, and the target position's estimated value is obtained using the matrix QR decomposition iteration algorithm. Finally, a TDOA pseudo-linear observation equation is established using the target estimate obtained in the DOA stage, and the coordinated positioning result is obtained using the matrix QR decomposition iteration algorithm again. Simulation results demonstrated that the proposed method had a stronger anti-multipath capability than traditional methods, and it improved the coordinated positioning accuracy of the DOA and TDOA. Measured data were used to validate the proposed method.

Keywords: shortwave; coordinated positioning; anti-multipath; prior information; deep learning

1. Introduction

Shortwave communication is an important means of long-distance communication and it has the characteristics of high destructibility and strong mobility. Shortwave signal source positioning has wide applications in both the military and civilian fields [1,2]. Target positioning based on position parameters is the main method for shortwave positioning. The position parameters mainly include direction of arrival (DOA), time of arrival, time difference of arrival (TDOA), and frequency difference of arrival. Usually, a single positioning parameter can be used to estimate the target position; however, a combination of multiple parameters can effectively improve the positioning accuracy [3]. DOA positioning and TDOA positioning are the two most representative shortwave positioning approaches. A combination of DOA and TDOA can significantly improve the accuracy of shortwave positioning.

For propagation within the visual range, researchers have proposed many algorithms for DOA positioning [4–6], TDOA positioning [7–9], and DOA–TDOA coordinated positioning [10–12]. However, these algorithms cannot be used directly in beyond-visual-range (BVR) shortwave communication. At present, there are few positioning algorithms for BVR shortwave targets. In [13], the authors proposed a TDOA positioning method for BVR shortwave targets based on a grid search. Their method has high positioning accuracy; however, it is a complex computational process. In [14], the authors proposed a TDOA positioning algorithm with gradient projection based on an ionospheric quasi-parabolic (QP) model. However, using the QP model requires many ionospheric parameters, which are difficult to obtain in practice. Moreover, the QP model has the limitation of only considering specific ionospheric reflections. Compared with the QP model, the ionospheric virtual height reflection model [15,16] only requires the ionospheric reflection virtual height as an input parameter, and different reflection virtual heights correspond to different ionospheric reflections. In [17], the authors proposed a method to conduct pseudo-linearization on the DOA and TDOA observation equations based on an ionospheric virtual height reflection model. They obtained solutions using an iterative matrix QR decomposition algorithm with simple calculations. However, this method only considers a single transmission path without considering the shortwave multipath effect in the practical process. In [18], the authors proposed a DOA–TDOA coordinated positioning method based on an ionospheric virtual height reflection model using two propagation paths without the need for a known ionospheric virtual height. However, this method requires a large amount of computation and requires the accurate arrival elevation angles of two propagation paths, which are difficult to achieve in practice.

The multipath effect under the ionospheric influence leads to difficulties in shortwave positioning, which has a great impact on the estimation of positioning parameters such as the arrival angle, and thus, the positioning accuracy. In actual shortwave positioning, the measurement error of the azimuth angle can reach about 3°, while the measurement error of the elevation angle can reach about 5–10°. Therefore, the elevation angle is not usually used for positioning. At present, data fusion is used to handle multipath data after the signal is received in the main methods designed for multipath scenarios [16,19,20]. However, these cannot be applied directly to the DOA or TDOA positioning scenarios in this study.

To summarize, in this paper, shortwave anti-multipath coordinated positioning with DOA–TDOA is addressed. To accurately reflect the characteristics of ionospheric reflection, the arrival elevation angle was simulated as a random variable with a Gaussian mixture model (GMM). First, a convolutional neural network (CNN) was used to learn the three-dimensional spatial spectrogram to identify the modes of single-path propagation and multipath propagation. The corresponding probability distribution of the elevation angle was obtained and used to modify the estimated value of the elevation angle. Then, a two-dimensional DOA pseudo-linear observation equation was established based on the azimuth estimate and the elevation angle's modified estimate. The matrix QR decomposition iteration algorithm proposed in [17] was used to obtain the DOA positioning results. Finally, a TDOA pseudo-linear observation equation was established based on the DOA positioning results, and the final positioning results were obtained using the matrix QR decomposition iteration algorithm again.

The remainder of this paper is as follows: Section 2 introduces the impact caused by the multipath problem and the observation model. Section 3 proposes the improved method. Section 4 shows the results of the simulation experiments. Section 5 shows the results of the measured data, and Section 6 summarizes the paper.

2. Problem Statement and Observation Models

2.1. Multipath Problem

The ionospheric signal channel varies randomly with space, time, and frequency, which has an impact on transmitted signals such as multipath fading and polarization fading. The electromagnetic wave incident in the ionosphere is divided into an ordinary (O) mode and unusual (X) mode. These two modes correspond to different paths and generate multipath fading. Ionospheric passive oblique detection technology can be used to obtain the ionospheric information regarding the reflection points between the transmitting and receiving stations. The high frequency radio signal is transmitted by the transmitting station, and the delay in the propagation time is calculated by the signal received by the receiving station, so as to infer the characteristics of the ionosphere. Figure 1 shows the result of passive oblique detection. The abscissa in Figure 1 represents frequency and the ordinate represents the group path. The colors indicate the electron concentrations. It can be observed that when the frequency is between 10 and 12 MHz, there are two propagation modes that correspond to the O and X waves. Therefore, when the communication frequency falls within this frequency range, there are multiple propagation paths that affect signal reception.

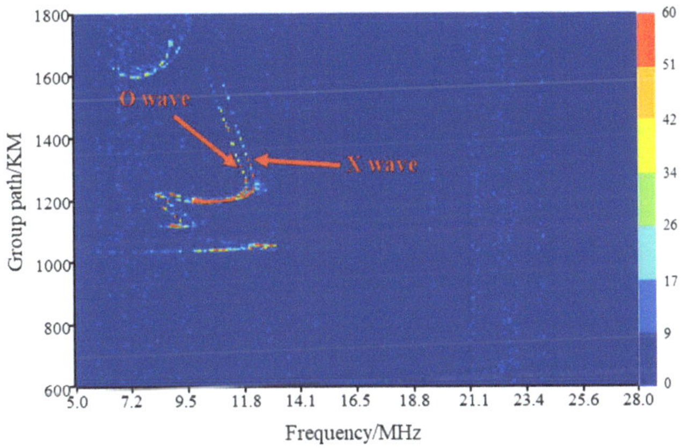

Figure 1. An example of an ionization map in ionospheric passive oblique detection.

Figure 2 shows a geometric illustration of multipath propagation. The electromagnetic wave reaches the receiving station after it is transmitted through the ionosphere. Because of the layered structure of the ionosphere, each layer corresponds to a different reflection height. The electromagnetic wave can reach the receiving station through different reflection heights that correspond to different paths. The elevation angles of the signals that reach the receiving station through different paths are different [15]. Target positioning under two paths was investigated in this study.

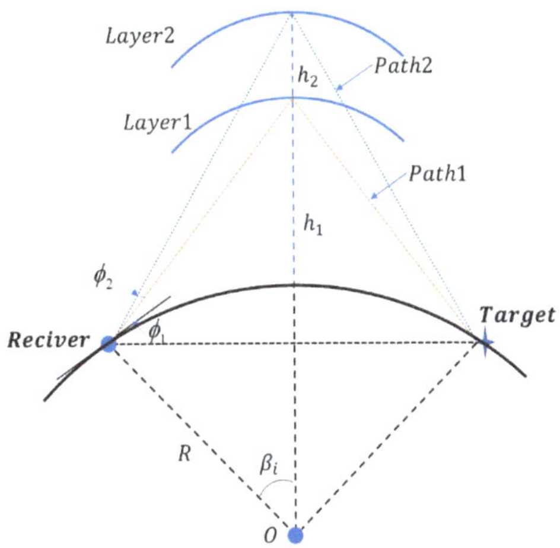

Figure 2. Geometric illustration of multipath propagation.

Multipath propagation in the spatial spectrogram is manifested by multiple spectral peaks at the same azimuth. Therefore, angle estimation ambiguity may occur [16]. Figure 3 shows spatial spectrogram examples of a single path and two paths. In the case of two paths, the spectral peaks are formed at two elevation angles, and the two spectral peaks are not easy to separate. Therefore, the angle estimation accuracy is affected.

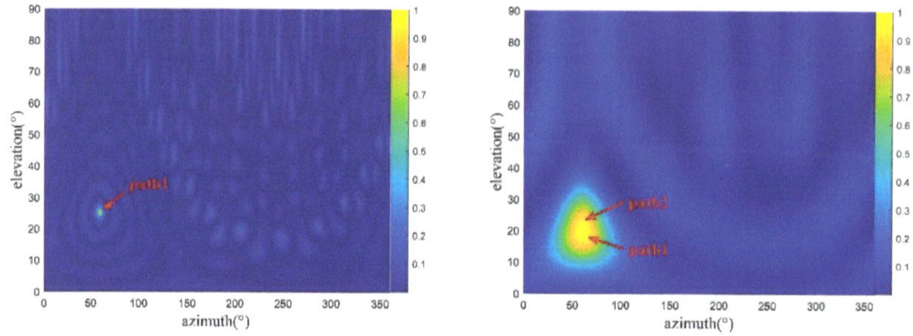

(a) (b)

Figure 3. Spatial spectrogram examples of a single path (**a**) and two paths (**b**) (top view).

Remark 1. *Only one single path and two paths are taken as examples to illustrate the influence of multipath on the spatial spectrogram. When the number of paths increases, the spectral peak of spatial spectrum will increase correspondingly; however, the case for a larger number of paths is very rare [20].*

When the fading amplitude follows the Rayleigh distribution, the probability distribution of the multipath time delay follows a normal distribution [21]. Inspired by this phenomenon, in this study, the authors modeled the single-path arrival elevation angle as a Gaussian distribution, with the true value as the mean. Figure 2 shows that the reflection height corresponds to the elevation angle. Because the ionosphere is time varying, the reflection height also varies, which results in a change in the elevation angle. Therefore, a Gaussian distribution model of the elevation angle can reflect the ionospheric change. When the elevation angles under the two paths are simulated as a GMM, their mean values are the true values at the corresponding reflection heights. The multipath elevation angles are modeled as a GMM to determine target positioning with multipath.

2.2. Array Receiving Signal Model

For an assumed circular array with p elements, there are q signals that have incident angles $\Theta_1, \Theta_2, \ldots, \Theta_q$ in the array, where $\Theta_i = (\theta_i, \phi_i)$, and θ_i Lining ϕ_i are the arrival angle and elevation angle of the ith signal, respectively. The arrival azimuth refers to the angle between the incident direction of the signal and the north direction of the local observation station. The elevation angle refers to the angle between the incident direction of the signal and the surface plane of the observation station. The output of the kth element in the antenna array can be expressed as

$$\boldsymbol{x}_k(t) = \sum_{i=1}^{q} g_{ki}\boldsymbol{s}_i(t-\tau_{ki}) + \boldsymbol{n}_k(t) \tag{1}$$

where $\boldsymbol{s}_i(t)$ denotes the ith incident signal in the array, g_{ki} denotes the complex gain of the kth array element in the ith signal array, $\boldsymbol{n}_k(t)$ denotes the additive noise in the kth array element, and τ_{ki} denotes the time delay of the signal reaching the array element relative to the reference point.

It is assumed that the array elements are isotropic and there are no impacts, such as channel inconsistency and mutual coupling. Therefore, $g_{ki} = 1$, and the received signal model of the array is given by

$$\mathbf{X} = \mathbf{A}(\Theta)\mathbf{S} + \mathbf{N} \tag{2}$$

In Equation (2), $\mathbf{X} = [x_1(t), \ldots, x_p(t)]^T \in \mathbf{C}^{p\times 1}$ is the array output vector. $\mathbf{N} = [n_1(t), \ldots, n_p(t)]^T \in \mathbf{C}^{p\times 1}$ is the additive noise vector of the array. $\mathbf{S} = [s_1(t), \ldots, s_q(t)]^T \in \mathbf{C}^{q\times 1}$ is the signal source vector. $\mathbf{A}(\Theta) = [\boldsymbol{a}(\Theta_1), \ldots, \boldsymbol{a}(\Theta_q)] \in \mathbf{C}^{p\times q}$ is an array flow pattern matrix, where $\boldsymbol{a}(\Theta_i) = [e^{-j\omega_o\tau_{1i}}, \ldots, e^{-j\omega_o\tau_{1p}}]^T \in \mathbf{C}^{p\times 1}$ is the array direction vector, and ω_o is the carrier frequency of the signal.

Remark 2. *$\mathbf{A}$ is the array manifold matrix, which is related to the shape of the array and the direction of the signal. In practice, the shape of the antenna array will not change once it is fixed, so $\mathbf{A}$ is closely related to the direction of the signal. In practice, the self-adjustment of the antenna array can ensure $\mathbf{A}$ can be used for angle estimation, and there will be no ill-conditioned matrix. $\mathbf{N}$ stands for additive noise. The larger it is, the lower the accuracy of angle estimation. When it reaches a certain level, angle estimation cannot be carried out. In addition, $\mathbf{N}$ does not have a substantial impact on the structure of matrix $\mathbf{A}$, and $\mathbf{N}$ only has an impact on the estimation accuracy.*

Under the narrowband model defined in Equation (2), the DOA estimation based on the multiple signal classification (MUSIC) algorithm is given by

$$\Theta_{MUSIC} = \underset{\Theta}{\operatorname{argmax}}(1/(\boldsymbol{a}^H(\Theta)\mathbf{U}_N\mathbf{U}_N^H\boldsymbol{a}(\Theta))) \tag{3}$$

In Equation (3), $\mathbf{U}_N \in \mathbf{C}^{p\times(p-q)}$ is the noise subspace in the MUSIC algorithm.

2.3. MUSIC Algorithm

The MUSIC algorithm is a classical direction of arrival estimation algorithm, and the main steps are as follows:

(1) Calculate the covariance matrix of array output data based on Equation (2);
(2) Eigenvalue decomposition is performed on the covariance matrix obtained in step (1) to obtain the signal subspace $\mathbf{U}_S$ and noise subspace $\mathbf{U}_N$;
(3) Search the angle corresponding to the maximum value of Equation (3), which is the angle estimate value.

2.4. Positioning Solution Algorithm

The positioning scenario in this study is the same as that in [16], with N DOA positioning observation stations and M TDOA positioning observation stations, which are denoted as $u_1, u_2, \ldots, u_N, u_{N+1}, \ldots, u_{N+M}$.

The positioning algorithm is divided into two stages. In the first stage, the pseudo-linear observation equations of the azimuth and the elevation angle are established, as shown in Equation (4). According to the geographical constraints of the earth surface and the algebraic relationship between the auxiliary variables $||\boldsymbol{u}||_2^2$ and the target position $\boldsymbol{u} \in \mathbf{R}^{3\times 1}$, the equality constraint shown in Equation (5) can be obtained. $\boldsymbol{u}$ is the target position vector in the earth-centered earth-fixed coordinate system. The estimation criterion with a double quadratic equality can be obtained by combining Equations (4) and (5). Then, the DOA stage estimate can be obtained using the matrix QR decomposition iteration algorithm [16] as follows:

$$\begin{aligned} &\mathbf{B}_\theta \boldsymbol{u} = \boldsymbol{b}_\theta \\ &\mathbf{B}_\phi \begin{bmatrix} \boldsymbol{u} \\ ||\boldsymbol{u}||_2^2 \end{bmatrix} = \boldsymbol{b}_\phi \end{aligned} \tag{4}$$

$$\begin{cases} \boldsymbol{t}_u^{\mathrm{T}}\mathbf{A}_1\boldsymbol{t}_u^{\mathrm{T}} = R_e^2 \\ \boldsymbol{t}_u^{\mathrm{T}}\mathbf{A}_2\boldsymbol{t}_u^{\mathrm{T}} + \boldsymbol{c}_1^{\mathrm{T}}\boldsymbol{t}_u = 0 \end{cases} \tag{5}$$

where $\mathbf{B}_\theta \in \mathbf{R}^{N\times 3}$, $\mathbf{B}_\phi \in \mathbf{R}^{N\times 4}$, $t_u \in \mathbf{R}^{4\times 1}$ and R_e is the equatorial radius of the earth and e is the first eccentricity of the earth. In Equation (5)

$$\begin{aligned}
&\mathbf{B}_\theta[i,:] = [s_{i,1}\cos(\theta_i) - s_{i,2}\sin(\theta_i)]^{\mathrm{T}}, \boldsymbol{b}_\theta[i,:] = (s_{i,1}\cos(\theta_i) - s_{i,2}\sin(\theta_i))^{\mathrm{T}}\boldsymbol{u}_i \\
&\mathbf{B}_\phi[i,:] = [2\boldsymbol{u}_i^{\mathrm{T}}, -1]^{\mathrm{T}}, \boldsymbol{b}_\phi = ||\boldsymbol{u}_i||_2^2 - ||\boldsymbol{u} - \boldsymbol{u}_i||_2^2, i = 1,2,\ldots,N \\
&\mathbf{A}_1 = diag\{1,1,1/(1-e^2),0\}, \mathbf{A}_2 = diag\{1,1,1,0\} \\
&c_1 = [0,0,0,-1]^{\mathrm{T}} \\
&t_u = \begin{bmatrix} \boldsymbol{u} \\ ||\boldsymbol{u}||_2^2 \end{bmatrix}
\end{aligned} \quad (6)$$

$s_{i,2} = [-\cos(\omega_{i,1})\sin(\omega_{i,2}), -\sin(\omega_{i,1})\cos(\omega_{i,2}), \cos(\omega_{i,2})]^{\mathrm{T}}$, $s_{i,1} = [-\sin(\omega_{i,1}), \cos(\omega_{i,2}), 0]^{\mathrm{T}}$, $\omega_{i,1}, \omega_{i,2}$ are the longitude and latitude of the ith observation station, respectively.

In the second stage, the pseudo-linear equation of the time difference is established based on the DOA estimate, as shown in Equation (7). Similarly, Equations (5) and (7) are combined to construct an estimation criterion that contains the double quadratic equality. The matrix QR decomposition iteration algorithm is used to obtain the final positioning results [16].

$$\begin{aligned}
&\mathbf{B}_\tau \begin{bmatrix} \boldsymbol{u} \\ ||\boldsymbol{u}||_2^2 \end{bmatrix} = \boldsymbol{b}_\tau \\
&\mathbf{B}_\tau[m-1.:] = [2a_m^2\boldsymbol{u}_{N+m}^{\mathrm{T}}, -a_m^2], \boldsymbol{b}_\tau[m-1,:] = 4R_0^2\mu_m^2 + a_m^2(||\boldsymbol{u}_{N+m}||_2^2 - 4R_0^2) \\
&m = 2,3,\ldots,M
\end{aligned} \quad (7)$$

where $\mathbf{B}_\tau \in \mathbf{R}^{(M-1)\times 4}$, $a_m = 2R_0(R_0 + h_m)$, μ_m is defined in Equation (73) of [16]. R_0 is the average radius of the earth.

3. Improved Method

3.1. Path Number Discrimination and the Elevation Angle's Prior Distribution Learning Based on the CNN

The CNN is a type of feedforward neural network that contains convolutional computation and a deep structure [22,23]. It is a representative deep learning network and widely used in the field of image processing. Its learning technology has matured gradually. In this study, when the two paths are close to each other, the two spectral peaks are merged into one spectral peak in the spatial spectrogram, as shown in Figure 4. Traditional spectral peak search techniques cannot distinguish a single path from two paths; hence, only one spectral peak can be obtained rather than the elevation angles of two paths.

In this study, the CNN was used to detect and identify spatial spectrograms to discriminate between a single path and multiple paths. In the multipath case, the CNN was used to obtain the elevation angle of each path.

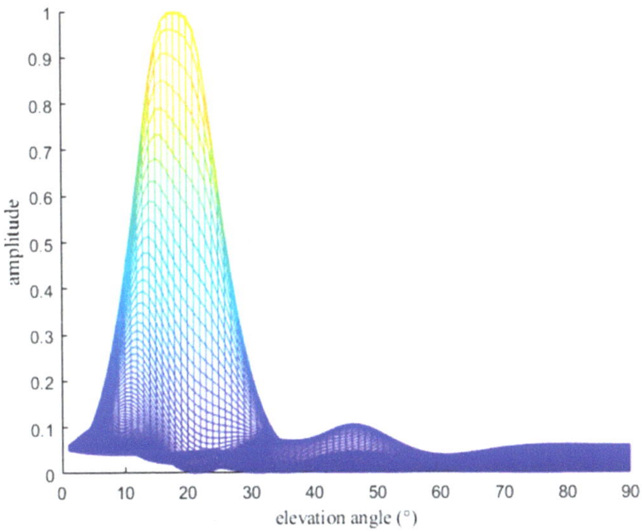

Figure 4. Illustration of spatial spectrograms with elevation angles of 16.4° and 20.5° under two paths (elevation view).

3.1.1. Path Number Discrimination

The problem of path number discrimination based on spatial spectrograms belongs to an image classification problem with supervised learning. In this study, 10,000 spatial spectrogram images with different numbers of paths were generated with an elevation angle range of 5° to 70°. The number ratio between the single path and the two paths was approximately 1:1. The spatial spectrogram contained RGB images of 256 × 256 pixels.

Remark 3. *When there are multiple signals, there are many possible combinations of signals and the number of paths, which will lead to the very complex training process of neural network and a large amount of training data is required. In addition, based on the theory of array signal processing [24], when the arrival direction of multiple signals is different, multiple signals can be reduced into a single signal by dimensionality reduction. Therefore, each signal can be processed in this way. By using this method, the accuracy will not decrease, but the computation can be greatly reduced.*

Remark 4. *The network used in this paper can be used to identify the number of paths under more paths, and only needs to increase the corresponding training data. However, the case for a larger number of paths is very rare [21] so this paper focuses on two common cases: single path and two paths.*

The CNN structure is shown in Figure 5. It contained four convolutional layers and two fully connected layers. The convolutional layer is used to extract spatial spectral features. The fully connected layer plays the role of mapping the learned feature representation to the label space of samples. The activation function was the ReLU function. The loss function was the BCELoss function. To prevent overfitting, an early stop mechanism was used in the training process. If the loss value of the validation set did not decrease for 10 consecutive times, the training was stopped.

The loss curves of the training set and validation set are shown in Figure 6. The loss values of the training and validation sets maintained the same decreasing trend, and there was no overfitting during the training process. A total of 2000 spatial spectrograms were used as the test set. The accuracy of path number discrimination was 99.8%. Therefore, we can use the trained network to determine the number of paths.

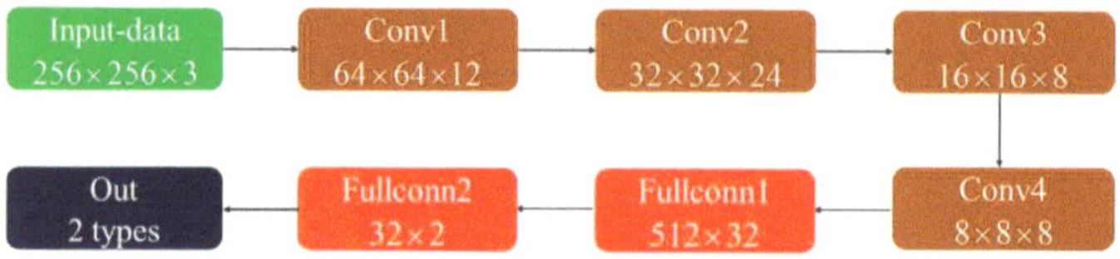

Figure 5. Structure of the CNN for path number discrimination.

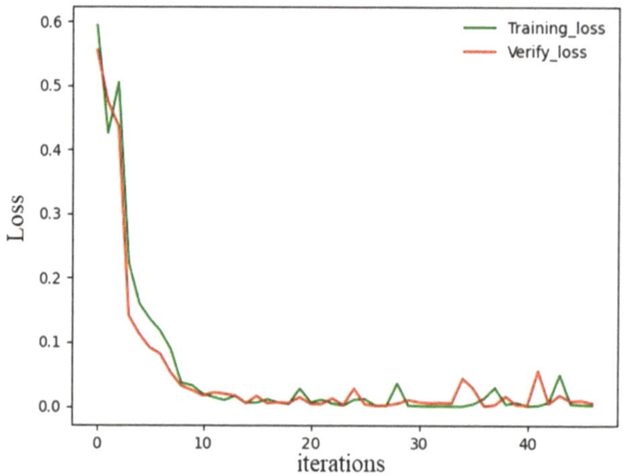

Figure 6. Loss curve of the path number discrimination model.

3.1.2. Search for Multipath Spectral Peaks

After the number of paths was determined, the search for the spectral peak in the spatial spectrogram was performed using a regression method based on the CNN. The regression network model structure is shown in Figure 7. The convolutional layer dimension was added in the model structure based on the path number discrimination model to improve the characteristic extraction capability of the model. The model output was modified from the classification result to the regression result of the elevation angle. The loss function was the MSELoss function. A total of 10,000 spatial spectral spectrogram images of the two paths were generated as the training data with an elevation angle range of 5° to 70°. The early stop mechanism was the same as the network shown in Figure 5. The loss curve in the training process is shown in Figure 8. The decreasing trend of the loss value in the training set was consistent with that in the validation set. There was no overfitting in the training process. A total of 1000 randomly generated spatial spectrogram images were used as the test set. When the angle error was within 1.5°, the prediction accuracy was acceptable. The prediction accuracy rate of the elevation angle is defined as the ratio of the number of correctly predicted spatial spectral spectrograms to the number of total spatial spectral spectrograms. It was 94.75%, which demonstrated satisfactory model performance that met the expected requirements.

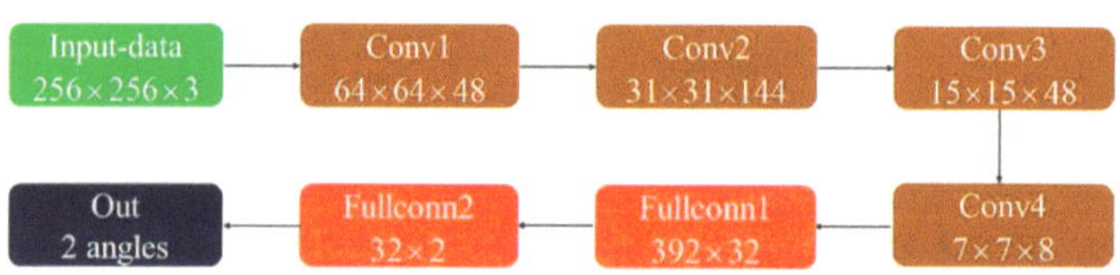

Figure 7. CNN structure used in the elevation angle calculation.

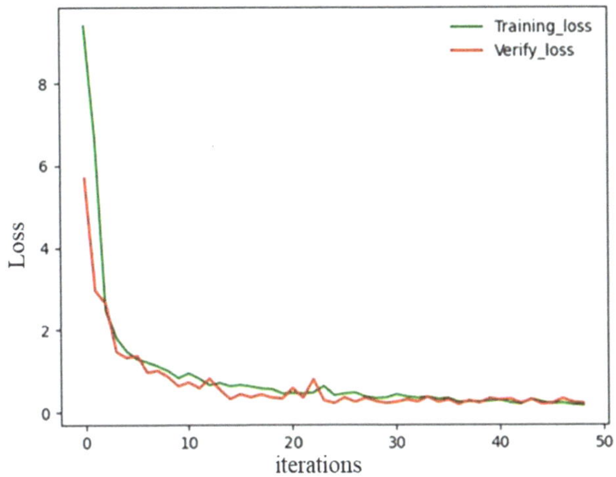

Figure 8. Loss curve of the elevation angle calculation model.

3.1.3. Elevation Angle's Prior Distribution Learning

A simulation was conducted to generate 10,000 spatial spectrogram images that included a single path and two paths for a specific region. The elevation angles of the two paths followed a GMM. First, the number of paths was determined by the path discrimination model to obtain the numbers of single-path and two-path spectrograms, which were 3823 and 6177, respectively. For the single-path spatial spectrograms, the elevation angle was obtained using a traditional search method for the spectral peak. For the two-path spatial spectrums, the network shown in Figure 7 was used to determine the elevation angle, and the probability distribution of the elevation angle was also collected. Figure 9 shows the result of the comparison between the actual probability distribution of the elevation angle and the probability distribution obtained by CNN learning. As shown in Figure 9, neural network learning helped us to obtain an important characteristic, that is, the elevation angles of the two paths followed a GMM. Therefore, the network met the expected requirements.

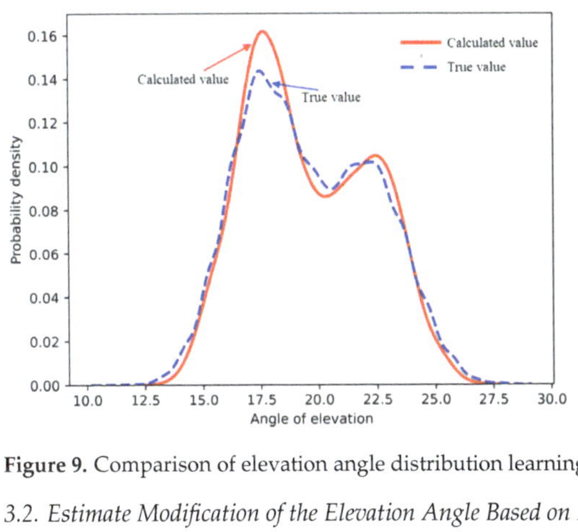

Figure 9. Comparison of elevation angle distribution learning.

3.2. *Estimate Modification of the Elevation Angle Based on Prior Information*

Under the influence of multipath fading, the elevation angle ϕ of the same target is no longer a fixed value but a random variable that varies with the direction measurement's time and site, denoted by $\phi(r, t)$, where r is the location of the measurement station and

t is time. Different from traditional methods in which the elevation angle is considered as a deterministic parameter for direction measurement, in this paper, Bayesian estimation theory was introduced and the elevation angle distribution under the ionospheric multi-layer reflection was considered as prior information to estimate the maximum posterior probability of the elevation angle as follows:

$$P(\phi|\mathbf{X}) = \frac{P(\mathbf{X}|\phi)P(\phi)}{P(\mathbf{X})} \quad (8)$$

where $\mathbf{X}$ is the array observation vector defined by Equation (2), $P(\phi|\mathbf{X})$ is the posterior probability, $P(\mathbf{X}|\phi)$ is the likelihood function of the elevation angle with respect to the observation vector, and $P(\phi)$ is the prior probability distribution of the elevation angle. Equation (8) can be simplified as follows:

$$P(\phi|\mathbf{X}) = \eta P(\mathbf{X}|\phi)P(\phi) \quad (9)$$

where

$$\eta = (P(\mathbf{X}))^{-1} = \frac{1}{\sum_{\mathbf{A}} P(\mathbf{X}|\phi)P(\phi)} \quad (10)$$

The prior distribution information of the elevation angle can be used to modify the posterior probability to obtain the modified posterior probability. This idea is important and provided by Equation (9). According to this idea, the prior distribution of the elevation angle can be used to modify the elevation angle's estimated value obtained using Equation (3).

$$\phi_{MUSIC} = \underset{\phi}{\operatorname{argmin}}(\boldsymbol{a}^H(\Theta)\mathbf{U}_N\mathbf{U}_N^H\boldsymbol{a}(\Theta))\ P(\phi) \quad (11)$$

It should be noted that the elevation angle's correction only involved multiplying the angle's estimate by a coefficient, so that the increase in the computational effort was limited. Although the prior probability distribution of the elevation angle requires a large amount of data analysis, the computation is an offline process. Therefore, only one deep learning calculation is needed to obtain the distributions for actual positioning.

As shown in Figure 2, the calculation method of the azimuth and the elevation angle of the signal that passes each path to arrive at the ith DOA observation station is as follows:

$$\theta_i = \arctan\left(\frac{u_x - u_{ix}}{u_y - u_{iy}}\right), \frac{R}{\sin\left(\frac{\pi}{2} - \phi_i - \beta_i\right)} = \frac{R + h_i}{\sin\left(\frac{\pi}{2} + \phi_i\right)}, i = 1, 2, \ldots, N \quad (12)$$

where R is the average radius of the earth, and $\boldsymbol{u}$ were $\boldsymbol{u}_i$ are the coordinates of the target and ith observation station, respectively, in the earth-centered earth-fixed coordinate system.

The modified estimates of the azimuth and the elevation angle were obtained using Equation (3) and (11). The estimates were substituted into Equation (4) to obtain the pseudo-linear equations of the azimuth and the elevation angle. The matrix QR decomposition iteration algorithm was used to obtain the DOA stage positioning results.

According to Figure 2, the propagation distance of each path can be expressed as follows:

$$d_m = 2\sqrt{(R\sin(\beta_m))^2 + (R - R\cos(\beta_m) + h_m)^2}\ m = 1, 2, \ldots, M \quad (13)$$

Therefore, the difference between the arrival time of the signal at the mth observation station and the arrival time at the first observation station can be calculated as

$$\tau_m = \frac{2}{c}(d_m - d_1) \quad (14)$$

Considering $\beta_m = \arcsin(||\boldsymbol{u} - \boldsymbol{u}_m||/(2R))$ and that this calculation is related to the target position, Equation (14) cannot be pseudo-linearized directly. In this case, the target estimate of the DOA in the first stage can be considered as a known value in the TDOA stage to obtain the pseudo-linear equation shown in Equation (7) and obtain the final results using the matrix QR decomposition iteration algorithm.

4. Simulation Studies

Simulations were conducted to verify the positioning performance of the proposed method. The default settings of the simulation parameters included the following. The longitude and latitude of the radiation source were 134° and 34°, respectively. There were seven shortwave observation stations, among which the first three had the DOA positioning system and the last four had the TDOA positioning system. Their longitudes, latitudes, and corresponding ionospheric reflection heights are shown in Table 1. The observation errors of the azimuth, elevation angle, TDOA, and ionospheric reflection height all followed Gaussian distributions with zero means and were independent of each other. Their covariance matrices were

$$\begin{aligned}\boldsymbol{\Omega}_{\Theta} &= blkdiag\{\boldsymbol{\Omega}_{\theta}, \boldsymbol{\Omega}_{\phi}\} = \sigma_{\theta}^2 \mathbf{I}_{2N} \\ \boldsymbol{\Omega}_{\mathcal{T}} &= \sigma_{\mathcal{T}}^2 \mathbf{R}_{M-1} \\ \boldsymbol{\Omega}_{h} &= \sigma_{h}^2 \mathbf{I}_{N+M}\end{aligned} \tag{15}$$

where the diagonal elements of the matrix $\mathbf{R}_{M-1}$ were 1, and all other elements were 0.5. $blkdiag\{\cdot\}$ is a block-diagonal matrix formed from the matrices or vectors. The root mean square error of positioning was the measurement standard of positioning accuracy. The calculation formula for the root mean square error of positioning is given by

$$RMSE = \sqrt{\sum_{i=1}^{K} \frac{||\hat{\boldsymbol{u}} - \boldsymbol{u}||_2^2}{K}} \tag{16}$$

where K denotes the number of Monte Carlo simulation runs. K takes 5000 in the simulation.

Table 1. Longitude, latitude, and ionospheric reflection virtual height of the observation station.

Observation Station	Longitude (°)	Latitude (°)	Ionospheric Reflection Virtual Height (km)
DOA-1	116.23	40.22	340.00
DOA-2	112.54	33.00	390.00
DOA-3	116.00	29.71	370.00
TDOA-1	123.47	41.80	310.00
TDOA-2	114.54	38.04	350.00
TDOA-3	114.03	30.58	385.00
DOA-1	116.23	40.22	340.00

Figure 10 shows that the results for the positioning accuracy vary with the standard deviation of the angle's error with a single path. Other parameters were set as $\sigma_{\mathcal{T}} = (0.3/c)/s$ and $\sigma_h = 2$km, where the elevation angle's search step was 0.01°. "Max" means that the spectral peak with the largest spatial spectrum was selected as the angle's estimate when the MUSIC algorithm was used. Figure 10 shows that the traditional positioning algorithm failed when the angle error was greater than 1.5°, which resulted in a sharply increasing positioning error. By contrast, the proposed method improved the estimation accuracy of the elevation angle; therefore, the failure threshold of the algorithm was raised. Figure 11 shows that the results for the positioning accuracy vary with the target's longitude under a single path. As shown in Figure 11, when the distance between the observation station and the target increased, the positioning accuracy decreased, mainly because DOA positioning was sensitive to the distance between the target and the observation station. However, the accuracy did not decrease greatly. Therefore, the proposed method has a certain generalization capability at the location of the radiation source.

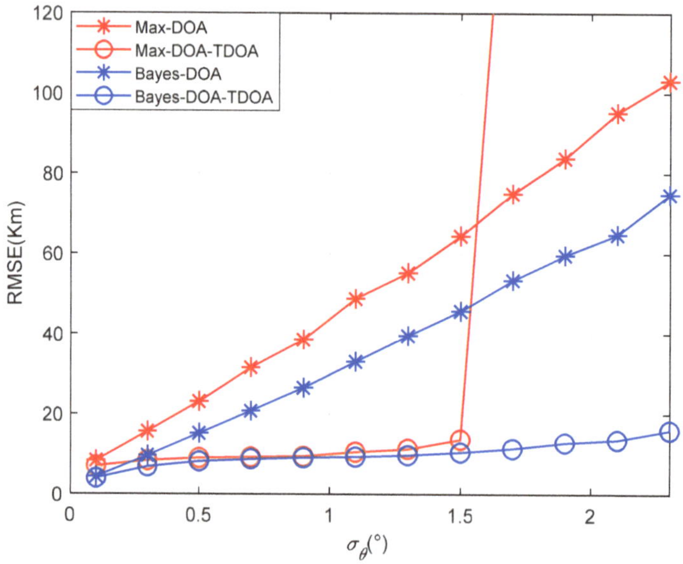

Figure 10. The positioning accuracy varies with the angle's standard deviation.

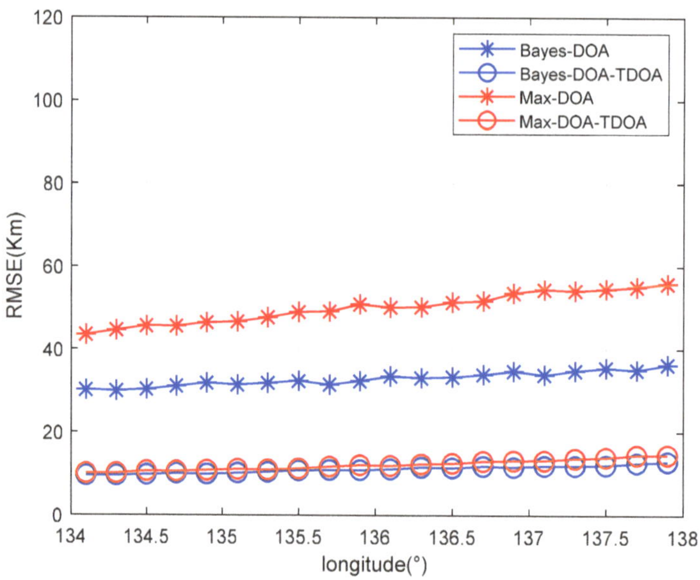

Figure 11. The positioning accuracy varies with the target's accuracy.

Table 1 shows the ionospheric reflection height; hence, the reflection elevation angle can be determined. The elevation angle of the second path was given by the elevation angle difference of 1°, 3°, 5°, 7°, and 10° in Table 1. In the case of two paths, two methods are usually used to estimate the angle, and these are called traditional methods. The first method selects the angle corresponding to the maximum peak of the two spectral peaks in the spatial spectrum as the angle's estimated value, denoted by "Max." The second method uses the average value of the angles corresponding to the two spectral peaks in the spatial spectrum as the angle's estimated value, denoted by "Average." The data given by the proposed method are denoted as "Bayes."

Figures 12 and 13 show the root mean square errors of the estimated values of the elevation angles given by the traditional methods and proposed method when the elevation angles were different. Figure 12 shows that when the difference between the elevation angles of the two paths was greater than 5°, the angle's estimation accuracy was less than 3°, and the positioning algorithm failed. Figure 13 shows that the elevation angle's estimation accuracies for the proposed method were all below 2° under various angle differences, which indicates that the proposed method had a certain anti-multipath effect.

Figures 14 and 15 show that the results for the positioning accuracy of the traditional methods and proposed method varied with the standard deviation of the angle's error when the differences between the elevation angles of the two paths were 1° and 3°, respectively. As the angles' difference between the two paths increased, the positioning error of the traditional methods gradually increased, and the threshold value was reduced. This shows that when the difference between the angles of the two paths was large, the traditional method could not position accurately. Moreover, when the DOA positioning error was large in the first stage, the accuracy of the proposed coordinated positioning was even lower than that of the standalone DOA positioning. This indicates that when the DOA accuracy was low, coordinated positioning was not meaningful. In this case, the accuracy of standalone positioning was better than that of coordinated positioning. It also indicates that the accuracy of coordinated positioning was not always better than that of standalone positioning.

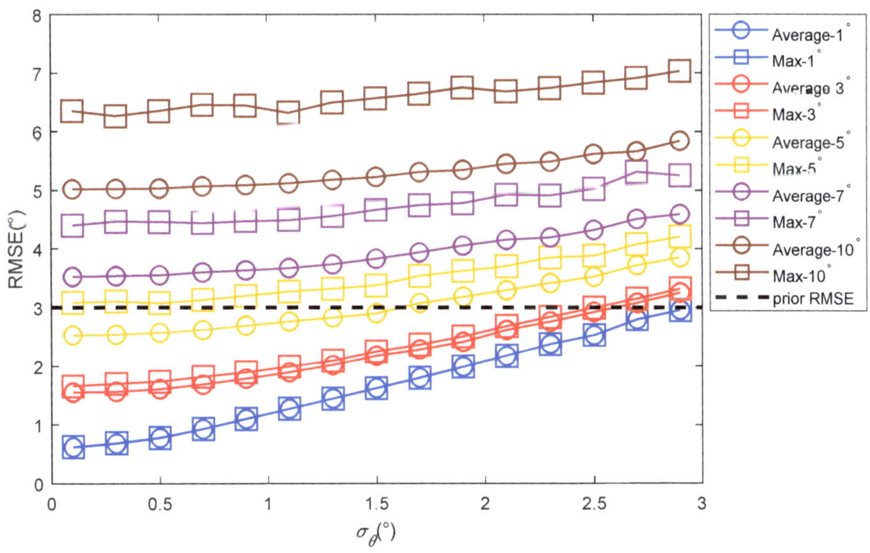

Figure 12. The elevation angle's estimate accuracy varies with the angle's standard deviation for the traditional methods.

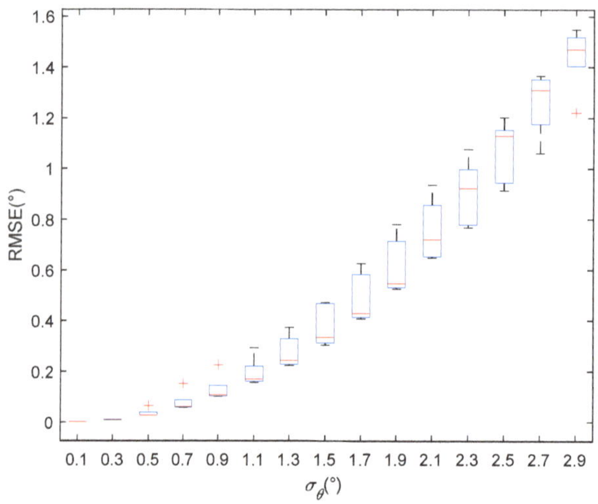

Figure 13. The elevation angle's estimate accuracy varies with the standard deviation of the angle's error for the proposed method.

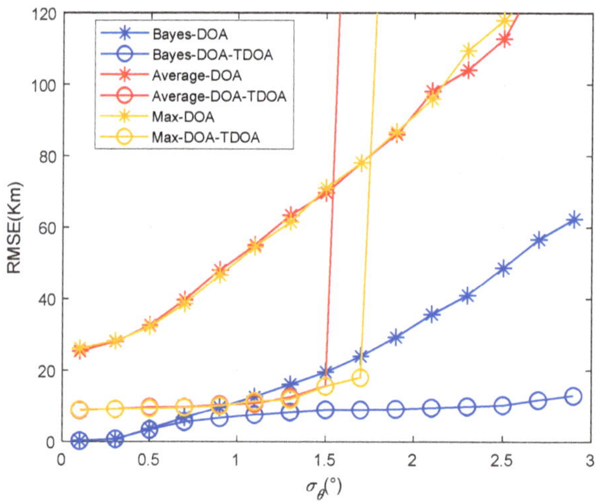

Figure 14. The positioning accuracy varied with the standard deviation of the angle's error when the multipath elevation angle difference was 1°.

Figure 16 shows the positioning accuracy varied with the standard deviation of the angle's error when the elevation angles' differences were 5°, 7°, and 10° for the proposed method. Because the traditional methods were not effective at this point, they are not shown in the figure.

Figures 14–16 show that as the angle difference increased, the proposed method maintained high accuracy. Therefore, the proposed method had a strong anti-multipath function. This is because the prior information for the elevation angle is related to the position and time of the observation station, but is irrelevant to the path's state. Therefore, prior information regarding the elevation angle can be used to correct the elevation angles under different angle differences.

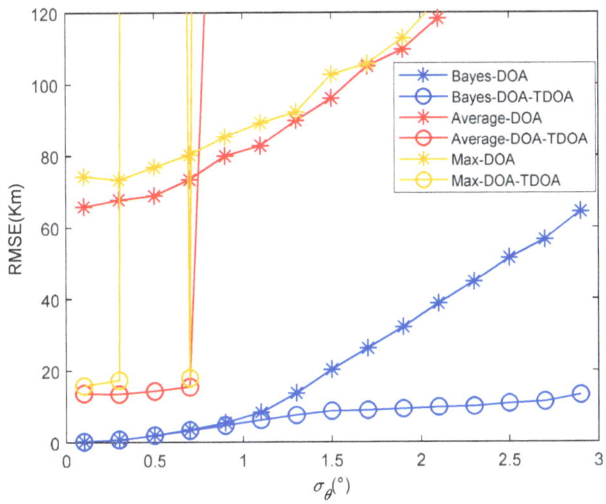

Figure 15. The positioning accuracy varied with the standard deviation of the angle's error when the multipath elevation angle difference was 3°.

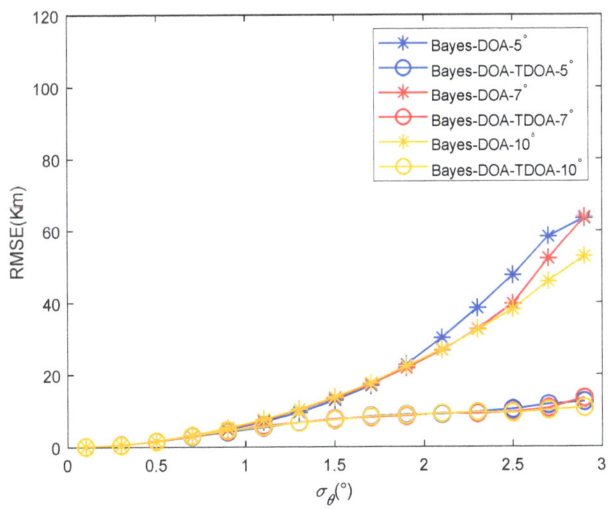

Figure 16. The positioning accuracy varied with the standard deviation of the angle's error when the multipath elevation angles' differences were 5°, 7°, and 10°.

5. Measured Data

The ionospheric multipath reflection model and its variation trends with respect to time were validated using measured data. The transmitting station used in the experiment was a radio station in Urumqi (87° E, 43° N), Xinjiang, China. The receiving station was a 20-channel shortwave direction measurement array located in Zhengzhou (113° E, 34° N), Henan Province, China. The experimental parameters are shown in Table 2.

Table 2. List of experimental parameters.

Parameter	Signal Modulation Type	Distance	Azimuth	Radius-to-Wavelength Ratio
Value	AM	2447 km	294.1°	2.5

Figure 17 shows an illustration of the arrival angle's estimate of the transmitting station obtained using the receiving array. In the signal direction, the elevation angle varied greatly, and the multipath effect existed. Based on the figure, the elevation angles of the two paths could not be separated accurately. Therefore, the estimation accuracy of the elevation angle was reduced because of the existence of the multipath effect.

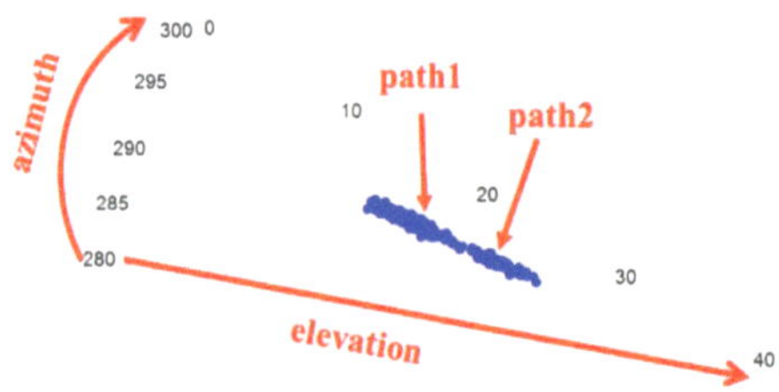

Figure 17. Local amplification of signal directions.

Figure 18 shows the time-varying results of the elevation angle of the signal. The difference between the elevation angles of the two paths was small at some moments, whereas the difference was large at other moments. Figure 19 shows the distribution of the signal's elevation angle. The elevation angle distribution of the two paths was approximately a GMM, thus verifying the previous analysis and the validity of this study.

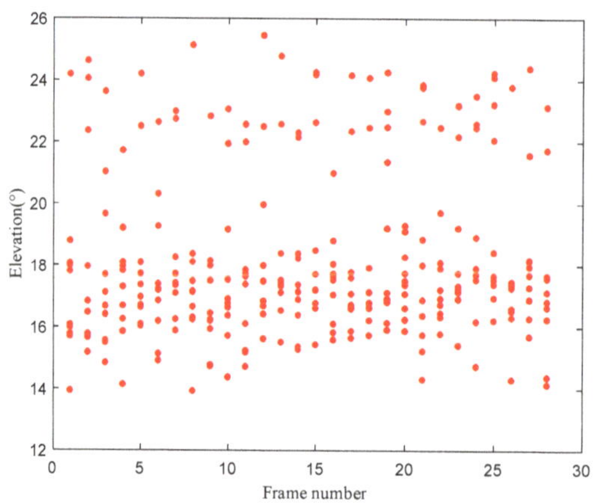

Figure 18. Illustration of the signal's elevation angle varying with time.

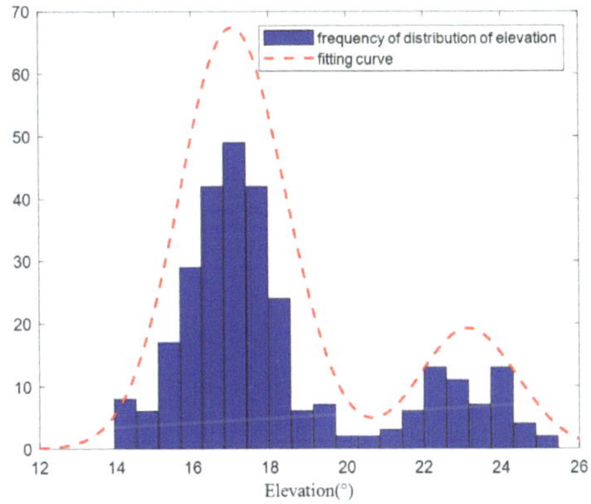

Figure 19. Statistical results for the elevation angle.

6. Conclusions

In this paper, DOA–TODA coordinated positioning for shortwave radiation sources was addressed based on prior information regarding the distribution of the elevation angles. First, a CNN was used to determine the number of single and multiple paths. The probability distribution of the elevation angle with the corresponding number of paths was learned. Then, the DOA and TDOA observation models were constructed according to the ionospheric virtual reflection model. The elevation angle's estimated value was modified using the elevation angle's prior information. An optimization model with double quadratic equality constraints was constructed according to the pseudo-linear equations of the azimuth, the elevation angle, and TDOA. The matrix QR decomposition iteration algorithm was used to solve the model.

Simulations were conducted for the single-path and two-path cases. The simulation results demonstrated that compared with traditional methods, the proposed method achieved better positioning accuracy when the angle error was large, and anti-noise performance improved with strong anti-multipath performance. Moreover, the proposed method generalized the target position. Finally, the measured data were used to validate the proposed method. It should be noted that in this paper, only DOA–TDOA coordinated positioning for stationary targets was addressed. A future study needs to be conducted for moving targets.

Author Contributions: Data curation, T.T.; Methodology, T.T.; Software, P.Z.; Supervision, P.Z. and N.-e.Z.; Writing—original draft, L.J.; Writing—review and editing, L.J. All authors have read and agreed to the published version of the manuscript.

Funding: This research was funded by the National Natural Science Foundation of China, grant number 62171469.

Institutional Review Board Statement: Not applicable.

Informed Consent Statement: Not applicable.

Data Availability Statement: Not applicable.

Conflicts of Interest: The authors declare no conflict of interest.

References

1. Huang, D.; Zhao, Y.; Hu, D.; Zhao, Y. 3D TDOA/AOA Localization in MIMO Passive Radar with Transmitter and Receiver Position Errors. *Math. Probl. Eng.* **2022**, *2022*, 9964332. [CrossRef]
2. Fresno, J.M.; Robles, G.; Martínez-Tarifa, J.M.; Stewart, B.G. Survey on the Performance of Source Localization Algorithms. *Sensors* **2017**, *17*, 2666. [CrossRef] [PubMed]

3. Zhang, Z.; Lin, Y.; Jin, B. Underwater TDOA/AOA joint localization algorithm based on hybrid invasive weed optimization algorithm. *IET Commun.* **2021**, *15*, 2376–2389. [CrossRef]
4. Shao, H.-J.; Zhang, X.-P.; Wang, Z. Efficient Closed-Form Algorithms for AOA Based Self-Localization of Sensor Nodes Using Auxiliary Variables. *IEEE Trans. Signal Process.* **2014**, *62*, 2580–2594. [CrossRef]
5. Badriasl, L.; Dogancay, K. Three-dimensional target motion analysis using azimuth/elevation angles. *IEEE Trans. Aerosp. Electron. Syst.* **2014**, *50*, 3178–3194. [CrossRef]
6. Chen, X.; Wang, G.; Ho, K. Semidefinite relaxation method for unified near-Field and far-Field localization by AOA. *Signal Process.* **2021**, *181*, 107916. [CrossRef]
7. Zhang, L.; Zhang, T.; Shin, H.S. An Efficient Constrained Weighted Least Squares Method with Bias Reduction for TDOA-Based Loacalization. *IEEE Sens. J.* **2021**, *21*, 10122–10131. [CrossRef]
8. Chan, Y.T.; Ho, K.C. A simple and efficient estimator for hyperbolic location. *IEEE Trans. Signal Process.* **1994**, *42*, 1905–1915. [CrossRef]
9. Yang, K.; An, J.; Bu, X.; Sun, G. Constrained Total Least-Squares Location Algorithm Using Time-Difference-of-Arrival Measurements. *IEEE Trans. Veh. Technol.* **2010**, *59*, 1558–1562. [CrossRef]
10. Yin, J.; Wan, Q.; Yang, S.; Ho, K.C. A Simple and Accurate TDOA-AOA Localization Method Using Two Stations. *IEEE Signal Process. Lett.* **2016**, *23*, 144–148. [CrossRef]
11. Zhang, F.; Sun, Y.; Wan, Q. Calibrating the error from sensor position uncertainty in TDOA-AOA localization. *Signal Process.* **2020**, *166*, 107213. [CrossRef]
12. Jiexin, Y.; Ding, W.; Xin, Y.; Peng, W. Combination of Land-Based and Satellite-Based OTH Geolocations Using Differentiable Exact Penalty Method. *IEEE Trans. Aerosp. Electron. Syst.* **2021**, *58*, 2363–2382. [CrossRef]
13. Zhang, T.; Mao, X.; Zhao, C.; Liu, J. A novel grid selection method for sky-wave time difference of arrival localisation. *IET Radar Sonar Navig.* **2019**, *13*, 538–549. [CrossRef]
14. Huang, S.; Pun, Y.-M.; So, A.M.-C.; Yang, K. A Provably Convergent Projected Gradient-Type Algorithm for TDOA-Based Geolocation under the Quasi-Parabolic Ionosphere Model. *IEEE Signal Process. Lett.* **2020**, *27*, 1335–1339. [CrossRef]
15. Jain, A.; Pagani, P.; Fleury, R.; Ney, M.M.; Pajusco, P. Efficient time domain HF geolocation using multiple distributed receivers. In Proceedings of the European Conference on Antennas and Propagation (EURAAP), Paris, France, 19–24 March 2017.
16. Fabrizio, G.; Heitmann, A. A multipath-driven approach to HF geolocation. *Signal Process.* **2013**, *93*, 3487–3503. [CrossRef]
17. Ding, W.; Yin, J.; Zhu, Z. A novel cooperative localization method of over-the-horizon shortwave emitter based on direction-of-arrival and time-difference-ofarrival measurements. *China Sci. Inf. Sci.* **2022**.
18. Wang, G.; Jiang, X.; Razul, S.G.; See, C.M.; Lin, Z. Passive TDOA and DOA based HF geolocation without ionosphere information. In Proceedings of the 2015 10th International Conference on Information, Communications and Signal Processing (ICICS), Singapore, 2–4 December 2015; pp. 1–5. [CrossRef]
19. Yan, L.; Mao, X.; Chen, M. Direct Localization of Shortwave Emitters in Multipath Ionosphere Channel. In Proceedings of the 2020 IEEE Radar Conference, Florence, Italy, 21–25 September 2020.
20. Rawer, K. *Wave Propagation in the Ionosphere*; Kluwer Academic: London, UK, 1993.
21. Wang, X.; Jiang, C. Distribution of Multiple Time Delay and Doppler Shift for HF Communication on Ionospheric Channels. *Acta Electron. Sin.* **1993**, *21*, 85–88.
22. Fan, J.; Chen, S.; Luo, X.; Zhang, Y.; Li, G.Y. A Machine Learning Approach for Hierarchical Localization Based on Multipath MIMO Fingerprints. *IEEE Commun. Lett.* **2019**, *23*, 1765–1768. [CrossRef]
23. Qiu, S.; Sava, K.; Guo, W. Robust satellite antenna fingerprinting under degradation using recurrent neural network. *Mod. Phys. Lett. B* **2022**, *36*, 2250043. [CrossRef]
24. Tadaion, A.A.; Derakhtian, M.; Gazor, S.; Aref, M.R. A Fast Multiple-Source Detection and Localization Array Signal Processing Algorithm Using the Spatial Filtering and ML Approach. *IEEE Trans. Signal Process.* **2007**, *55*, 1815–1827. [CrossRef]

sensors

Communication

Efficient Multi-Sound Source Localization Algorithm for Transformer Faults Based on Polyphase Filters

Hualiang Zhou [1,2,*], Zhantao Su [1,2], Yuxuan Huang [3], Lu Lu [1,2] and Mingwei Shen [3]

[1] NARI Group Corporation (State Grid Electric Power Research Institute), Nanjing 211106, China; suzhantao@sgepri.sgcc.com.cn (Z.S.); lulu1@sgepri.sgcc.com.cn (L.L.)
[2] NARI Technology Nanjing Control System Co., Ltd., Nanjing 211106, China
[3] College of Information Science and Engineering, Hohai University, Nanjing 210098, China; yx_huang@hhu.edu.cn (Y.H.); smw_nuaa@hhu.edu.cn (M.S.)
* Correspondence: zhouhualiang@sgepri.sgcc.com.cn; Tel.: +86-139-1381-3031

Abstract: Power transformers play a critical role in power systems, and the early detection of their faults and defects, accounting for over 30%, can be achieved through abnormal sound analysis. Sound source localization based on microphone arrays has proven effective in focusing on the troubleshooting scope, preventing potential severe hazards caused by delays in fault removal, and significantly reducing operational and maintenance difficulties and costs. However, existing microphone array-based sound source localization algorithms face challenges in maintaining both accuracy and simplicity and especially suffer from a sharp decrease in performance when dealing with multiple sound sources. This paper presents a multi-sound source localization algorithm for transformer faults based on polyphase filters, integrating the sum-difference monopulse angle measurement technique into the microphone array. Firstly, the signals received from the transformers are divided into multiple subbands using polyphase filters, allowing for multi-source separation and reducing the sampling rate of each subband. Next, the time-domain signals in subbands subject to noise suppression are processed into sum and difference beams. The resulting beam outputs are transformed into frequency-domain signals using the Fast Fourier Transform (FFT), effectively enhancing the signal-to-noise ratio (SNR) for separate sound sources. Finally, each subband undergoes sum-difference monopulse angle measurement in the frequency domain to achieve the high-precision localization of specific faults. The proposed algorithm has been demonstrated to be effective in achieving higher localization accuracy and reducing computational complexity in the presence of actual amplitude-phase errors in microphone arrays. These advantages can facilitate its practical applications. By enabling early targeting of fault sources when abnormalities occur, this algorithm provides valuable assistance to operation and maintenance personnel, thereby enhancing the maintenance efficiency of transformers.

Keywords: transformer; polyphase filter; sum-difference monopulse; multi-source separation; multi-source localization

1. Introduction

The rapid growth in power demand has led to the constant expansion of grid capacity and the continuous improvement of voltage levels. As a result, higher and stricter requirements for safe operation and reliable power supply have been imposed on power systems. Power transformers serve as core equipment in power systems, and their operating conditions have a direct impact on the overall functioning of power grids [1–3]. It is essential to enhance the operational reliability of transformers as well as promptly identify and localize any faults that occur during their operation, which is a crucial task for ensuring the safe and reliable operation of power grids. At present, some typical faults of power transformers, such as partial discharge, short-circuit impulse, and DC magnetic bias, can be distinguished from the voiceprint signal. For example, when a partial discharge occurs in a power transformer, air bubbles and impurities will appear inside, leading to partial damage to the dielectric and a "squeak" or "crackle" sound. When a power transformer encounters a short-circuit shock, the short-circuit current will surge, producing a "gurgling" boiling sound inside the transformer [4]. Therefore, in engineering design, when facing voiceprint information, it is necessary to first detect and identify the faulty voiceprint, and then locate the identified faulty voiceprint. However, due to various factors such as the presence of interference sounds in their operating environments, complex structures, and diverse operating conditions, existing localization techniques can only target large components, falling short of achieving precise localization. In recent years, microphone array-based sound source localization technology has gained significant attention due to its convenience and efficiency [5–9].

Many scholars, both in China and abroad, have conducted extensive studies on microphone array-based sound source localization. For instance, Hahn et al. proposed a sound source localization method based on steerable beamforming [10]. However, this method relies on prior knowledge of sound sources and ambient noises, which is often challenging to obtain in real-world scenarios. Another method presented by Dibiase et al. is the guided response power-phase conversion approach [11]. However, this technique necessitates a global search and a subsequent heavy computational workload. Schmidt introduced the multiple signal classification (MUSIC) algorithm [12], which is only applicable in scenarios where the number of sources is known and the sources are non-coherent. This algorithm has subsequently sparked various improvements by other researchers. For instance, Reference [13] suggests using a windowing function to filter and process coherent signals for estimating their direction of arrival (DOA). Reference [14] proposes a joint diagonalization matrix-based algorithm to construct a cost function, enabling DOA estimation even in scenarios with unknown numbers of mixed signals. Although these improved versions of the MUSIC algorithm have made progress in overcoming the limitations of the original method, they still face challenges in real-time processing due to the high computational complexity associated with covariance matrix estimation and matrix inversion.

While these algorithms demonstrate effectiveness in localizing sound sources under specific scenarios, applying them directly in complex multi-source transformer environments is impractical. In such scenarios, a feasible approach to localizing each sound source involves using a filter bank for separate extraction of sources, followed by direction finding. However, traditional analysis filter banks perform filtering before sampling, which can lead to increased hardware resource requirements when a high sampling rate is necessary. In contrast, the utilization of polyphase filters (PFs) [15–17] offers advantages in improving resource efficiency by sharing a low-pass filter among different branches and leveraging the sampling rate transformation theorem that allows for downsampling before filtering.

Among the various angle measurement algorithms, the sum-difference monopulse (MP) method stands out due to its high accuracy, robustness, and easy applicability, and has been extensively applied in radar systems [18–20]. However, there are a limited number of studies exploring its application in the field of sound source localization. To summarize, this paper presents a sound source localization algorithm for detecting faults in power transformers, combining polyphase filters with the MP method, to give a solution for achieving high-precision localization of sound sources in transformers within complex environments. This research on spatial localization technology for abnormal sound sources facilitates the precise localization of zones containing abnormal sound sources within transformers and improves maintenance efficiency by narrowing down the troubleshooting scope.

2. Signal-Receiving Model of Microphone Array

The model depicted in Figure 1 illustrates the signal received by microphones in a uniform linear array with M elements. In this model, the normal direction represents the direction perpendicular to the array. On the assumption of a far-field model, signals from different sound sources propagate as plane waves coming from various incidence directions. Considering the leftmost signal as the reference element, the received signals of the array can be expressed as [21]

$$X = a_1 x_1 + \ldots + a_Q x_Q + N = \mathrm{Re}(\sum_{n=1}^{Q} a_n x_n + N) \quad (1)$$

$$a_n = [1 \ldots \exp(\frac{j2\pi f_n (M-1) d \sin\theta_n}{c})]^T \quad (2)$$

where d denotes the element spacing; Q denotes the number of sound sources; $x(n)$ denotes the complex amplitude of the nth sound signal; f_n denotes the frequency of the nth sound signal; θ_n denotes the angle of incidence of the nth sound signal; c denotes the velocity of sound; N denotes the M $\times$ 1-dimensional noise vector; a_n denotes the steering vector of the array response; Re($\cdot$) denotes the operation to extract the real part; and T denotes the transpose operator.

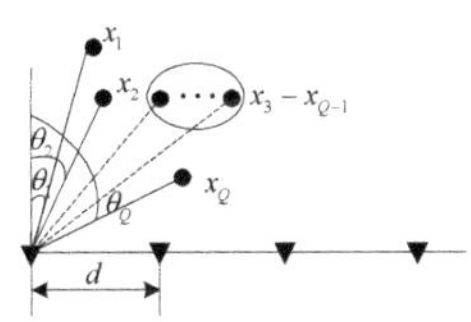

Figure 1. Linear array model of signal receiving.

3. Multi-Source Separation Using Polyphase Filter

The sound emitted by running transformers primarily originates from the vibrations of various components, such as windings, iron cores, cooling fans, and others. These components generate vibrations at different frequencies. To localize fault sound sources separately and eliminate noise interference in complex multi-source environments, an analysis filter bank is employed to divide the sound signals received from the transformers into multiple subbands, allowing for the separation of fault sound sources.

Traditional analysis filter banks typically first divide the wideband signals x_n into K subbands using a modulation process and then shift these subbands to the baseband while decimating them to reduce the sampling rate. The theoretical formulas can be found in [22]. This process necessitates numerous unnecessary calculations, leading to a waste of hardware resources. On the contrary, analysis filter banks based on a polyphase structure share a low-pass filter among branches to improve resource utilization.

An N-point FIR filter system function is assumed as

$$H(z) = \sum_{l=0}^{N-1} h(n) \quad (3)$$

Taking N as a multiple of K and letting $L = N/K$, we obtain

$$E_l(Z^K) = \sum_{n=0}^{L-1} h(nK+l)\left(z^K\right)^{-n} \quad (4)$$

as the polyphase component of the filter, and then the filter system function can be expressed as

$$H(z) = \sum_{l=0}^{M-1} z^{-l} E_l(z^K) \quad (5)$$

For traditional analysis filter banks, the kth (k = 0, 1, ..., $K-1$) output can be expressed as

$$v_k(n) = [x(n) * h_k(n)] e^{-jw_k n} \quad (6)$$

$$y_k(m) = v_k(n)|_{n=Km} = \sum_{i=0}^{N-1} x(Km-i)\left[h_0(i)e^{j\omega_k(i-Km)}\right] \tag{7}$$

where $x(n)$ denotes the input signal, $h_k(n)$ denotes the kth filter response, with $h_k(n) = h_0(n)e^{j\omega_k n}$, and $h_0(n)$ denotes the low-pass prototype filter.

Letting

$$i = lK + r \tag{8}$$

Equation (7) can be expressed as

$$y_k(m) = \sum_{r=0}^{K-1} e^{j\omega_k r} \sum_{l=0}^{L-1} x_r(m-l)e_r(l)e^{j\omega_k K(l-m)} \tag{9}$$

In Equation (9),

$$x_r(m-l) = x(Km - Kl - r) \tag{10}$$

$$e_r(l) = h_0(Kl + r) \tag{11}$$

In odd-channel layout scenarios, the center frequency of the kth subband is expressed as

$$w_k = -\pi + \frac{2\pi k}{K} \tag{12}$$

Substituting it into Equation (9), we obtain

$$y_k(m) = K(IDFT[(x_r(m) * e_r(m))(-1)^r]) \tag{13}$$

Figure 2 illustrates a channelized polyphase filter. In the polyphase filter (PF) approach, it is important to note that the sequence length of all subbands is $1/K$ of the original sequence length. Specifically, the 0th and $(K/2)$th subbands carry real signals, while the remaining subbands carry complex signals. Additionally, the information in the 1st to $(K/2-1)$th subbands is the same as that in the $(K/2+1)$th to $(K-1)$th subbands. When using PFs for signals received by each element, the output in the kth channel of the array is expressed as

$$\mathbf{Y}_k(m) = \boldsymbol{a} * (K(IDFT[(x_r(m) * e_r(m)(-1)^r)])) \tag{14}$$

where $\boldsymbol{a}$ denotes the steering vector of the array.

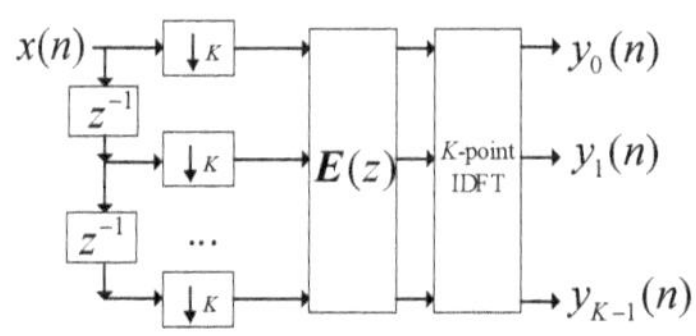

Figure 2. Channelized polyphase filter.

The incorporation of polyphase filters allows for the separation of multiple fault sound sources. Moreover, the sequence length of the subbands containing fault sound sources is reduced to $1/K$ of the original length by downsampling before filtering, which helps reduce the computational workload of subsequent operations, contributing to enhanced calculation efficiency. In addition, sharing a low-pass filter among the polyphase filter branches can greatly improve the utilization rate of system resources.

4. Multi-Source Localization Based on Sum-Difference Monopulse in Frequency Domain

4.1. Principle of Sum-Difference Monopulse Angle Measurement

Current sound source localization algorithms based on microphone arrays often demand extensive calculations to achieve high localization accuracy, which imposes a high cost for real-world implementation. Therefore, the exploration of a microphone array-based sound source localization technology that combines both accuracy and simplicity holds paramount significance in engineering practice. Considering the extensive applications of the sum-difference MP angle measurement in radar systems, which boasts a simple structure, low calculation complexity, and high accuracy, this section presents a multi-source localization algorithm based on microphone arrays, which aims to strike a balance between accuracy and simplicity, emphasizing the integration of the sum-difference MP technique with the previously mentioned PF in the microphone array.

The sum-difference monopulse angle measurement is achieved by aligning the main lobe of the sum beam with the desired direction and forming the null of the difference beam in the desired direction. We adjust the distance between the microphone array and the transformer and weight the amplitude of each subband to ensure that the transformer's sound surface is located within the main lobe of the beam. Through the output ratio calculation between the sum and difference beams, a specific value is derived. This value represents the extent of deviation of the beams in the target direction from the beam center, leading to a precise estimation of the target location. Detailed theoretical derivations can be found in reference [23].

Assuming the element spacing of an M-element uniform linear array as d and the beam pointing as θ_0, the sum beam weight $\boldsymbol{w}_\Sigma$ can be taken as

$$\boldsymbol{w}_\Sigma = \boldsymbol{a}(\theta_0) \tag{15}$$

Based on the symmetry of uniform linear arrays, the difference in beam weight $\boldsymbol{w}_\Delta$ can be taken as

$$\boldsymbol{w}_\Delta = [\overbrace{-1,\ldots,-1}^{M/2},\overbrace{1,\ldots,1}^{M/2}]^T \odot \boldsymbol{a}(\theta_0) \tag{16}$$

In Equation (16), $\odot$ represents the Hadamard product. Then, the outputs of the sum and difference beams in the kth subband can be expressed by Equations (17) and (18), respectively.

$$\Sigma(\theta) = \boldsymbol{w}_\Sigma^H \mathbf{Y}_k(m) \tag{17}$$

$$\Delta(\theta) = \boldsymbol{w}_\Delta^H \mathbf{Y}_k(m) \tag{18}$$

The in-band signaling angle can be estimated from the outputs of the sum and difference beams as follows:

$$\hat{\theta} = \theta_0 + \frac{1}{k'}\mathrm{Re}(\frac{\Delta(\theta)}{\Sigma(\theta)}) \tag{19}$$

In Equation (19), k' denotes the monopulse ratio slope [24,25].

$$k' = j\frac{\pi M d}{2\lambda} \tag{20}$$

It can be seen from Equation (20) that the value of k' is determined solely by the number of elements (M), element spacing (d), and signal wavelength (λ). The angle measurement accuracy is correlated with the value of k' and SNR. Notably, higher values of k' and SNR contribute to improved accuracy in angle measurement.

4.2. Multi-Source Localization in Subbands

To enhance the accuracy of sound source localization, the complex variational mode decomposition (CVMD) [26] method is used to suppress noise. CVMD is applied within the subbands that contain faulty sound sources. It enables the transformation of the outputs of the sum and difference beams to the frequency domain, thereby providing an enhanced SNR.

$$\hat{\sum}(\theta) = \mathrm{FFT}(\boldsymbol{w}_{\Sigma}^{H}\boldsymbol{Y}_k(m)) \quad (21)$$

$$\hat{\Delta}(\theta) = \mathrm{FFT}(\boldsymbol{w}_{\Delta}^{H}\boldsymbol{Y}_k(m)) \quad (22)$$

where FFT denotes the Fast Fourier Transform, and $\hat{\sum}(\theta)$ and $\hat{\Delta}(\theta)$ represent the frequency-domain outputs of $\hat{\sum}(\theta)$ and $\hat{\Delta}(\theta)$, respectively.

In conclusion, the proposed algorithm can be explained using a flow chart, as shown in Figure 3. This algorithm, integrating PF and frequency-domain MP angle measurement, can be used for multi-source sound localization of faults in power transformers.

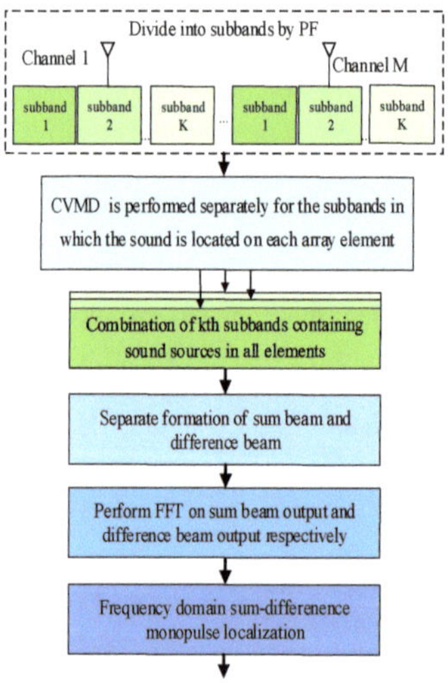

Figure 3. Flow chart of the algorithm for multi-sound source localization.

5. Analysis of Algorithm Performance

The performance of the algorithm was verified using the experimental data, which are detailed below. The microphone array used in the experiment was an eight-element line array, with each array element having a center spacing of 4.25 cm and a sampling rate of 96 KHz. Figure 4 shows the microphone array used in the experiment. The first transformer sound source to be estimated emitted electromagnetic sound, while the second sound source emitted a whistling sound. Figure 5 shows a sketch of the experimental setup. The system parameters are shown in Table 1.

Figure 4. Microphone array.

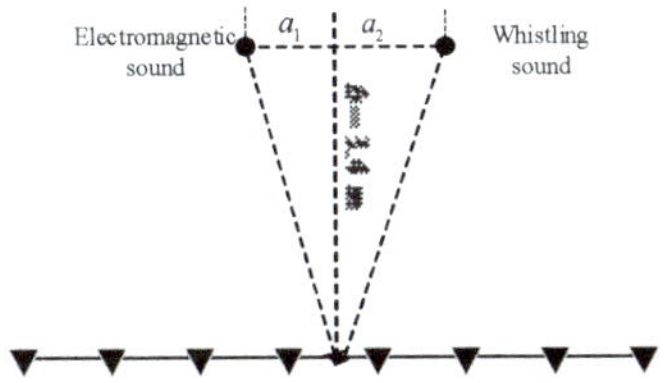

Figure 5. Diagram for experimental data collection.

Table 1. Parameters of experimental system.

Parameter	Symbol	Value
Sampling rate	f_s	96,000 Hz
Number of elements	M	8
Element spacing	d	0.0425 m
Center frequency of electromagnetic sound	f_1	293.5 Hz
Center frequency of whistling sound	f_2	1404 Hz
True offset value of electromagnetic sound	a_1	−0.3 m
True offset value of whistling sound	a_2	0.36 m
Velocity of sound	c	340 m/s
Number of subbands	K	data

The entire Nyquist spectrum was divided into 640 subbands using PFs, with each subband having a bandwidth of 150 Hz. In engineering applications, the transformer's voiceprint needs to be detected and identified before localization. In fault identification, the frequency domain information corresponding to different faulty voiceprints can be determined, and this can be used to determine the subbands in which the faulty voiceprint is located. The electromagnetic sound was localized in the 323rd subband, while the whistling sound was in the 330th subband. Figures 6 and 7 illustrate the frequency spectrograms of the subbands containing these two sound sources in their respective elements.

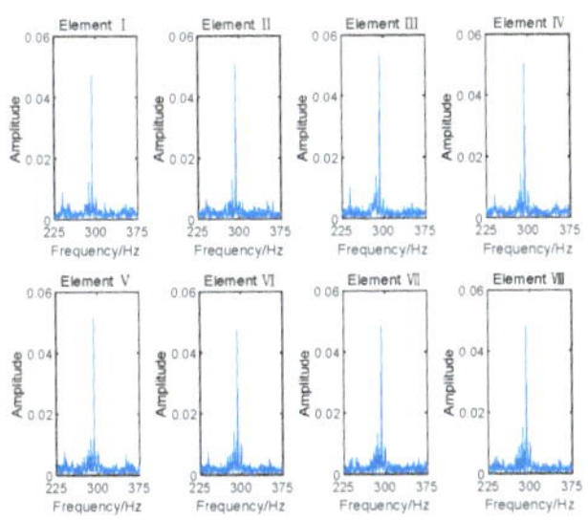

Figure 6. Subband spectrum of array elements for electromagnetic sound.

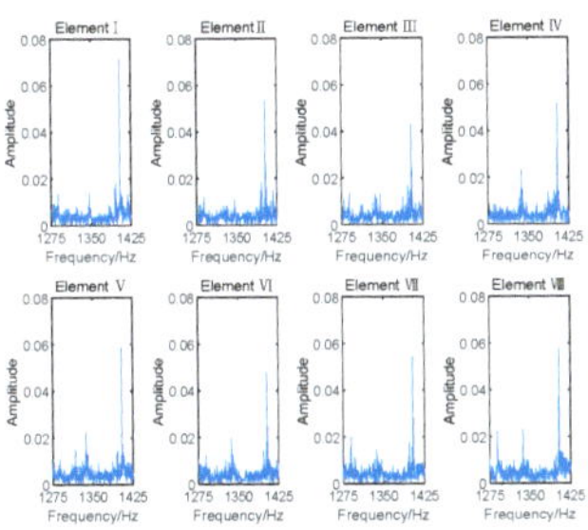

Figure 7. Subband spectrum of array elements for whistling sound.

Figures 8 and 9 show the antenna patterns for the sum and difference beams corresponding to f_1 and f_2, along with their monopulse response curves. These illustrations demonstrate that the main lobe of the beams is aligned at an angle of 0 degrees, indicating a notch in the difference beams. It is important to note that the sum-difference monopulse angle measurement exploits the linear interval of the monopulse response curve near the main lobe. Consequently, as the sound source frequency increases, the main lobe narrows, resulting in a larger value of k' and higher accuracy in angle measurement. Typically, the range of the 3dB main lobe attenuation is considered the linear interval for angle measurement. Thus, in Figure 8, the angle measurement range for the sum and difference beams of the electromagnetic sound is approximately -30° to 30°, while for the whistling sound, it is approximately -15° to 15°.

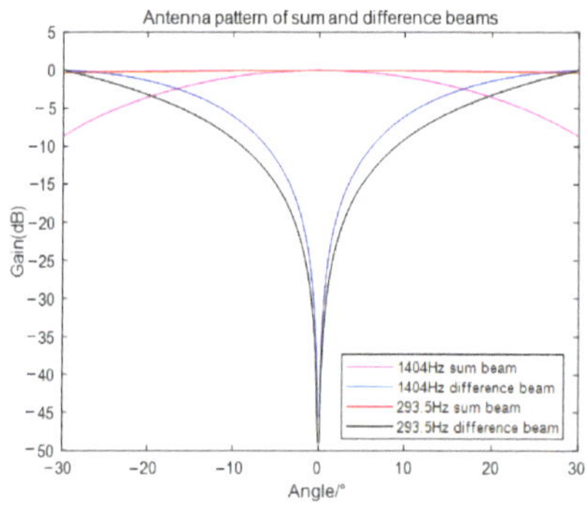

Figure 8. Antenna pattern of sum and difference beams.

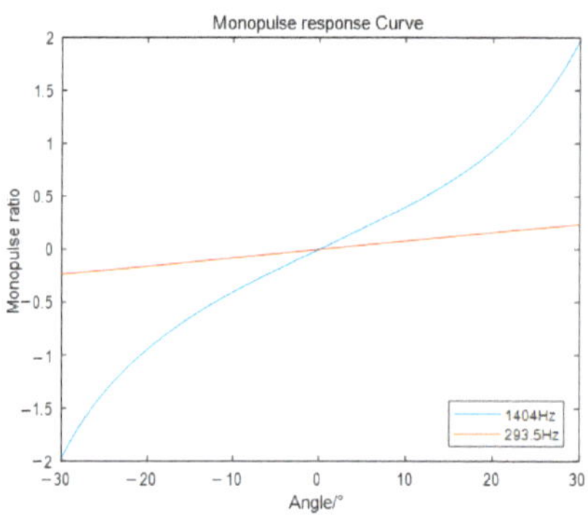

Figure 9. Monopulse response curves.

Figures 10 and 11 show the FFT outputs of the sum and difference beams, respectively, for the electromagnetic sound and the whistling sound. In the validation of the actual measurement data, we used two different spectra of electromagnetic and whistling sounds corresponding to the two kinds of faulty sound patterns for localization. Table 2 shows the offsets in the conversion results obtained from angle estimation using MP and MUSIC algorithms. Through a comparison of these results, it can be concluded that the algorithm proposed in this paper demonstrates significantly improved localization accuracy, both at high and low frequencies.

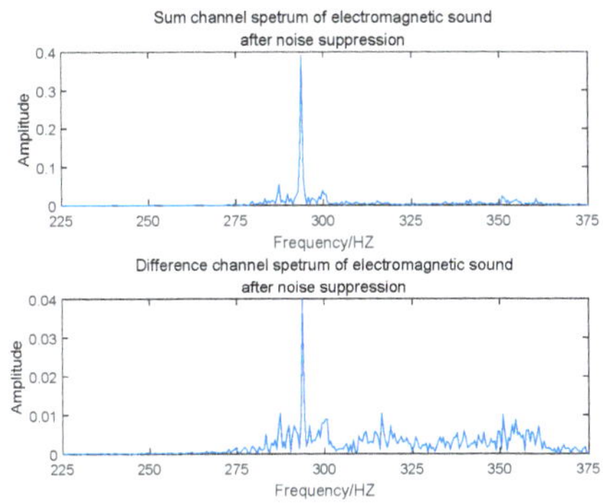

Figure 10. FFT outputs of sum and difference beams of electromagnetic sound.

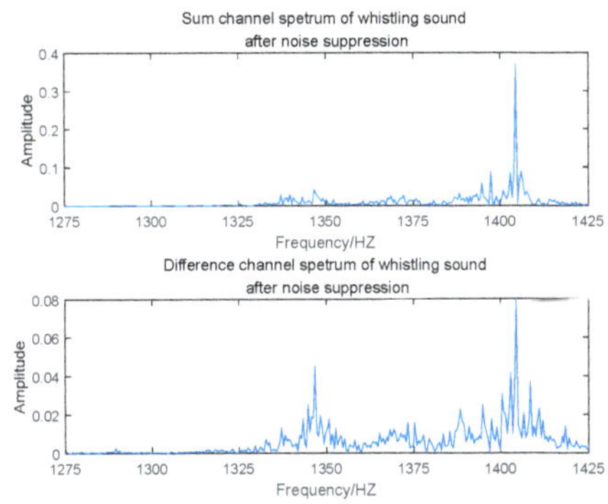

Figure 11. FFT outputs of sum and difference beams of whistling sound.

Table 2. Angle measurement results of different algorithms.

Algorithm	f_1 Estimation Offset/m	f_1 Estimation Error/m	f_2 Estimation Offset/m	f_2 Estimation Error/m
MP	−0.6039	−0.3039	0.3322	−0.0278
MUSIC	−0.6850	−0.3850	0.2601	−0.0999

6. Conclusions

To tackle the challenges faced by classical microphone array-based sound source localization algorithms, which often struggle to balance accuracy and simplicity and, in particular, encounter significant performance degradation when addressing transformer faults in environments with multiple sound sources, this paper investigates the algorithm for multi-sound source localization of power transformer faults based on polyphase filters. The algorithm incorporates PFs to first divide the received signal into different subbands, enabling the downsampling of signals in the time-domain subbands and the separate formation of sum and difference beams. The outputs of these sum and difference beams are transformed into the frequency domain, and their ratio is utilized to achieve high-precision localization of fault sound sources simultaneously. In this paper, the performance of the proposed algorithm is evaluated by comparing its angle measurement results with those obtained using the MUSIC algorithm. The experimental data demonstrate that the proposed algorithm outperforms the MUSIC algorithm in terms of localization accuracy when faced with array amplitude and phase errors. Moreover, the proposed algorithm offers practical advantages due to its lower computational complexity, making it suitable for real-world applications. This algorithm has proven effective in achieving high-precision localization of sound sources for transformer faults. It helps narrow down the troubleshooting scope and assists operation and inspection personnel in targeting fault sources when abnormalities initially emerge. These benefits significantly enhance the efficiency of operation and inspection, thereby demonstrating substantial application potential. The present study also establishes a strong foundation for future investigations concentrating on ways to further suppress interference and enhance the localization accuracy of this algorithm in complex environments where interference and noise coexist.

Author Contributions: Conceptualization, H.Z. and L.L.; methodology, H.Z., L.L. and M.S.; software, M.S. and Y.H.; validation, H.Z., L.L., Z.S. and M.S.; formal analysis, H.Z. and Z.S.; investigation, H.Z. and Z.S.; resources, H.Z., L.L. and Z.S.; data curation, H.Z. and L.L.; writing—original draft preparation, H.Z.; writing—review and editing, L.L., M.S., Z.S. and Y.H.; visualization, H.Z.; supervision, H.Z. and L.L.; project administration, H.Z. and L.L.; funding acquisition, H.Z., L.L., M.S., Z.S. and Y.H. All authors have read and agreed to the published version of the manuscript.

Funding: This research was funded by the Science and Technology Project of NARI Technology Nanjing Control System Co., Ltd. (No. 524608220231) and the Project of Jiangsu Key Laboratory of Power Transmission and Distribution Equipment Technology (No. 2023JSSPD03).

Institutional Review Board Statement: Not applicable.

Informed Consent Statement: Not applicable.

Data Availability Statement: The data used to support the findings of this study are available from the corresponding author upon request.

Acknowledgments: We sincerely thank all the reviewers for their patient reviews and valuable suggestions on this article.

Conflicts of Interest: Authors Hualiang Zhou, Zhantao Su and Lu Lu were employed by the companies NARI Group Corporation (State Grid Electric Power Research Institute) and NARI Technology Nanjing Control System Co., Ltd. The remaining authors declare that the research was conducted in the absence of any commercial or financial relationships that could be construed as a potential conflict of interest.

References

1. Wang, W.Z.; Wang, C.; Ao, X. Fruit fly optimization algorithm based fault localization in distribution networks. *Power Syst. Prot. Control* **2019**, *47*, 108–114.
2. Zheng, C.; Zhou, H.F.; Zhen, D.Q.; Lin, Z.H.; Zhang, X.J. Research on active distribution network fault localization method based on improved multivariate universe algorithm. *Power Syst. Prot. Control* **2023**, *51*, 169–179.
3. Li, H.; Zhang, L.L.; Luan, Y.F.; Shan, H.T.; Mo, C. Distribution network fault localization method based on subsystem division and injected current ratio. *Power Syst. Prot. Control* **2023**, *51*, 63–72.
4. Wang, J.; Zhao, Z.; Zhu, J.; Li, X.; Dong, F.; Wan, S. Improved Support Vector Machine for Voiceprint Diagnosis of Typical Faults in Power Transformers. *Machines* **2023**, *11*, 539. [CrossRef]
5. Wang, H.; Liu, J.; Xu, F.; Liu, Y. 3-D sound source localization with a ternary microphone array based on TDOA-ILD algorithm. *IEEE Sens. J.* **2022**, *22*, 19826–19834. [CrossRef]
6. Zhang, X.; Sun, H.; Wang, S.; Xu, J. A new regional localization method for indoor sound source based on convolutional neural network. *IEEE Access* **2018**, *6*, 72073–72082. [CrossRef]
7. Zhang, C.; Chen, J.D.; Luo, L.E.; Sheng, G.H.; Jiang, X.C. Localization method of abnormal acoustic source in substation based on maximum likelihood estimation. In Proceedings of the 2021 4th International Conference on Energy, Electrical and Power Engineering (CEEPE), Chongqing, China, 23–25 April 2021.
8. Zhang, C.; Chen, J.D.; Luo, L.E.; Sheng, G.H.; Jiang, X.C. A method for acoustic source orientation in substation based on non redundant fourth order cumulants. *High Volt. Technol.* **2022**, *48*, 75–83.
9. Qian, X.; Tang, X.; Wang, Z.; Meng, Q. A narrowband acoustic source localization method and system based on a non-uniform linear microphone array. In Proceedings of the IEEE 22nd International Conference on Communication Technology (ICCT), Nanjing, China, 11–14 November 2022.
10. Hahn, W.; Tretter, S. Optimum Processing for delay-vector estimation in passive signal arrays. *IEEE Trans. Inf. Theory* **1973**, *19*, 608–614. [CrossRef]
11. Dibiase, J.H. A High-Accuracy, Low-Latency technique for talker localization in reverberant environments using microphone arrays. *Eur. J. Biochem.* **2000**, *216*, 281–291.
12. Schmidt, R.O. Multiple Emitter location and signal parameter estimation. *IEEE Trans. Antennas Propag.* **1986**, *34*, 276–280. [CrossRef]

13. Zhang, H.M.; Zhang, H.Y. Research on DOA estimation method of sonar radar target based on MUSIC algorithm. *J. Phys. Conf. Series. IOP Publ.* **2019**, *1176*, 32001–32005. [CrossRef]
14. Wang, Y.M.; Liu, S.; Jin, M. Localization of coherent signals based on toeplitz matrix reconstruction in spatially colored nosie. In Proceedings of the 2019 IEEE 2nd International Conference on Electronic Information and Communication Technology (ICEICT), Harbin, China, 20–22 January 2019.
15. Vaidyanathan, P.P. Multirate digital filters, filter banks, polyphase networks, and applications: A tutorial. *Proc. IEEE* **1990**, *78*, 56–93. [CrossRef]
16. Li, H.Q.; Zhang, H.B. Design of a Digital Channelized Receiver Based on Multiphase Filtering. *Inf. Commun.* **2014**, *31*, 47–49.
17. Zhao, W.C.; Huang, W.; Li, X.; Hou, G.L. Software Radio Receiver Based on Multiphase Filtering. *Commun. Technol.* **2021**, *54*, 1514–1520.
18. Zhang, X.; Li, Y.; Yang, X.; Zheng, L.; Long, T.; Baker, C.J. A novel constrained monopulse technique for adaptive phased arrays in the presence of interference. In Proceedings of the 2016 CIE International Conference on Radar (RADAR), Guangzhou, China, 10–13 October 2016.
19. Yu, K.B.; Fernández, M.F. Robust adaptive monopulse processing for multiple observations with applications to TS-MIMO radar. In Proceedings of the 2020 IEEE 11th Sensor Array and Multichannel Signal Processing Workshop (SAM), Hangzhou, China, 8–11 June 2020.
20. Zhou, G.L.; Xu, J.; Gao, Y.; Mao, M.Y.; Yu, M. High precision secondary radar tracking algorithm based on monopulse tracking technology. *Technol. Perspect.* **2020**, *22*, 25–27.
21. Wang, C.H. Research on Radar Angle Measurement Method. Master's Thesis, Xi'an University of Electronic Science and Technology, Xi'an, China, 2014.
22. Zhou, W. Research and Implementation of Broadband Channelized Digital Receiver Based on FPGA. Master's Thesis, University of Electronic Science and Technology of China, Chengdu, China, 2012.
23. Zhu, L. Research and Implementation of Single Pulse Angle Measurement and Differential Beamforming Algorithm. Master's Thesis, Nanjing University of Science and Technology, Nanjing, China, 2018.
24. Takahashi, R.; Inaba, T.; Takahashi, T.; Tasaki, H. Digital monopulse beamforming for achieving the CRLB for angle accuracy. *IEEE Trans. Aerosp. Electron. Syst.* **2018**, *54*, 315–323. [CrossRef]
25. Fu, M.; Gao, C.; Li, Y.; Deng, Z.; Chen, D. Monopulse-radar angle estimation of multiple targets using multiple observations. *IEEE Trans. Aerosp. Electron. Syst.* **2021**, *57*, 968–983. [CrossRef]
26. Yu, S.; Ma, J. Complex Variational mode decomposition for slop-preserving denoising. *IEEE Trans. Geosci. Remote Sens.* **2018**, *56*, 586–597. [CrossRef]

sensors

MDPI

Article

Direct Position Determination of Non-Gaussian Sources for Multiple Nested Arrays: Discrete Fourier Transform and Taylor Compensation Algorithm

Hao Hu [1,2,3], Meng Yang [1,2], Qi Yuan [1,2], Mingyi You [4,5], Xinlei Shi [1,2,*] and Yuxin Sun [1,2]

[1] College of Electronic Information Engineering, Nanjing University of Aeronautics and Astronautics, Nanjing 211106, China; 18061368794@163.com (H.H.); yangmeng19861213@126.com (M.Y.); yuanqi@nuaa.edu.cn (Q.Y.); sunyuxin@nuaa.edu.cn (Y.S.)
[2] Key Laboratory of Dynamic Cognitive System of Electromagnetic Spectrum Space, Nanjing University of Aeronautics and Astronautics, Ministry of Industry and Information Technology, Nanjing 211106, China
[3] Jiangsu Automation Research Institute, Lianyungang 222000, China
[4] China Electronics Technology Group Corporation No.36 Research Institute, Jiaxing 314033, China; youmingyi@126.com
[5] National Key Laboratory of Electromagnetic Space Security, Jiaxing 314033, China
* Correspondence: lincoln@nuaa.edu.cn

Abstract: This paper delves into the problem of direct position determination (DPD) for non-Gaussian sources. Existing DPD algorithms are hindered by their high computational complexity from exhaustive grid searches and a disregard for the received signal characteristics by multiple nested arrays (MNAs). To address these issues, the paper proposes a novel DPD algorithm for non-Gaussian sources with MNAs: the Discrete Fourier Transform (DFT) and Taylor compensation algorithm. Initially, the fourth-order cumulant matrix of the received signal is computed, and the vectorizing method is applied. Subsequently, a computationally efficient DPD cost function is proposed by leveraging a normalized DFT matrix to reduce complexity. Finally, first-order Taylor compensation is utilized to enhance the accuracy of the localization results. The superiority of the proposed algorithm is demonstrated through numerical simulation results.

Keywords: non-Gaussian signal; multiple nested arrays; direct position determination; Discrete Fourier Transform; Taylor compensation

1. Introduction

The advancement of wireless positioning technology has led to its widespread application in various sectors, including military defense, emergency rescue [1,2], resource exploration, intelligent transportation, and more [3–5]. Precisely locating enemy radiation sources is crucial for victory in battlefield scenarios. Wireless localization technology is categorized into traditional two-step positioning and direct position determination (DPD) technology [6–8].

The two-step positioning framework can be further classified into angle of arrival (AOA) [9], time of arrival (TOA) [10], time difference of arrival (TDOA) [11], received signal strength (RSS) [12], and other categories based on intermediate parameters estimation. While this technology estimates intermediate parameters from received signals and then solves spatial geometry problems to determine radiation source positions, it suffers from information loss between steps and suboptimal accuracy due to errors in parameter matching processes [13].

A DPD algorithm is suggested in [14] as a solution to address the challenges mentioned. This algorithm has garnered attention for its ability to achieve superior localization accuracy compared to the two-step algorithm, particularly in low signal-to-noise ratio (SNR) environments. The DPD algorithm directly determines the source position from the original received data, eliminating the need for estimating signal parameters and thereby avoiding estimation errors associated with intermediate parameters in the two-step approach. Moreover, the DPD algorithm does not require a matching procedure. While no intermediate parameters are essential in the DPD framework, certain signal parameters must still be taken into account in the algorithm model. The maximum likelihood (ML) DPD estimator in [15] incorporates joint information from angle of arrival (AOA) and time difference of arrival (TDOA) to achieve high localization accuracy through exhaustive search, albeit at the cost of increased complexity in scenarios with multiple sources. To mitigate this complexity, the subspace data fusion (SDF) technology [16] based on multiple signal classification (MUSIC) [17] is proposed. Furthermore, the Capon DPD algorithm, which avoids eigenvalue decomposition (EVD), is introduced for situations with multiple sources [18]. However, all the aforementioned algorithms treat the source signal as arbitrary, whereas studies have proved that better positioning accuracy can be achieved when the property of the source signal is a priori known [19,20]. A localization algorithm tailored for orthogonal frequency division multiplexing (OFDM) signals is detailed in [21]. Due to the high computational demands of ML algorithms, an extended SDF DPD algorithm is proposed in [22]. Additionally, features of non-circular signals are harnessed to enhance positioning accuracy and increase the degree of freedom (DOF) as evidenced by studies such as [23–25].

In the realm of localization research, it is commonly presumed that a signal adheres to a Gaussian distribution, with the second-order cumulant (SOC) [26] being utilized to derive a probability density function that encapsulates all signal information. However, practical scenarios often involve signals that deviate from Gaussian distribution, rendering lower-order cumulants insufficient in capturing all signal details. Therefore, the fourth-order cumulant (FOC) introduced in [27] is employed for signal analysis. Unlike SOC, FOC is adept at disregarding Gaussian noise and expanding array elements, leading to enhanced parameter estimation precision as evidenced in [28–30]. Moreover, prevalent position estimation techniques rely on uniform linear arrays (ULAs), which suffer from densely positioned elements, inducing heightened mutual coupling. Proposals to mitigate this challenge include sparse arrays with larger apertures and reduced mutual coupling, exemplified by classic coprime arrays and nested arrays (NAs) as put forth in [31–33].

In this paper, we propose a novel DPD algorithm utilizing multiple nested arrays (MNAs) for non-Gaussian sources. The main contributions are summarized as follows:

- The property of non-Gaussian sources is fully exploited to suppress Gaussian noise and augment the virtual array aperture, which benefits the available degrees of freedom (DOFs).
- We propose a novel low-complexity DPD algorithm of non-Gaussian sources utilizing MNAs. We deploy the Discrete Fourier Transform (DFT) method to construct a computationally efficient DPD cost function to reduce the high computational complexity caused by exhaustive grid search.
- We utilize the Taylor compensation method to improve the localization accuracy at the expense of calculating the position estimation bias. It should be emphasized that even when the source position does not fall on the preset grid, the proposed algorithm can still estimate the position of sources accurately.
- Complexity analysis and extensive numerical results are presented to verify the superiority of the proposed algorithm in terms of location accuracy, resolution capability, and computational complexity.

The following structure of the article is, the MNAs localization model is introduced in Section 2. The proposed algorithm is derived in Section 3. Performance analysis, numerical results, and conclusions are drawn in Sections 4, 5 and 6, respectively.

Notations: Vectors and matrices are lower-case bold and upper-case bold, respectively. $(\cdot)^T$, $(\cdot)^*$, and $(\cdot)^H$ denote transposition, conjugate, and Hermitian transpose, respectively. $\mathbf{x}(n)$ extracts the nth element of vector $\mathbf{x}$; $E(\cdot)$ represents the expectation operator; $\otimes$ is the Kronecker product; $\text{vec}(\cdot)$ stacks the columns of a matrix into a vector; $\partial(x)/\partial(y)$ denotes the derivation of x with respect to y; diag($\mathbf{x}$) turns vector $\mathbf{x}$ into a diagonal matrix; and $\|\cdot\|$ denotes the l_2 norm.

2. Model Formulation

In the context depicted in Figure 1, we analyze a scenario where K incoherent non-Gaussian sources emit narrow-band stationary signals in a two-dimensional localization setting, with unknown positions. It is assumed that the quantity of sources K is pre-determined, and various established techniques can be employed to ascertain it [34,35]. To pinpoint these unidentified sources, we deploy L spatially distributed sensor arrays, each furnished with a NA comprising M array elements. The positions of the sources and sensor arrays are represented as $\mathbf{p}_k = [x_k, y_k]^T (k = 1, \cdots, K)$ and $\mathbf{u}_l = [x_l, y_l]^T (l = 1, \cdots, L)$, respectively. The signal received by the lth sensor array at the tth $(t = 1, \cdots, \mathrm{T})$ snapshot can be formulated as per [36]:

$$\mathbf{x}_l(t) = \mathbf{A}_l(\mathbf{p})\mathbf{s}(t) + \mathbf{n}_l(t) \tag{1}$$

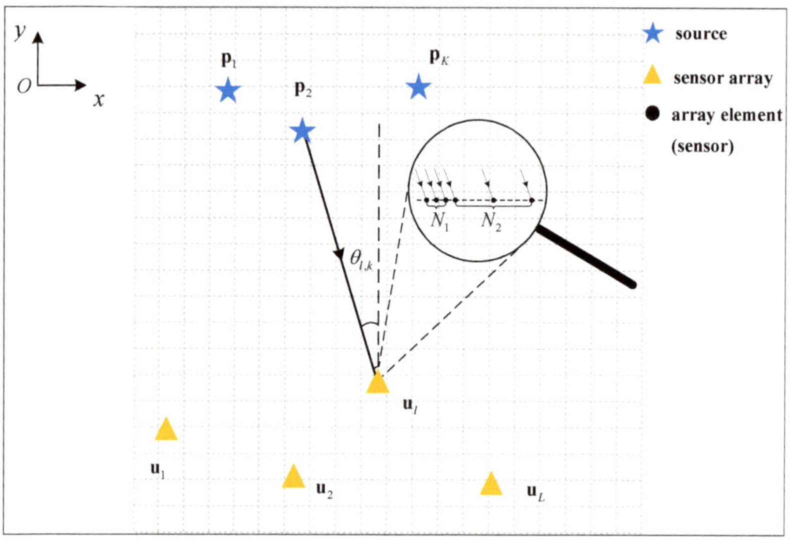

Figure 1. Geometry of multiple nested arrays localization.

with the following notational definitions:

- $\mathbf{p} = \left[\mathbf{p}_1^T, \cdots, \mathbf{p}_K^T\right]^T \in \mathbb{C}^{2K\times 1}$ is the position vector of K sources;
- $\mathbf{A}_l(\mathbf{p}) = [\mathbf{a}_l(\mathbf{p}_1), \mathbf{a}_l(\mathbf{p}_2), \cdots, \mathbf{a}_l(\mathbf{p}_K)] \in \mathbb{C}^{M\times K}$ is the array manifold of the lth NA, and $\mathbf{a}_l(\mathbf{p}_k) = \left[e^{j2\pi d_1 \frac{\sin\varphi_{l,k}}{\lambda}}, e^{j2\pi d_2 \frac{\sin\varphi_{l,k}}{\lambda}}, \cdots, e^{j2\pi d_M \frac{\sin\varphi_{l,k}}{\lambda}}\right]^T \in \mathbb{C}^{M\times 1}$ denotes the array steering vector, where $\sin\varphi_{l,k} = \frac{\mathbf{p}_k(1)-\mathbf{u}_l(1)}{\|\mathbf{u}_l-\mathbf{p}_k\|}$. Array element locations $d_1, d_2, \cdots, d_M$ are given by the set

$$\mathbb{S} = \{d, 2d, \cdots, N_1 d, (N_1+1)d, 2(N_1+1)d, \cdots, N_2(N_1+1)d\} \tag{2}$$

  where $M = N_1 + N_2$, and d denotes the space between array elements, which is usually set as half of wavelength $\lambda/2$;
- $\mathbf{s}(t) = [s_1(t), s_2(t), \cdots, s_K(t)]^T \in \mathbb{C}^{K\times 1}$ is the signal vector transmitted from the kth source at time t, where $k = 1, 2, \cdots, K$, $t = 1, 2, \cdots, \mathrm{T}$;
- $\mathbf{n}_l(t)$ denotes the independent additive white Gaussian noise vector of the lth sensor array.

3. Proposed Algorithm

Define the FOC of a stationary stochastic process x as [37]:

$$Cum\left(x_{k_1}, x_{k_2}, x_{k_3}^*, x_{k_4}^*\right) = E\left(x_{k_1}x_{k_2}x_{k_3}^*x_{k_4}^*\right) - E(x_{k_1}x_{k_2})E\left(x_{k_3}^*x_{k_4}^*\right) - E\left(x_{k_1}x_{k_3}^*\right)E\left(x_{k_2}x_{k_4}^*\right) - E\left(x_{k_1}x_{k_4}^*\right)E\left(x_{k_2}x_{k_3}^*\right) \quad (3)$$

where $k_1, k_2, k_3, k_4, 1 \leq k_1, k_2, k_3, k_4 \leq \mathrm{T}$ are integers. And the FOC matrix of the received signal $\mathbf{x}_l(t)$ can be expressed as [38,39]:

$$\mathbf{R}_{l,4} = \sum_{k=1}^{K} c_{s_k,4}[\mathbf{a}_l(\mathbf{p}_k) \otimes \mathbf{a}_l^*(\mathbf{p}_k)][\mathbf{a}_l(\mathbf{p}_k) \otimes \mathbf{a}_l^*(\mathbf{p}_k)]^H = \sum_{k=1}^{K} c_{s_k,4}\mathbf{a}_{l,4\mathbf{x}}(\mathbf{p}_k)\mathbf{a}_{l,4\mathbf{x}}^H(\mathbf{p}_k) \quad (4)$$

where $\mathbf{a}_{l,4\mathbf{x}}(\mathbf{p}_k) = \mathbf{a}_l(\mathbf{p}_k) \otimes \mathbf{a}_l^*(\mathbf{p}_k), 1 \leq k \leq K$, $c_{s_k,4} = Cum(s_k(t), s_k(t), s_k^*(t), s_{l,k}^*(t))$ denotes the FOC of the kth signal and is calculated using (3). Notably, the elements of $\mathbf{a}_{l,4\mathbf{x}}(\mathbf{p}_k)$ can be constructed with $e^{j2\pi(d_i-d_j)\frac{\sin\varphi_{l,k}}{\lambda}}$, $d_i, d_j \in \mathbb{S}$. In practice, $\mathbf{a}_{l,4\mathbf{x}}(\mathbf{p}_k)$ can be considered the steering vector of $\mathbf{R}_{l,4}$, which expands from $\mathbf{a}_l(\mathbf{p}_k)$. To further improve the accuracy of localization, we vectorize $\mathbf{R}_{l,4}$ by column [40]:

$$\tilde{\mathbf{z}}_l = \mathrm{vec}(\mathbf{R}_{l,4}) = \tilde{\mathbf{A}}_l(\mathbf{p})\mathbf{c} \quad (5)$$

where $\tilde{\mathbf{A}}_l(\mathbf{p})$ takes the form:

$$\tilde{\mathbf{A}}_l(\mathbf{p}) = \left[\mathbf{a}_{l,4\mathbf{x}}^*(\mathbf{p}_1) \otimes \mathbf{a}_{l,4\mathbf{x}}(\mathbf{p}_1), \cdots, \mathbf{a}_{l,4\mathbf{x}}^*(\mathbf{p}_K) \otimes \mathbf{a}_{l,4\mathbf{x}}(\mathbf{p}_K)\right] - [\mathbf{a}_{l,vec}(\mathbf{p}_1), \cdots, \mathbf{a}_{l,vec}(\mathbf{p}_K)] \quad (6)$$

and

$$\mathbf{c} = [c_{s_1,4}, c_{s_2,4}, \cdots, c_{s_K,4}]^T \quad (7)$$

Without loss of generality, suppose that the number of sensors of two subarrays of each NA is equal, i.e., $N_1 = N_2 = M/2$. Then, the closed-form expression of the $\mathbf{a}_{l,vec}(\mathbf{p}_k)$ location can be stacked as [40]

$$\mathbb{S}_{vec} = \{-M_{vec}d, -(M_{vec}-1)d, \cdots, (M_{vec}-1)d, M_{vec}d | M_{vec} = 2N_1(N_1+1) - 2\} \quad (8)$$

with many redundant elements. According to [41], matrix $\mathbf{A}_{l,sort}(\mathbf{p}) \in \mathbb{C}^{(2M_{vec}+1)\times K}$ is formed by removing the redundancy from $\mathbf{a}_{l,vec}(\mathbf{p}_k)$. Meanwhile, the corresponding virtual signal vector $\mathbf{z}_{l,sort}$ is given by

$$\mathbf{z}_{l,sort} = \mathbf{A}_{l,sort}(\mathbf{p})\mathbf{c} \quad (9)$$

Then, according to the geometry relationship showed in Figure 1, for a random grid $\mathbf{p} = [x, y]^T$, we have

$$\sin\varphi_l = \frac{x - x_l}{\sqrt{(x-x_l)^2 + (y-y_l)^2}} \quad (10)$$

Let $q_l = \frac{J}{2}\sin\varphi_l$, where $J = 2M_{vec} + 1$, yielding

$$q_l = \frac{J(x-x_l)}{2\sqrt{(x-x_l)^2 + (y-y_l)^2}} \quad (11)$$

Define DFT vector $\mathbf{v}_{l,\mathbf{p}} \in \mathbb{C}^{J\times 1}$, the τth element of which is

$$\mathbf{v}_{l,\mathbf{p}}(\tau) = \left(e^{-j2\pi q_l/J}\right)^{\tau} \quad (12)$$

Substitute (11) into (12), yielding

$$\mathbf{v}_{l,\mathbf{p}}(\tau) = e^{\frac{-j\pi\tau(x-x_l)}{\sqrt{(x-x_l)^2+(y-y_l)^2}}} \quad (13)$$

where $1 \leq \tau \leq J$. Combining the information of MNAs, we can construct the following DPD problem:

$$\hat{\mathbf{p}}_k^{\text{ini}} = \arg\max_{\mathbf{p}} \sum_{l=1}^{L} \sum_{\tau=1}^{J} \mathbf{v}_{l,\mathbf{p}}(\tau)\mathbf{z}_{l,sort}(\tau) \tag{14}$$

where the estimated location $\hat{\mathbf{p}}_k^{\text{ini}} = \left[\hat{x}_k^{\text{ini}}, \hat{y}_k^{\text{ini}}\right]^T, k = 1, 2, \cdots, K$ is based on the assumption that the position of the source falls on the preset grid. When the assumption is not met, the off-grid error will always exist. To overcome this issue, we apply the Taylor compensation method to the initial position estimation.

Define vectors

$$\mathbf{p}_x = [x_1, x_2, \cdots, x_K]^T \tag{15}$$

$$\mathbf{p}_y = [y_1, y_2, \cdots, y_K]^T \tag{16}$$

$$\hat{\mathbf{p}}_x^{\text{ini}} = \left[\hat{x}_1^{\text{ini}}, \hat{x}_2^{\text{ini}}, \cdots, \hat{x}_K^{\text{ini}}\right]^T \tag{17}$$

$$\hat{\mathbf{p}}_y^{\text{ini}} = \left[\hat{y}_1^{\text{ini}}, \hat{y}_2^{\text{ini}}, \cdots, \hat{y}_K^{\text{ini}}\right]^T \tag{18}$$

$$\mathbf{z}_{sort} = \left[\mathbf{z}_{1,sort}^T, \mathbf{z}_{2,sort}^T, \cdots, \mathbf{z}_{L,sort}^T\right]^T \tag{19}$$

and matrix

$$\mathbf{A}_{sort}(\mathbf{p}) = \left[\mathbf{A}_{1,sort}^T(\mathbf{p}), \mathbf{A}_{2,sort}^T(\mathbf{p}), \cdots, \mathbf{A}_{L,sort}^T(\mathbf{p})\right]^T \tag{20}$$

According to (9), we have

$$\mathbf{z}_{sort} = \mathbf{A}_{sort}(\mathbf{p})\mathbf{c} \tag{21}$$

Performing the first-order Taylor expansion of $\mathbf{A}_{sort}(\mathbf{p})$ at $\hat{\mathbf{p}}^{\text{ini}} = \left[\left(\hat{\mathbf{p}}_1^{\text{ini}}\right)^T, \cdots, \left(\hat{\mathbf{p}}_K^{\text{ini}}\right)^T\right]^T$, and ignoring the second-order and higher-order terms, we have:

$$\mathbf{A}_{sort}(\mathbf{p}) \approx \mathbf{A}_{sort}\left(\hat{\mathbf{p}}^{\text{ini}}\right) + \sum_{i=1}^{2K} \left(\mathbf{p}(i) - \hat{\mathbf{p}}^{\text{ini}}(i)\right) \frac{\partial \mathbf{A}_{sort}(\mathbf{p})}{\partial \mathbf{p}(i)}\bigg|_{\mathbf{p}=\hat{\mathbf{p}}^{\text{ini}}} \tag{22}$$

where $\mathbf{p}(i)$ and $\hat{\mathbf{p}}^{\text{ini}}(i)$ extract the ith element of $\mathbf{p}$ and $\hat{\mathbf{p}}^{\text{ini}}$, respectively. Substitute (15)–(18) into (22), yielding

$$\mathbf{A}_{sort}(\mathbf{p}) \approx \mathbf{A}_{sort}\left(\hat{\mathbf{p}}^{\text{ini}}\right) + \frac{\partial \mathbf{A}_{sort}(\mathbf{p})}{\partial \mathbf{p}_x^T}\bigg|_{\mathbf{p}_x=\hat{\mathbf{p}}_x^{\text{ini}}} \mathbf{\Lambda}_x + \frac{\partial \mathbf{A}_{sort}(\mathbf{p})}{\partial \mathbf{p}_y^T}\bigg|_{\mathbf{p}_y=\hat{\mathbf{p}}_y^{\text{ini}}} \mathbf{\Lambda}_y \tag{23}$$

where $\mathbf{\Lambda}_x = \text{diag}(\delta_x)$, $\mathbf{\Lambda}_y = \text{diag}(\delta_y)$ with $\delta_x = \mathbf{p}_x - \hat{\mathbf{p}}_x^{\text{ini}}$, $\delta_y = \mathbf{p}_y - \hat{\mathbf{p}}_y^{\text{ini}}$. Substitute (23) into (21), yielding

$$\hat{\mathbf{z}}_{sort} \approx \left(\mathbf{A}_{sort}\left(\hat{\mathbf{p}}^{\text{ini}}\right) + \frac{\partial \mathbf{A}_{sort}(\mathbf{p})}{\partial \mathbf{p}_x^T}\bigg|_{\mathbf{p}_x=\hat{\mathbf{p}}_x^{\text{ini}}} \mathbf{\Lambda}_x + \frac{\partial \mathbf{A}_{sort}(\mathbf{p})}{\partial \mathbf{p}_y^T}\bigg|_{\mathbf{p}_y=\hat{\mathbf{p}}_y^{\text{ini}}} \mathbf{\Lambda}_y\right)\mathbf{c} \tag{24}$$

Rewrite (24) as

$$\hat{\mathbf{z}}_{sort} \approx \left[\mathbf{A}_{sort}\left(\hat{\mathbf{p}}^{\text{ini}}\right), \frac{\partial \mathbf{A}_{sort}(\mathbf{p})}{\partial \mathbf{p}_x^T}\bigg|_{\mathbf{p}_x=\hat{\mathbf{p}}_x^{\text{ini}}}, \frac{\partial \mathbf{A}_{sort}(\mathbf{p})}{\partial \mathbf{p}_y^T}\bigg|_{\mathbf{p}_y=\hat{\mathbf{p}}_y^{\text{ini}}}\right]\begin{bmatrix} \mathbf{c} \\ \omega_x \\ \omega_y \end{bmatrix} \tag{25}$$

where $\boldsymbol{\omega}_x = \boldsymbol{\Lambda}_x \mathbf{c}$ and $\boldsymbol{\omega}_y = \boldsymbol{\Lambda}_y \mathbf{c}$. Let $\mathbf{B} = \left[\mathbf{A}_{sort}(\hat{\mathbf{p}}^{\text{ini}}), \left. \frac{\partial \mathbf{A}_{sort}(\mathbf{p})}{\partial \mathbf{p}_x^T} \right|_{\mathbf{p}_x = \hat{\mathbf{p}}_x^{\text{ini}}}, \left. \frac{\partial \mathbf{A}_{sort}(\mathbf{p})}{\partial \mathbf{p}_y^T} \right|_{\mathbf{p}_y = \hat{\mathbf{p}}_y^{\text{ini}}} \right]$, $\mathbf{y} = \begin{bmatrix} \mathbf{c} \\ \boldsymbol{\omega}_x \\ \boldsymbol{\omega}_y \end{bmatrix}$, the least square solution of $\mathbf{y}$ is readily given by

$$\hat{\mathbf{y}} = \left(\mathbf{B}^H \mathbf{B} \right)^{-1} \mathbf{B}^H \hat{\mathbf{z}}_{sort} \tag{26}$$

It should be noted that the estimated value $\hat{\mathbf{c}}$ is composed of the first to Kth elements of $\hat{\mathbf{y}}$, $\hat{\omega}_x$ is composed of the $(K+1)$th to $2K$th elements of $\hat{\mathbf{y}}$, and $\hat{\omega}_y$ is composed of the $(2K+1)$th to $3K$th elements of $\hat{\mathbf{y}}$. Therefore, the accurate position estimation after Taylor compensation is given by

$$\hat{\mathbf{p}}_x = \hat{\mathbf{p}}_x^{\text{ini}} + \hat{\omega}_x./\hat{\mathbf{c}} \tag{27}$$

$$\hat{\mathbf{p}}_y = \hat{\mathbf{p}}_y^{\text{ini}} + \hat{\omega}_y./\hat{\mathbf{c}} \tag{28}$$

The main steps of the proposed algorithm are summarized as in Algorithm 1.

Algorithm 1: Main Steps of the FOC-DFT-Taylor Algorithm

Input: $\{\mathbf{x}_l(t)\}_{t=1}^{\mathrm{T}}, K, \{\mathbf{u}_l\}_{l=1}^{L}, \lambda, d, M, N_1, N_2$.

1. Calculate the FOC matrix $\hat{\mathbf{R}}_{l,4}$ according to (4);
2. Vectorize the matrix $\hat{\mathbf{R}}_{l,4}$ and remove the redundancy to obtain the virtual signal vector $\mathbf{z}_{l,sort}$ utilizing (5) and (8);
3. Construct the normalized DFT vector according to (12) and obtain initial estimated values of the source position $\hat{\mathbf{p}}_k^{\text{ini}} = [\hat{x}_k, \hat{y}_k]^T, k = 1, 2, \cdots, K$ using (14);
4. Perform the first-order Taylor expansion of $\mathbf{A}_{sort}(\mathbf{p})$ at $\hat{\mathbf{p}}^{\text{ini}}$ and construct the least square constraint utilizing (23) and (26);
5. Obtain accurate position estimation after compensation according to (27) and (28).

Output: $\hat{\mathbf{p}}_x, \hat{\mathbf{p}}_y$

4. Performance Discussion

4.1. Complexity

Here, the main computational complexity of different algorithms is compared in terms of complex number multiplication times. The parameters used are listed as follows: M is the number of array elements; K denotes the number of sources; L is the number of NAs; T represents the number of snapshot; $J = 2M_{vec} + 1$ is the length of virtual signal vector $\mathbf{z}_{l,sort}$; and we denote the number of search grids for x and y directions as recorded as $\mathbb{D}_x$ and $\mathbb{D}_y$. The main complexity of the proposed algorithm lies in the calculation of the FOC matrix, $O(LTM^4)$; grid search, $O(LJ\mathbb{D}_x\mathbb{D}_y)$; complex differentiation, $O(2LKJ)$; and compensation, $O(27K^3 + 9K^2 + 9K^2LJ + 3KLJ)$. For the sake of comparison, the complexity of the proposed algorithm (termed as FOC-DFT-Taylor), the classic SDF DPD algorithm with spatial smoothing technology (termed as FOC-SS-SDF-DPD) [22,42], the Capon DPD algorithm with spatial smoothing technology (termed as FOC-SS-Capon-DPD) [18], and the proposed DFT DPD algorithm with SOC (termed as SOC-DFT-DPD) are given in Table 1, where $N_a = (J+1)/2$ denotes the length of the smoothing window.

Figure 2 shows the computational complexity of four different algorithms with the parameters setting as below: $M = 8$, $K = 12$, $N_1 = N_2 = 4$, $\mathrm{T} = 500$, $J = 77$, $N_a = 39$, $\mathbb{D}_x = \mathbb{D}_y = 400$, and the number of NAs is changed from 3 to 10. As we can see, compared to the FOC-SS-SDF-DPD and FOC-SS-Capon-DPD algorithm, the proposed FOC-DFT-DPD and FOC-DFT-Taylor algorithms reduce the computational complexity thanks to the computationally efficient cost function. The complexity of the proposed FOC-DFT-Taylor is a bit heavier than that of the FOC-DFT-DPD and SOC-DFT-DPD algorithms, while it exhibits higher localization accuracy, which will be shown in Section 5.

Table 1. Comparison of computational complexity.

Methods	Computational Complexity
SOC-DFT-DPD	$O\left(LTM^2 + LM_{vec}\mathbb{D}_x\mathbb{D}_y\right)$
FOC-DFT-DPD	$O\left(LTM^4 + LJ\mathbb{D}_x\mathbb{D}_y\right)$
FOC-DFT-Taylor	$O\left(LTM^4 + LJ\mathbb{D}_x\mathbb{D}_y + 27K^3 + 9K^2 + 9K^2LJ + 5KLJ\right)$
FOC-SS-SDF-DPD [22]	$O\left(LTM^4 + 2LN_a^3 + L\mathbb{D}_x\mathbb{D}_yN_a(N_a - K)\right)$
FOC-SS-Capon-DPD [18]	$O\left(LTM^4 + LN_a^3 + L\mathbb{D}_x\mathbb{D}_y\left(N_a^2 + N_a\right)\right)$

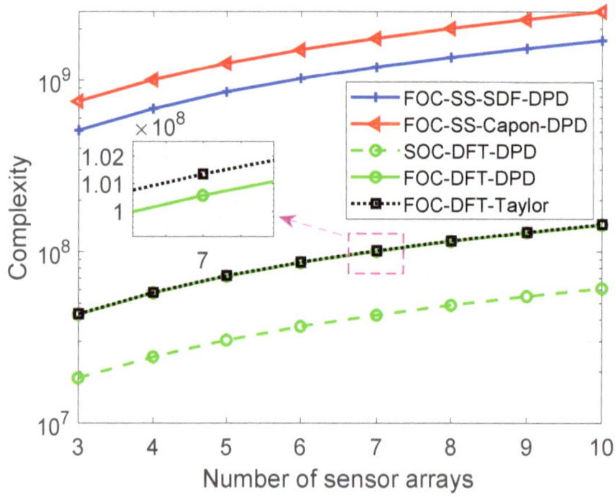

Figure 2. Comparison of complexity versus L.

Figure 3 shows the computational complexity of four different algorithms with the parameters setting as below: $K = 8$, $N_1 = N_2 = M/2$, $\mathrm{T} = 300$, $\mathbb{D}_x = \mathbb{D}_y = 500$, and the number of array elements is changed from 6 to 20. As we can see, compared to the FOC-SS-SDF-DPD and FOC-SS-Capon-DPD algorithm, the proposed FOC-DFT-DPD and FOC-DFT-Taylor algorithms reduce the computational complexity, as they are less sensitive to the number of array elements M. The complexity of the proposed FOC-DFT-Taylor algorithm is a bit heavier than that of FOC-DFT-DPD and SOC-DFT-DPD algorithms.

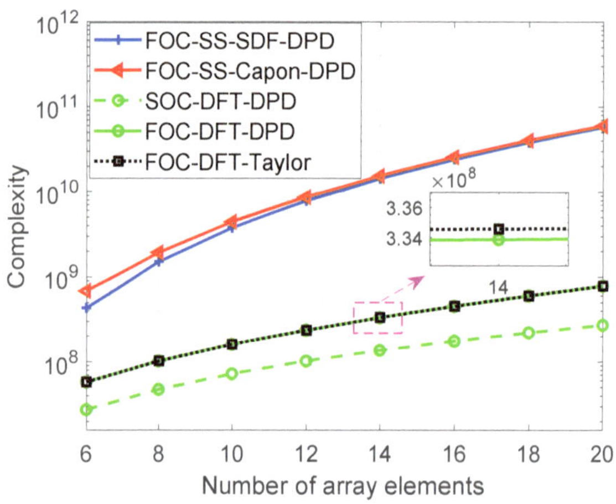

Figure 3. Comparison of complexity versus M.

4.2. Advantages

1. *Low Complexity:* Compared to the FOC-SS-SDF-DPD and FOC-SS-Capon-DPD algorithms, the proposed algorithm reduces much of the computational complexity thanks to the computationally efficient cost function.
2. *High DOF:* Compared to the FOC-SS-SDF-DPD and FOC-SS-Capon-DPD algorithms, the proposed algorithm does not need SS technology for decoherence, which is caused by vectorizing the FOC matrix. Thus, the proposed algorithm can estimate more sources.
3. *Suppress Gaussian Noise:* The proposed algorithm takes full advantage of the characteristics of non-Gaussian signals, and the Gaussian noise is suppressed during the process of calculating the FOC matrix.
4. *High Accuracy:* When the sources do not fall on the preset grids, i.e., the off-grid error exists, the proposed algorithm can still estimate the position of sources accurately thanks to the Taylor compensation.
5. *High-Resolution Capability:* Compared to FOC-SS-SDF-DPD and FOC-SS-Capon-DPD methods, the proposed method has higher resolution capability.

5. Numerical Analysis

We conduct numerical simulations in this section to evaluate various algorithms and demonstrate the effectiveness of the proposed approach. The superiority of the proposed method is established through comparisons. Our evaluation includes assessing localization accuracy using root mean square error (RMSE) across different algorithms, which is defined as

$$\text{RMSE} = \frac{1}{K}\sum_{k=1}^{K}\sqrt{\sum_{m_c=1}^{M_c}||\hat{\mathbf{p}}_{k,m_c} - \mathbf{p}_k||^2/M_c} \tag{29}$$

here, M_c denotes the number of Monte Carlo runs, and $\hat{\mathbf{p}}_{k,m_c}$ is the estimated position of the kth source in the m_cth Monte Carlo experiment. The resolution capability of the algorithms is assessed by the definition of the effective estimate rate (EER), which is calculated by

$$\text{EER} = \frac{1}{M_c}\sum_{m_c}^{M_c}\alpha_{m_c} \tag{30}$$

with

$$\alpha_{m_c} = \begin{cases} 1, \text{ if } \frac{1}{K}\sum_{k=1}^{K}||\hat{\mathbf{p}}_{k,m_c} - \mathbf{p}_k|| < \varepsilon_{\text{error}} \\ 0, \text{ otherwise} \end{cases} \tag{31}$$

where $\varepsilon_{\text{error}}$ represents the error threshold.

5.1. Effectiveness Analysis

Figure 4 shows scatter diagrams of the proposed FOC-DFT-Taylor algorithm, where the number of Monte Carlo runs is 300, $K = 2$, and $L = 4$. The location of sources and SAs are set as $\mathbf{u}_1 = [-900\text{ m}, 1000\text{ m}]^T$, $\mathbf{u}_2 = [-300\text{ m}, 500\text{ m}]^T$, $\mathbf{u}_3 = [300\text{ m}, -180\text{ m}]^T$, $\mathbf{u}_4 = [-800\text{ m}, 900\text{ m}]^T$, $\mathbf{p}_1 = [100.5\text{ m}, 900.2\text{ m}]^T$, $\mathbf{p}_2 = [900.6\text{ m}, 200.7\text{ m}]^T$. The results presented in Figure 4a demonstrate the successful estimation of sources by the FOC-DFT-DPD algorithm. Furthermore, analysis of Figure 4b reveals that the estimated positions closely align with the true positions following Taylor compensation, providing a compelling clue for the efficacy of the proposed FOC-DFT-Taylor algorithm.

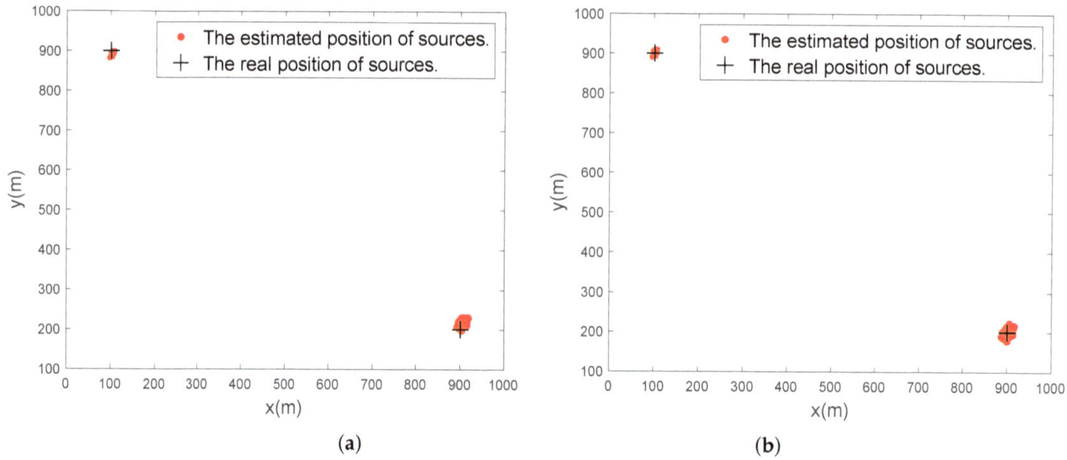

Figure 4. Scatter diagrams of the proposed algorithm when $M = 4$, $N_1 = N_2 = 2$, $\mathrm{T} = 300$, $\mathbb{D}_x = \mathbb{D}_y = 600$, $\mathbf{u}_1 = [-900\ \mathrm{m}, 1000\ \mathrm{m}]^T$, $\mathbf{u}_2 = [-300\ \mathrm{m}, 500\ \mathrm{m}]^T$, $\mathbf{u}_3 = [300\ \mathrm{m}, -180\ \mathrm{m}]^T$, $\mathbf{u}_4 = [-800\ \mathrm{m}, 900\ \mathrm{m}]^T$, $\mathbf{p}_1 = [100.5\ \mathrm{m}, 900.2\ \mathrm{m}]^T$, $\mathbf{p}_2 = [900.6\ \mathrm{m}, 200.7\ \mathrm{m}]^T$, SNR $= 5$ dB. (**a**) Initial estimation. (**b**) After Taylor compensation.

5.2. RMSE Results

Figure 5 depicts the RMSE of four algorithms versus SNR when the number of Monte Carlo runs is 600, $M = 6$, $N_1 = N_2 = 3$, $\mathrm{T} = 500$, $\mathbb{D}_x = \mathbb{D}_y = 600$, $L = 4$, $K = 2$, $\mathbf{u}_1 = [-900\ \mathrm{m}, -1200\ \mathrm{m}]^T$, $\mathbf{u}_2 = [-300\ \mathrm{m}, -1100\ \mathrm{m}]^T$, $\mathbf{u}_3 = [300\ \mathrm{m}, -1000\ \mathrm{m}]^T$, $\mathbf{u}_4 = [-800\ \mathrm{m}, -900\ \mathrm{m}]^T$, $\mathbf{p}_1 = [200.5\ \mathrm{m}, 623.5\ \mathrm{m}]^T$, $\mathbf{p}_2 = [920.5\ \mathrm{m}, -174.5\ \mathrm{m}]^T$, SNR is changed from -5 dB to 30 dB, and the gap of search grids is 1 m. Observe that the RMSE values of the four algorithms exhibit close proximity except for the SOC-DFT-DPD algorithm when SNR is below 10 dB, thanks to the property of non-Gaussian sources. The proposed FOC-DFT-Taylor algorithm demonstrates superior performance over the others at SNR levels exceeding 10 dB, showcasing its exceptional capabilities. Moreover, the RMSE associated with the proposed FOC-DFT-Taylor algorithm consistently remains below that of the FOC-DFT-DPD algorithm at the expense of performing Taylor compensation.

Figure 6 shows the RMSE of four algorithms versus T when the number of Monte Carlo runs is 600, $M = 6$, $N_1 = N_2 = 3$, SNR $= 15$ dB, $\mathbb{D}_x = \mathbb{D}_y = 600$, $L = 4$, $K = 2$, $\mathbf{u}_1 = [-900\ \mathrm{m}, -1200\ \mathrm{m}]^T$, $\mathbf{u}_2 = [-300\ \mathrm{m}, -1100\ \mathrm{m}]^T$, $\mathbf{u}_3 = [300\ \mathrm{m}, -1000\ \mathrm{m}]^T$, $\mathbf{u}_4 = [-800\ \mathrm{m}, -900\ \mathrm{m}]^T$, $\mathbf{p}_1 = [100.5\ \mathrm{m}, 815.5\ \mathrm{m}]^T$, $\mathbf{p}_2 = [970.5\ \mathrm{m}, 54.5\ \mathrm{m}]^T$, the number of snapshots is changed from 200 to 900, and the gap of the search grids is 1 m. The localization performance of the FOC-DFT-DPD and SOC-DFT-DPD remains consistent, as they do not operate as super resolution algorithms and are constrained by M. Thanks to the property of non-Gaussian sources, the localization accuracy of the FOC-DFT-DPD is higher than that of SOC-DFT-DPD. Moreover, the RMSE values of the FOC-DFT-Taylor algorithm consistently outperform the other three algorithms due to its compensation method, showcasing its superior accuracy in localization.

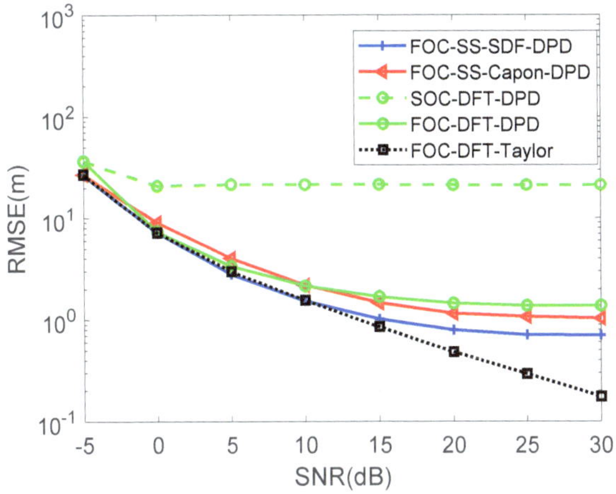

Figure 5. Comparison of RMSE versus SNR.

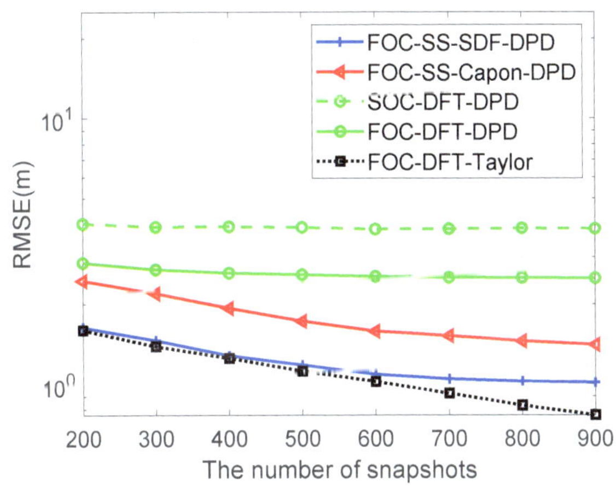

Figure 6. Comparison of RMSE versus T.

Figure 7 shows the RMSE performance of different algorithms versus the number of array elements M over 500 Monte Carlo runs, when $N_1 = N_2 = M/2$, $\mathbb{D}_x = \mathbb{D}_y = 600$, $L = 4$, $K = 2$, $\mathbf{u}_1 = [-900\,\text{m}, -1200\,\text{m}]^T$, $\mathbf{u}_2 = [-300\,\text{m}, -1100\,\text{m}]^T$, $\mathbf{u}_3 = [300\,\text{m}, -1000\,\text{m}]^T$, $\mathbf{u}_4 = [-800\,\text{m}, -900\,\text{m}]^T$, $\mathbf{p}_1 = [100.5\,\text{m}, 815.5\,\text{m}]^T$, $\mathbf{p}_2 = [970.5\,\text{m}, 54.5\,\text{m}]^T$, $\text{T} = 500$, SNR = 20 dB, the number of array elements is changed from 4 to 18, and the gap of the search grids is 1 m. Observe that when the array size exceeds 8, the RMSEs of FOC-SS-SDF-DPD, FOC-SS-Capon-DPD, and FOC-DFT-DPD algorithms stabilize around 0.7 m due to their inability to address off-grid errors. In comparison, the RMSE values of the proposed FOC-DFT-Taylor algorithm consistently outperform the other algorithms thanks to its compensation method, demonstrating superior accuracy in localization. Unfortunately, the SOC-DFT-DPD performs poorly, as it fails to take full advantage of non-Gaussian sources. Furthermore, the RMSE values of FOC-DFT-DPD decrease as the array size increases, indicating that localization accuracy is constrained by the number of array elements.

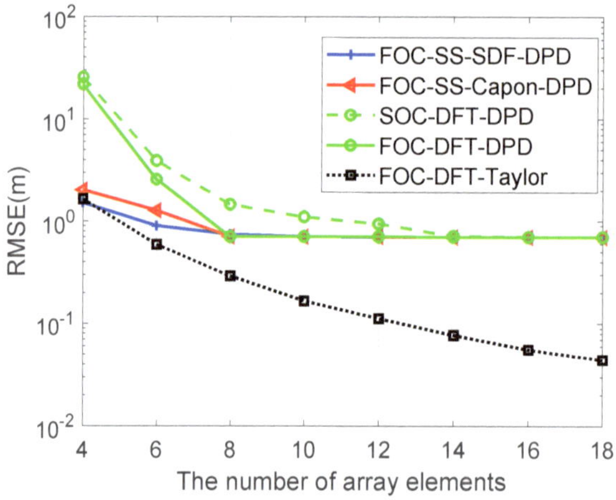

Figure 7. Comparison of RMSE versus M.

5.3. EER Analysis

Figure 8 shows the EER of four algorithms versus SNR when the number of Monte Carlo runs is 600, $M = 8$, $N_1 = N_2 = 4$, $\mathrm{T} = 500$, $\mathbb{D}_x = \mathbb{D}_y = 600$, $L = 4$, $K = 2$, $\mathbf{u}_1 = [-900\ \mathrm{m}, -1200\ \mathrm{m}]^T$, $\mathbf{u}_2 = [-300\ \mathrm{m}, -1100\ \mathrm{m}]^T$, $\mathbf{u}_3 = [300\ \mathrm{m}, -1000\ \mathrm{m}]^T$, $\mathbf{u}_4 = [-800\ \mathrm{m}, -900\ \mathrm{m}]^T$, $\mathbf{p}_1 = [100.5\ \mathrm{m}, 900.2\ \mathrm{m}]^T$, $\mathbf{p}_2 = [900.6\ \mathrm{m}, 200.7\ \mathrm{m}]^T$, SNR is changed from 6 dB to 20 dB, the gap of search grids is 1 m, and the error threshold $\varepsilon_{\mathrm{error}}$ is set as 0.75 m. The EER achieved by the proposed FOC-DFT-Taylor algorithm surpasses 90%, contrasting with the EERs below 90% for the other four algorithms when SNR reaches a level of 14 dB. Unfortunately, the SOC-DFT-DPD performs poorly, as it fails to take full advantage of non-Gaussian sources. Notably, the proposed FOC-DFT-Taylor algorithm exhibits superior resolution capabilities compared to the remaining algorithms in the study.

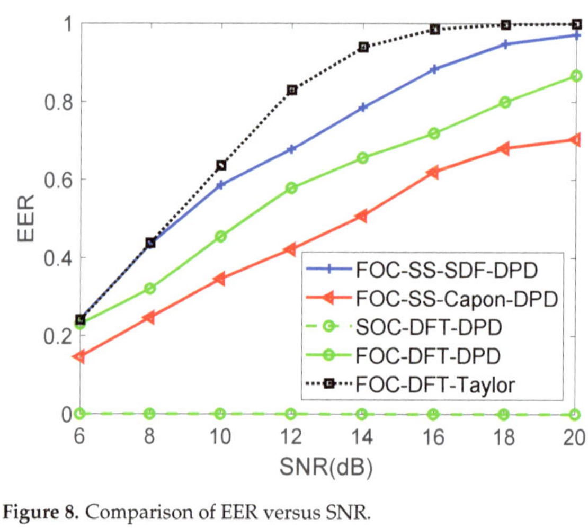

Figure 8. Comparison of EER versus SNR.

Figure 9 shows the EER of four algorithms versus T, when the number of Monte Carlo runs is 600, $M = 6$, $N_1 = N_2 = 3$, $\mathrm{SNR} = 20$ dB, $\mathbb{D}_x = \mathbb{D}_y = 600$, $L = 4$, $K = 2$, $\mathbf{u}_1 = [-900\ \mathrm{m}, -1200\ \mathrm{m}]^T$, $\mathbf{u}_2 = [-300\ \mathrm{m}, -1100\ \mathrm{m}]^T$, $\mathbf{u}_3 = [300\ \mathrm{m}, -1000\ \mathrm{m}]^T$,

$\mathbf{u}_4 = [-800\ \text{m}, -900\ \text{m}]^T$, $\mathbf{p}_1 = [200.5\ \text{m}, 623.5\ \text{m}]^T$, $\mathbf{p}_2 = [920.5\ \text{m}, -174.5\ \text{m}]^T$, the number of snapshots is changed from 20 to 600, the gap of search grids is 1 m, and the error threshold $\varepsilon_{\text{error}}$ is set as 1 m. The EER of the proposed FOC-DFT-Taylor algorithm shows a slight increase compared to the FOC-SS-SDF-DPD algorithm, yet with a significant reduction in complexity. Unfortunately, the SOC-DFT-DPD performs poorly, as it fails to take full advantage of non-Gaussian sources. Additionally, when compared to the FOC-SS-Capon-DPD and FOC-DFT-DPD algorithms, the EER of the FOC-DFT-Taylor algorithm is higher, underscoring its superior resolution capabilities.

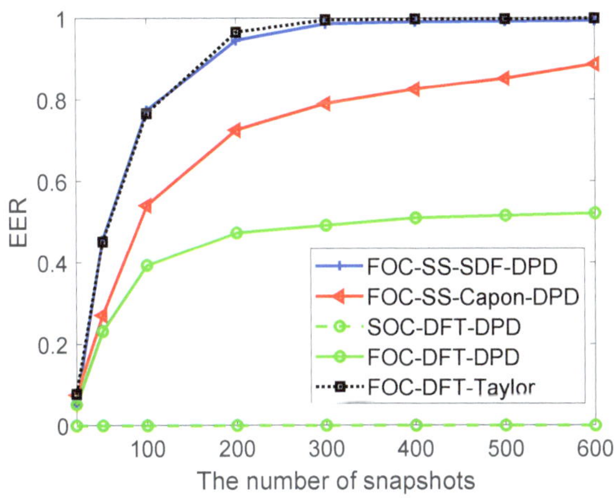

Figure 9. Comparison of EER versus T.

6. Conclusions

In this article, the focus is on discussing the direct position determination (DPD) of non-Gaussian sources and proposing a new algorithm called the FOC-DFT-Taylor algorithm with multiple nested arrays (MNAs). This novel algorithm is designed to create an efficient DPD cost function that significantly reduces complexity when compared to conventional FOC-SS-SDF-DPD and FOC-SS-Capon-DPD algorithms. By leveraging the characteristics of non-Gaussian signals, this proposed algorithm offers a greater number of achievable degrees of freedom (DOFs). Additionally, due to the utilization of Taylor compensation, this new algorithm demonstrates improved localization accuracy and resolution capability over the traditional FOC-SS-SDF-DPD and FOC-SS-Capon-DPD algorithms.

Author Contributions: Conceptualization, H.H. and M.Y. (Meng Yang); methodology, Q.Y. and M.Y. (Mingyi You); software, X.S.; validation, H.H., M.Y. (Meng Yang), Q.Y., M.Y. (Mingyi You), X.S. and Y.S.; formal analysis, Y.S.; investigation, H.H.; resources, M.Y. (Meng Yang); data curation, Q.Y.; writing—original draft preparation, M.Y. (Mingyi You); writing—review and editing, H.H.; visualization, M.Y. (Meng Yang); supervision, Q.Y.; project administration, X.S.; funding acquisition, X.S. All authors have read and agreed to the published version of the manuscript.

Funding: This research was funded by the National science foundation of China under Grants 62371225 and 62371227.

Institutional Review Board Statement: Not applicable.

Informed Consent Statement: Not applicable.

Data Availability Statement: The data used to support the findings of this study are included within the article.

Conflicts of Interest: The authors declare no conflicts of interest.

Abbreviations

The following abbreviations are used in this manuscript:

DPD	Direct Position Determination
MNA	Multiple Nested Array
DFT	Discrete Fourier Transform
AOA	Angle of Arrival
TDOA	Time Difference of Arrival
TOA	Time of Arrival
RSS	Received Signal Strength
ML	Maximum Likelihood
SNR	Signal-to-noise Ratio
SDF	Subspace Data Fusion
MUSIC	Multiple Signal Classification
EVD	Eigenvalue Decomposition
OFDM	Orthogonal Frequency Division Multiplexing
SOC	Second-order Cumulant
FOC	Fourth-order Cumulant
ULA	Uniform Linear Array
NA	Nested Array
DOF	Degree of Freedom
SS	Spatial Smoothing
RMSE	Root Mean Square Error
EER	Effective Estimate Rate

References

1. Coluccia, A.; Fascista, A. A review of advanced localization techniques for crowdsensing wireless sensor networks. *Sensors* **2019**, *19*, 988. [CrossRef]
2. Sibanyoni, S.V.; Ramotsoela, D.T.; Silva, B.J.; Hancke, G.P. A 2-D acoustic source localization system for drones in search and rescue missions. *IEEE Sens. J.* **2018**, *19*, 332–341. [CrossRef]
3. Zhao, K.; Zhao, T.; Zheng, Z.; Yu, C.; Ma, D.; Rabie, K.; Kharel, R. Optimization of time synchronization and algorithms with TDOA based indoor positioning technique for Internet of Things. *Sensors* **2020**, *20*, 6513. [CrossRef]
4. Zhang, R.; Cheng, L.; Wang, S.; Lou, Y.; Gao, Y.; Wu, W.; Ng, D.W.K. Integrated Sensing and Communication with Massive MIMO: A Unified Tensor Approach for Channel and Target Parameter Estimation. *IEEE Trans. Wirel. Commun.* 2024, *early access* .
5. Ma, Y.; Miao, C.; Long, W.; Zhang, R.; Chen, Q.; Zhang, J.; Wu, W. Time-Modulated Arrays in Scanning Mode Using Wideband Signals for Range-Doppler Estimation with Time-Frequency Filtering and Fusion. *IEEE Trans. Aerosp. Electron. Syst.* **2023**, *60*, 980–990. [CrossRef]
6. Wang, Y.; Ho, K. Unified near-field and far-field localization for AOA and hybrid AOA-TDOA positionings. *IEEE Trans. Wirel. Commun.* **2017**, *17*, 1242–1254. [CrossRef]
7. Wu, M.; Xiong, N.; Tan, L. Adaptive range-based target localization using diffusion Gauss–Newton method in industrial environments. *IEEE Trans. Ind. Inform.* **2019**, *15*, 5919–5930. [CrossRef]
8. Weiss, A.J.; Amar, A. Direct position determination of multiple radio signals. *EURASIP J. Adv. Signal Process.* **2005**, *2005*, 653549. [CrossRef]
9. Lo, T.K.; Leung, H.; Litva, J. Artificial neural network for AOA estimation in a multipath environment over the sea. *IEEE J. Ocean. Eng.* **1994**, *19*, 555–562. [CrossRef]
10. Yu, K.; Guo, Y.J. Statistical NLOS identification based on AOA, TOA, and signal strength. *IEEE Trans. Veh. Technol.* **2008**, *58*, 274–286. [CrossRef]
11. Wen, F.; Shi, J.; Zhang, Z. Joint 2D-DOD, 2D-DOA, and polarization angles estimation for bistatic EMVS-MIMO radar via PARAFAC analysis. *IEEE Trans. Veh. Technol.* **2019**, *69*, 1626–1638. [CrossRef]
12. Kulas, L. RSS-based DOA estimation using ESPAR antennas and interpolated radiation patterns. *IEEE Antennas Wirel. Propag. Lett.* **2017**, *17*, 25–28. [CrossRef]
13. Zeng, H.; Shi, S.; Yue, H.; Li, J. Coherent signal localization with multiple arrays: A successive perspective. *Multidimens. Syst. Signal Process.* **2023**, *34*, 321–340. [CrossRef]
14. Weiss, A.J. Direct position determination of narrowband radio frequency transmitters. *IEEE Signal Process. Lett.* **2004**, *11*, 513–516. [CrossRef]
15. Tirer, T.; Weiss, A.J. High resolution direct position determination of radio frequency sources. *IEEE Signal Process. Lett.* **2015**, *23*, 192–196. [CrossRef]

16. Yin, J.; Wu, Y.; Wang, D. Direct position determination of multiple noncircular sources with a moving array. *Circuits Syst. Signal Process.* **2017**, *36*, 4050–4076. [CrossRef]
17. Vallet, P.; Mestre, X.; Loubaton, P. Performance analysis of an improved MUSIC DOA estimator. *IEEE Trans. Signal Process.* **2015**, *63*, 6407–6422. [CrossRef]
18. Tzafri, L.; Weiss, A.J. Application of capon method to direct position determination. *ICT Express* **2016**, *2*, 5–9. [CrossRef]
19. Shu, T.; Li, L.; He, J. Near-field localization for non-circular sources in the presence of sensor phase uncertainties. *IEEE Wirel. Commun. Lett.* **2020**, *10*, 562–566. [CrossRef]
20. Wang, Q.; Wang, X.; Chen, H.; Zhu, X.; Liu, W.; Yan, W.; Wang, L. An effective localization method for mixed far-field and near-field strictly non-circular sources. *Digit. Signal Process.* **2019**, *94*, 125–136. [CrossRef]
21. Lin, Y.; Jin, S.; Matthaiou, M.; You, X. Channel estimation and user localization for IRS-assisted MIMO-OFDM systems. *IEEE Trans. Wirel. Commun.* **2021**, *21*, 2320–2335. [CrossRef]
22. Lu, Z.; Wang, J.; Ba, B.; Wang, D. A novel direct position determination algorithm for orthogonal frequency division multiplexing signals based on the time and angle of arrival. *IEEE Access* **2017**, *5*, 25312–25321. [CrossRef]
23. Xie, J.; Tao, H.; Rao, X.; Su, J. Efficient method of passive localization for near-field noncircular sources. *IEEE Antennas Wirel. Propag. Lett.* **2015**, *14*, 1223–1226. [CrossRef]
24. Zhang, Y.K.; Xu, H.Y.; Ba, B.; Wang, D.M.; Geng, W. Direct position determination of non-circular sources based on a Doppler-extended aperture with a moving coprime array. *IEEE Access* **2018**, *6*, 61014–61021. [CrossRef]
25. Shi, X.; Zhang, X.; Qian, Y.; Ahmad, M. Closed-form rectilinear emitters localization with multiple sensor arrays: Phase alignment and optimal weighted least-square method. *IEEE Sens. J.* **2023**, *23*, 7266–7278. [CrossRef]
26. Wang, H.; Zhao, J.; Qian, L. A study based on fourth-second order normalized cumulant in TDOA estimation. In Proceedings of the IEEE 10th International Conference on Signal Processing Proceedings, Beijing, China, 24–28 October 2010; pp. 215–218.
27. Hou, C.; Wang, D. The Fourth Order Cumulant-Based MUSIC Algorithm for Harmonic Retrieval in Colored Noise. In Proceedings of the 2009 WRI International Conference on Communications and Mobile Computing, Kunming, China, 6–8 January 2009; Volume 2, pp. 564–567.
28. He, Q.; Cheng, Z.; Wang, Z.; He, Z. Mixed far-field and near-field source separation and localization based on FOC matrix differencing. *Digit. Signal Process.* **2022**, *131*, 103753. [CrossRef]
29. Liu, C.; Ye, Z.; Zhang, Y. DOA estimation based on fourth-order cumulants with unknown mutual coupling. *Signal Process.* **2009**, *89*, 1839–1843. [CrossRef]
30. An, L.; Shu, W. Propagator method for DOA estimation using fourth-order cumulant. In Proceedings of the 2011 7th International Conference on Wireless Communications, Networking and Mobile Computing, Wuhan, China, 23–25 September 2011; pp. 1–4.
31. Alinezhad, P.; Seydnejad, S.R.; Abbasi-Moghadam, D. DOA estimation in conformal arrays based on the nested array principles. *Digit. Signal Process.* **2016**, *50*, 191–202. [CrossRef]
32. Peng, Z.; Ding, Y.; Ren, S.; Wu, H.; Wang, W. Coprime nested arrays for DOA estimation: Exploiting the nesting property of coprime array. *IEEE Signal Process. Lett.* **2021**, *29*, 444–448. [CrossRef]
33. Qin, G.; Zhang, Y.D.; Amin, M.G. DOA estimation exploiting moving dilated nested arrays. *IEEE Signal Process. Lett.* **2019**, *26*, 490–494. [CrossRef]
34. Han, K.; Nehorai, A. Improved source number detection and direction estimation with nested arrays and ULAs using jackknifing. *IEEE Trans. Signal Process.* **2013**, *61*, 6118–6128. [CrossRef]
35. Chen, Y.; Wang, W.; Wang, Z.; Xia, B. A source counting method using acoustic vector sensor based on sparse modeling of DOA histogram. *IEEE Signal Process. Lett.* **2018**, *26*, 69–73. [CrossRef]
36. Li, J.; He, Y.; Zhang, X.; Wu, Q. Simultaneous localization of multiple unknown emitters based on UAV monitoring big data. *IEEE Trans. Ind. Inform.* **2021**, *17*, 6303–6313. [CrossRef]
37. Ye, Z.; Zhang, Y. DOA estimation for non-Gaussian signals using fourth-order cumulants. *IET Microwaves Antennas Propag.* **2009**, *3*, 1069–1078. [CrossRef]
38. Dogan, M.C.; Mendel, J.M. Applications of cumulants to array processing. I. Aperture extension and array calibration. *IEEE Trans. Signal Process.* **1995**, *43*, 1200–1216. [CrossRef]
39. Yuen, N.; Friedlander, B. DOA estimation in multipath: An approach using fourth-order cumulants. *IEEE Trans. Signal Process.* **1997**, *45*, 1253–1263. [CrossRef]
40. Ye, C.; Chen, W.; Zhu, B.; Tang, L. Direction of arrival estimation of non-Gaussian signals for nested arrays: Applying fourth-order difference co-array and the successive method. *ETRI J.* **2021**, *43*, 869–880. [CrossRef]
41. Pal, P.; Vaidyanathan, P.P. Coprime sampling and the MUSIC algorithm. In Proceedings of the 2011 Digital Signal Processing and Signal Processing Education Meeting (DSP/SPE), Sedona, AZ, USA, 4–7 January 2011; pp. 289–294.
42. Pannert, W. Spatial smoothing for localized correlated sources–Its effect on different localization methods in the nearfield. *Appl. Acoust.* **2011**, *72*, 873–883. [CrossRef]

MDPI AG
Grosspeteranlage 5
4052 Basel
Switzerland
Tel.: +41 61 683 77 34

Sensors Editorial Office
E-mail: sensors@mdpi.com
www.mdpi.com/journal/sensors

www.ingramcontent.com/pod-product-compliance
Lightning Source LLC
Chambersburg PA
CBHW041354210926
51850CB00013B/3074
* 9 7 8 3 7 2 5 8 4 3 9 3 0 *